Diagnosis and Fault Tolerance of Electrical Machines, Power Electronics and Drives

Edited by
Antonio J. Marques Cardoso

The Institution of Engineering and Technology

Published by The Institution of Engineering and Technology, London, United Kingdom

The Institution of Engineering and Technology is registered as a Charity in England & Wales (no. 211014) and Scotland (no. SC038698).

First published 2018

The Institution of Engineering and Technology
Michael Faraday House
Six Hills Way, Stevenage
Herts, SG1 2AY, United Kingdom

www.theiet.org

British Library Cataloguing in Publication Data
A catalogue record for this product is available from the British Library

ISBN 978-1-78561-531-3 (hardback)
ISBN 978-1-78561-532-0 (PDF)

Typeset in India by MPS Limited

Contents

About the authors

Adérito N. Alcaso was born in 1966. He received the diploma in Electrical and Computer Engineering from the Technical University of Lisbon, Lisbon, Portugal, in 1990. He received the MSc degree in Systems and Automation, and the PhD degree in Electrical Engineering from the University of Coimbra, Coimbra, Portugal, in 1995 and 2005, respectively. Since 1996 he is Adjunct Professor at the Polytechnic of Guarda, Guarda, Portugal, where he has been director of the Mechanical Engineering Department and member of the Pedagogical and Scientific Councils of the School of Technology and Management. He is also a Researcher of CISE – Electromechatronic Systems Research Centre. He has published several papers in technical journals and conference proceedings. His current research interests are focused in renewable energy systems, particularly of the co-generation type, and in electric mobility systems, exploring the application of low-cost microcontrollers and internet of things for monitoring and optimizing the operation of these systems.

Acácio M. R. Amaral was born in Luso, Angola, in 1974. He received the Electrical Engineering Diploma, the MSc degree and the PhD degree from the University of Coimbra, Coimbra, Portugal in 1998, 2005 and 2010, respectively. Since 1998 he has been with the Polytechnic Institute of Coimbra, where he is currently an Adjunct Professor in the Department of Informatics and Systems. He is also a Researcher of CISE – Electromechatronic Systems Research Centre. He is the author of three books entitled: *Circuit Analysis and Electronic Devices* (Porto, Portugal, Publindustria, 2013, in Portuguese); *Digital Systems: Principles, Analysis and Projects* (Lisboa, Portugal, Edições Sílabo, 2014, in Portuguese); and *Analog Electronics: Principles, Analysis and Projects* (Lisboa, Portugal, Edições Sílabo, 2017, in Portuguese). He has also published more than 40 papers in technical journals and conference proceedings. His research activities include fault diagnosis and design of linear and switch-mode power supplies, with emphasis on the consequences of aging of electrolytic and film capacitors, as well as the development of solutions to this problem.

Fernando Bento received both the BSc and MSc degrees in Electric and Computer Engineering from the University of Beira Interior, Covilhã, Portugal, in 2014 and 2016, respectively. Currently, he is a PhD student in Electric and Computer Engineering at the University of Beira Interior, and PhD student of CISE – Electromechatronic Systems Research Centre. His scientific research activities focus on energy efficiency analysis, fault diagnostic and fault tolerance in electronic power converters, namely DC-DC converters.

Antonio J. Marques Cardoso received the Dipl. Eng., Dr. Eng., and Habilitation degrees from the University of Coimbra, Coimbra, Portugal, in 1985, 1995 and 2008, respectively, all in Electrical Engineering. From 1985 to 2011, he was with the University of Coimbra, Coimbra, Portugal, where he was the director of the Electrical Machines Laboratory. Since 2011 he has been with the University of Beira Interior (UBI), Covilhã, Portugal, where he is Full Professor at the Department of Electromechanical Engineering and director of CISE – Electromechatronic Systems Research Centre (http://cise.ubi.pt). He was Vice-Rector of UBI (2013–2014). His current research interests are in fault diagnosis and fault tolerance in electrical machines, power electronics and drives. He is the author of a book entitled *Fault Diagnosis in Three-Phase Induction Motors* (Coimbra, Portugal: Coimbra Editora, 1991), (in Portuguese) and he is also the author of around 500 papers published in technical journals and conference proceedings. He currently serves as an associate editor for the *IEEE Transactions on Industry Applications*, *IEEE Transactions on Industrial Electronics*, *IEEE Transactions on Power Electronics*, *IEEE Journal of Emerging and Selected Topics in Power Electronics*, and also for the Springer *International Journal of Systems Assurance Engineering and Management*.

Jorge O. Estima was born in Aveiro, Portugal, in 1984. He received the Dipl. Eng. and the Dr. Eng. degree from the University of Coimbra, Coimbra, Portugal, in 2007 and 2012, respectively. From 2012 to 2016, he was a postdoctoral researcher at CISE – Electromechatronic Systems Research Centre (http://cise.ubi.pt), University of Beira Interior (UBI), Covilhã, Portugal. Since 2016 he has been with the UBI where he is an Invited Assistant Professor at the Department of Electromechanical Engineering and Researcher of CISE. He has also been with the company Enging where he is R&D Manager. His research interests are focused on condition monitoring and diagnostics of electric machines, power electronics, fault-tolerant variable speed drives and energy efficiency in motor drive systems.

Davide S. B. Fonseca was born in Castelo Branco, Portugal, on December 23, 1972. He received the Electrical Engineering Diploma in 1996 from the University of Coimbra, Coimbra, Portugal, and the PhD in electrical engineering from the University of Beira Interior, Covilhã, Portugal, in 2008. He has been with the University of Beira Interior since 1997, where he is currently an Assistant Professor in the Department of Electromechanical Engineering, and the Coordinator of the Electrical Machines and Power Electronics Laboratory. He is also Researcher of CISE – Electromechatronic Systems Research Centre. His research interests are focused on reluctance machines design and fault analysis. He has published more than 40 papers in technical journals and conference proceedings.

Natália S. Gameiro received the Electrical Engineering Diploma, the MSc degree in Electrical Engineering, and the PhD degree in Electrical Engineering from the University of Coimbra, Coimbra, Portugal, in 1997, 2004 and 2014, respectively. Since 1997, she has been with the Polytechnic Institute of Leiria, Leiria, Portugal, where she is currently an Adjunct Professor with the Department of Electrical Engineering. She is also Researcher of CISE – Electromechatronic Systems Research Centre. Her teaching interests cover electrical machines, control systems and basic electric network analysis, and her research interests also include electrical machines and drives, control of variable electric drives, fault diagnosis and fault-tolerant control. Currently, she is mainly focused on the analysis and development of fault-tolerant solutions, based on inverter and/or control reconfigurations applied to switched reluctance motor drives.

Konstantinos N. Gyftakis was born in Patras, Greece, in May 1984. He received the Diploma in Electrical and Computer Engineering from the University of Patras, Patras, Greece, in 2010. He pursued a PhD in the same institution in the area of electrical machines condition monitoring and fault diagnosis (2010–2014). Then he worked as a Post-Doctoral Research Assistant in the Department of Engineering Science, University of Oxford, UK (2014–2015). In 2015, he was appointed Lecturer on Electrical and Electronic Engineering, School of

Computing, Electronics and Mathematics, Faculty of Engineering, Environment and Computing, Coventry University, UK. Moreover, he is a member of the Research Institute for Future Transport and Cities, Coventry University, UK. Additionally, he is a member of CISE – Electromechatronic Systems Research Centre, Portugal. Finally, he is an IEEE member as well as a member of the IEEE Industry Applications Society and IEEE Industrial Electronics Society.

His research interests focus in the fault diagnosis, condition monitoring and degradation of electrical machines. He has authored/co-authored more than 60 papers in international scientific journals and conferences.

Eunice Ribeiro holds a PhD degree in Electrical Engineering from the University of Coimbra (Portugal) awarded with 'The Joseph J. Suozzi INTELEC Award Fellowship in Power Electronics'. As part of her studies and involvement in research projects, she has a considerable experience on power electronic converters covering a wide range of applications, such as energy management systems, electric vehicles, renewable energies power conditioning, hybrid energy storage solutions, fault diagnostic methods and fault tolerance strategies. She has published more than 20 scientific papers in peer-reviewed conferences and journals related to power electronic converters, renewable energies, energy storage and energy systems. Previously, Eunice Ribeiro was H2020 National Contact Point, National Representative in H2020 Programme Committees and National Delegate for the European Strategic Energy Technology Plan (SET-Plan). Currently, she is the EU Programmes Manager at Ubiwhere and a Researcher at CISE – Electromechatronic Systems Research Centre.

Mohamed Sahraoui was born in Biskra, Algeria, on May 26, 1978. He received the Engineer and Magister diploma and the PhD degree in Electrical Engineering from the University of Biskra, in 2001, 2004 and 2010, respectively. From 2005 to 2012, he was an Assistant Professor with the University of Constantine, Constantine, Algeria. Since 2012, he has been with the University of Biskra, Biskra, Algeria, where he is an Assistant Professor at the Department of Electrical Engineering and a member of the LGEB Laboratory. Dr Sahraoui is also a PhD Researcher of CISE – Electromechatronic Systems Research Centre (http://cise.ubi.pt). His research interests are related to condition monitoring and fault diagnosis in power electronics systems and AC machines.

Alberto Tessarolo received the Laurea and PhD degrees in Electrical Engineering from the University of Trieste, Trieste, Italy, in 2000 and 2011, respectively. Before joining the University, he worked in the design and development of large innovative motors, generators and drives with NIDEC-ASI (formerly Ansaldo Sistemi Industriali). Since 2006, he has been with the Department of Engineering and Architecture, University of Trieste, where he teaches the course of Electric Machine Design. He holds the scientific responsibility for several funded research projects in coordination with leading companies and institutions. He has authored more than 150 international technical papers in the area of electrical machine and drive modelling and design. He serves as an editor for the *IEEE Transactions on Energy Conversion* and an associate editor for the *IEEE Transactions on Industry Applications* and *IET Electric Power Applications*. He received the Electric Machinery Committee 2012 Prize Paper Award of the IEEE Power and Energy Society and of various best paper awards for contributions presented at IEEE-sponsored or co-sponsored conferences. He is a senior member of the IEEE and a member of the Industry Applications, Power and Energy, Power Electronics, Industrial Electronics, and Magnetics and Reliability Societies of the IEEE.

Chapter 1

Introduction

Antonio J. Marques Cardoso[1]

Electrical machines, drives, and their associated power electronics, namely, converters and capacitors, play a key role in an ever increasingly technological society. Transportation electrification, renewable energies, and more efficient buildings are just some of the areas where the intensive application of these systems has been most noticed.

This book will address, in the next five chapters, voltage source inverter (VSI)-fed drives, switched reluctance machine (SRM) drives, high-power synchronous machine drives, capacitors, and DC–DC converters.

VSI-fed drives, SRM drives, and high-power synchronous machine drives are extensively used, namely, in the aforementioned areas of transportation electrification and renewable energies.

Electrolytic capacitors and metallized polypropylene film capacitors are commonly found in the DC-link of the power converters of such drives.

DC–DC converters are facing an exponential growth in the context of the ever increasing use of DC microgrids in the homes and businesses, driven by the fact that the vast majority of renewable energy sources, electrical appliances, and storage devices operate either in true DC mode or at least involve an intermediate DC-link bus.

In all these applications, efficiency and reliability are of major concern. Reliability is a major challenge in these systems design, operation, and maintenance. Unreliable systems are not only the cause of users frustration but they also drive up the cost, so diagnostics and fault tolerance become important to help maintain the systems and estimate their operational life.

The scope of the book encompasses the issues related to fault analysis, fault detection and isolation, diagnostics, prognostics, condition monitoring, post-fault reconfiguration, remedial operation, robust control, and fault tolerance of electromechatronic systems.

[1]CISE – Electromechatronic Systems Research Centre, Universidade da Beira Interior, Portugal

1.1 Electromechatronics

Electromechatronics, introduced in the 1980s by Professor Yuri P. Koskin of the Department of Electromechanics and Electromechatronics of the Saint Petersburg State Electrical Engineering University, Russia, corresponds to the integration of electromechanical and electronic areas in a single technical/scientific field of electrical engineering [1].

Electromechatronics, term formed by agglutination, according to the following expression [1]

$$\text{ELECTROMECHATRONICS} = \text{ELECTROMECHA}\textsc{nics} + \textsc{elec}\text{TRONICS} \tag{1.1}$$

brings together the areas of electrical machines, drives, and their associated power electronics, namely, converters and capacitors.

1.2 Fault diagnosis

A fault may be defined as the condition of an equipment, material, or system, characterized by the termination of the ability to fully perform the required functions.

The following main categories of faults are distinguished [2]:

- Catastrophic – Characterized by a sudden occurrence and involving the total and immediate stoppage of the functions performed until then.
- Evolutionary – Associated with a gradual development and affecting, at first only partially, the performance of the functions.
- Intentional – Deliberately caused and involving the interruption of the performance of the functions, regardless of the registered condition.

Evolutionary faults are, therefore, the most appropriate to the application of early diagnostic methods.

Similarly to the clinical diagnosis, the diagnosis of faults involves the characterization of equipment state condition through the consideration of the symptoms it manifests. Thus, four intrinsic aspects to the diagnostic process are distinguished, as shown in Figure 1.1.

The detection of faults is the primary objective of the monitoring of parameters indicative of the fault occurrence. However, consideration of the remaining aspects presented in Figure 1.1 is not possible through the exclusive use of monitoring. Indeed, the detailed analysis of the information contained in the various indicators of the occurrence of faults becomes crucial for the development of methods capable of providing a complete and reliable diagnosis. Additionally, fault analysis, making possible the understanding of the various phenomena associated to a fault occurrence, also becomes essential [2].

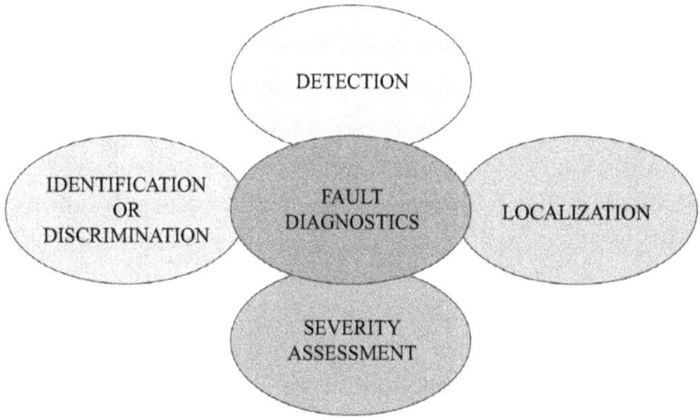

Figure 1.1 Intrinsic aspects to the fault diagnostics process

After fault detection, the identification or discrimination of faults is another important aspect to be considered in the diagnostics process. Depending on the particular type of fault identified, specific and more appropriate diagnostic approaches can be applied toward a complete and reliable diagnosis.

Fault localization is particularly important when it comes to the application of subsequent repair actions. Knowledge about the position of the fault eliminates the need to completely dismantle the whole equipment, thus reducing the repair time and costs. Therefore, a complete and reliable diagnosis should also provide information regarding the fault localization.

Fault severity assessment is another key aspect in the diagnostics process. Nowadays, risk analysis and decision support systems are widely recognized management tools that strongly rely on that piece of information.

1.2.1 Diagnostic methods

In accordance with the way in which they are applied, diagnostic methods may be grouped into

Off-line – Characterized by the need for the equipment to be out of service when they are applied, they even require, in most cases, that equipment should be disassembled, in order to make accessible some of its components.

Online – Diagnosis can be achieved without the need to resort to interrupting the operation of the equipment.

Obviously, online diagnostic methods are the most attractive. Among the online diagnostic methods, a further distinction can also be established between invasive and noninvasive methods. The former require that sensors have to be attached to the equipment structure or even inside, like accelerometers, search coils, thermistors, or thermocouples, while the latter are solely based on the information captured from sensors placed away from the equipment itself, like current probes or thermographic cameras.

1.3 Prognosis

Prognosis, or the anticipated knowledge, is the next step following diagnostics activities. It requires an accurate modeling of equipment-degradation mechanisms, and the manipulation of past and present condition related data, through suitable methods of analysis, in order to be able to predict equipment future condition, behavior, performance, or remaining useful life estimation.

It is therefore a scientific area where a deep knowledge of the equipment under analysis is required, together with the application of statistical techniques, estimation and identification techniques, numerical analysis, risk analysis, etc.

As far as electrical machines are concerned, prognosis has gained lately a focal research interest due to the importance of insulating materials' prognosis for motors used in transportation electrification, where reliability and safety are of major concern.

1.4 Fault tolerance

Similar to the fault diagnostics process, four aspects intrinsic to fault tolerance are also considered, as shown in Figure 1.2.

Fault diagnostics is the first step to be considered. Only after this, it is possible to isolate the faulty component(s) and define the most appropriate hardware/software reconfigurations to be adopted. Time-to-diagnostics and isolation is a critical aspect. Indeed, post-fault remedial operating strategies have to be implemented before a complete shutdown may occur. For that, suitable hardware and/or software reconfiguration approaches have to be almost instantly considered, always aiming at minimizing any additional hardware requirements. Indeed, the basic principle behind the fault tolerance concept is the guarantee of a continuous operation, although under an acceptable degraded mode, at the cost of minimum changes. Otherwise, the use of full-duplicated components can always be considered, but that is redundancy – i.e., the most primary form of fault tolerance.

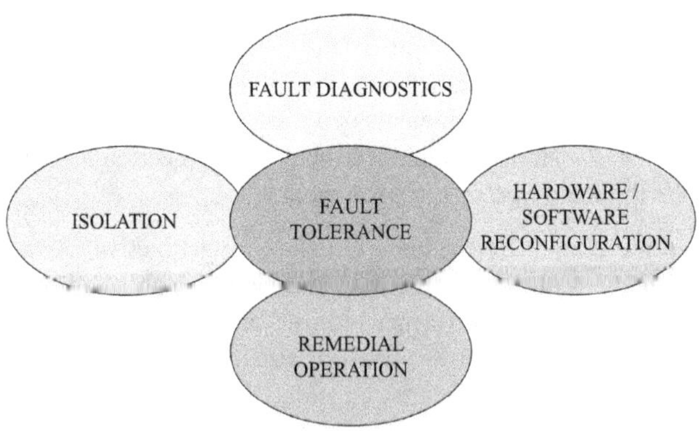

Figure 1.2 Intrinsic aspects to fault tolerance

The challenge is therefore twofold: reaching the maximum functionality at the cost of minimum changes.

1.5 Diagnosis and fault tolerance of electrical machines, power electronics, and drives

The next five chapters will address the issues related to diagnosis and fault tolerance of electrical machines, power electronics, and drives.

Chapter 2 focuses on VSI-fed drives. First, condition monitoring and fault diagnostics of electrical machines, particularly induction and permanent magnet machines, are considered. Eccentricity, inter-turn faults, broken rotor bars or end-rings, demagnetization of permanent magnets, and bearing faults are among the addressed types of machine faults. Fault prognosis is also considered. Second, fault diagnostic techniques applied to VSIs, particularly two-level VSIs, are addressed. Current-based fault diagnostic approaches and voltage-based fault diagnostic approaches are discussed. Fault-tolerant techniques applied to VSI-fed drives are also presented.

Chapter 3 is dedicated to SRM drives. First, the overall characteristics related to the constitution, operation, and control of SRM drives are introduced, followed by a comprehensive description of SRM drives fault analysis. Secondly, fault diagnostic techniques and fault-tolerant strategies, applied to SRM drives, are presented.

Chapter 4 addresses high-power synchronous machine drives. First of all, an overview is provided on the main technologies and design features which characterize large synchronous machines and the relevant supplying converters, also taking into account their field of application. Subsequently, the attention is focused on the major strategies intended to improve high-power synchronous machine drives fault tolerance, acting on the system-level drive architecture as well as on the design and operation of the individual components (electric motor, converter, control system). Finally, the main diagnostics and condition monitoring techniques for high-power synchronous machines drives is covered, describing the main methods to detect possible malfunctioning, anomalies, and faults in drive operation before they result in serious damages or hazards.

Chapter 5 deals with capacitors, one of the most vulnerable components of electromechatronic systems. Capacitors main technologies (electrolytic capacitors, film capacitors, and ceramic capacitors) are presented firstly. Subsequently, a particular emphasis is given to aluminum electrolytic capacitors and metalized polypropylene film capacitors, currently the most commonly used capacitors in the DC-link of power electronic converters. Capacitors diagnostic techniques are then introduced. Off-line, online, and quasi-online techniques are described in detail. At the end, some key ideas are presented, which synthesize the advantages and disadvantages of the discussed fault diagnostic techniques, and some envisaged advancements in this domain are also addressed.

Chapter 6 outlines the most important advances achieved in the development of fault diagnostic tools and fault-tolerant strategies aimed at DC–DC converters.

An exhaustive description of both signal-processing-based fault diagnostic algorithms and model-based-fault diagnostic algorithms is provided, as well as comprehensive summary of the most relevant features and limitations of the algorithms pertaining to each category. Next, the most relevant fault-tolerant architectures and control strategies developed to overcome the negative effects of the occurrence of faults in DC–DC converters are presented. Their applicability, main merits, and drawbacks are addressed and a comparative analysis of their main features is also provided.

Acknowledgment

This work was supported by the European Regional Development Fund (ERDF) through the Operational Programme for Competitiveness and Internationalization (COMPETE 2020), under Project POCI-01-0145-FEDER-029494, and by National Funds through the FCT – Portuguese Foundation for Science and Technology, under Projects PTDC/EEI-EEE/29494/2017 and UID/EEA/04131/2013.

References

[1] Koskin, Y. P.: 'The electromechatronics as the scientifical background of electromechanical converters and electronical components integration.' *Proceedings of the International Conference on Electrical Machines*, Vigo, Spain, 1996, Vol. III, pp. 513–518.

[2] Cardoso, A. J. M.: *Fault Diagnosis in Three-Phase Induction Motors* (in Portuguese), Coimbra, Editora, 1991, pp. 34–35.

Chapter 2
Voltage-source inverter-fed drives

Jorge Oliveira Estima[1] *and Konstantinos N. Gyftakis*[1,2]

2.1 Condition monitoring, fault diagnosis and prognosis of electrical machines

2.1.1 Introduction

Electrical machines have infiltrated and supported our everyday modern life. Electrical machines produce electric power working as generators or transform it into mechanical power operating as motors. Electrical machines are operating devices in power plants, wind farms, pumps, industry applications, cranes, conveyors, belts, mills, transportation and many other applications. So, it is to be expected that electrical machines are related to huge financial variables as well as safety and reliability.

Although electrical machines are robust devices, faults may appear and interrupt their working life cycle in many ways. Faults can be classified in three categories: stator related, rotor related and mechanical.

Stator faults include electrical failures which means short-/open-circuits, as well as inter-turn short-circuits which is a special case of short-circuits. Supply imbalance belongs in this category also. Other stator faults are iron core related ones.

Similarly, rotor faults may be of electrical nature when the rotor has windings and iron related ones. Other special faults are broken/cracked rotor bars/end-rings for cage induction motors, permanent magnet cracks or demagnetisation for permanent magnet motors and commutator/slip rings/brushes failures for all machines with rotor winding.

Mechanical faults mainly include bearing failures. However, in this category, we can also include issues with the cooling fan as well as irregularities with the connected mechanical load such as overloading and load oscillations.

Several surveys have been carried out in the past leading to percentages distributions of machine failures [1–3]. Representative results are shown in Figure 2.1. It is interesting that three out of four failures in low voltage motors are bearings related, while stator faults account for only 9% of total faults. The distribution is

[1]CISE – Electromechatronic Systems Research Centre, Universidade da Beira Interior, Portugal
[2]School of Computing, Electronics and Mathematics and the Research Institute for Future Transport and Cities, Coventry University, UK

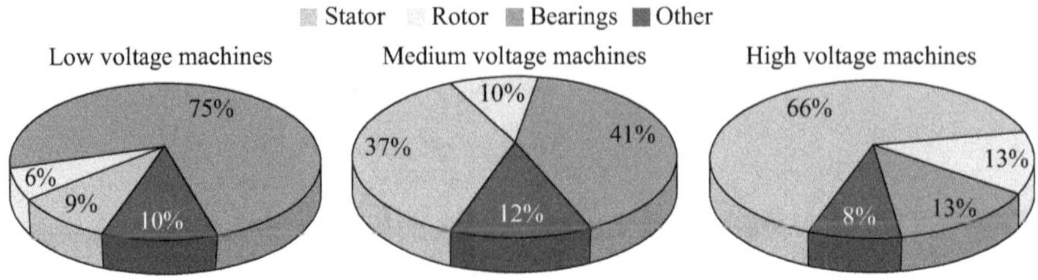

Figure 2.1 Distributions of electrical machine failures depending on the voltage supply level

exactly opposite when looking at high voltage motors where two out of three failures are stator related, while bearing faults account for 13% of total failures. This is due to the fact that large machines have sleeve bearings degradation of which is significantly slower than ball bearings. In medium voltage, motors stator and bearing faults are more or less of equal severity. Finally, in all cases, the rotor faults account for about 10% of total motor failures.

Due to the significance of the electrical machines uninterrupted operation and the negative impacts of failures, the area of electrical machines' condition monitoring has flourished during the last 30 years and has known tremendous development and progress. However, the plethora of machine sizes, geometries, components and applications have been reported to create unique and special diagnostic cases where misdiagnosis may happen. Moreover, new applications constantly appear, where electrical machines and drives are the key components and as such, their reliable operation is of high importance. Typically new applications require proper adjustment and reconfiguration of existing diagnostic procedures or even completely new diagnostic approaches, and this is one more reason for continuous active research in this field.

2.1.2 Condition monitoring, fault diagnosis and prognosis

Condition monitoring of electrical machines is a broad scientific area, the ultimate purpose of which is to ensure the safe, reliable and continuous operation of electrical machines. Condition monitoring can be divided into two sub-areas, namely fault prognosis and fault diagnosis.

Prognosis (in Greek: "Πργνωσις") is a complex Greek word from "pro" which means before and "gnosis" which means knowledge. Fault prognosis is the scientific area which aims to predict failures before they happen. The estimation of the remaining useful life (RUL) of a component or a device is the main goal of the fault prognosis area. For this reason, prognosis is strongly related to material science and degradation.

On the other hand, diagnosis (in Greek: "Διάγνωσις") is also a complex Greek word from "dia" which is the term for division and "gnosis" for knowledge. So, the term diagnosis is related to penetrating into the problem to get knowledge. Fault diagnosis assumes that a fault has already happened in a device, while the final goal is to detect the fault with an appropriate diagnostic procedure. In electrical

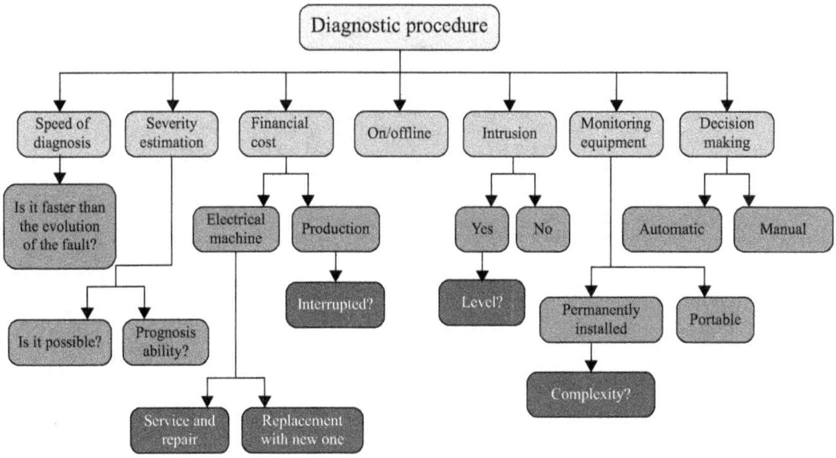

Figure 2.2 Characteristics of the diagnostic procedure

machines a fault will automatically create an asymmetry in the magnetic field. This asymmetry will pass on to various electromagnetic variables like the currents, voltages, magnetic flux, electric and mechanical power, torque and speed. So, the diagnostics engineer needs to monitor and analyse some of the above variables and detect any divergence from the expected healthy machine characteristics. Figure 2.2 illustrates the most important characteristics of the diagnosis procedure.

2.1.3 Fault diagnosis of electrical machines

As there are many different machines, drives and applications, it is only natural that there is also a plethora of electrical machines faults and diagnostic methods. Amongst them the most favourable is the motor current signature analysis (MCSA) which is the analysis of the stator current harmonic index [4,5]. Most define the MCSA as the monitoring and spectral analysis of the stator current at steady state. Despite the method's origins, the name is very generic and should include the analysis of the stator current spectra under transient operation also. Anyway, this method has become favourable due to its unique characteristics such as remote monitoring [6] (Figure 2.3), low implementation costs and equipment, and continuous and online monitoring capability. However, many other methods have been proposed and they rely on the monitoring of other variables such as the magnetic flux [7], torque [8], electric power [9], voltage [10], etc. As a priority, in this chapter, all MCSA formulas will be given for most common faults. However, other methods will be also discussed and analysed.

To facilitate reading and comprehension, this section will be organised in the following strategy. First, faults that are common in both induction and permanent magnet machines will be discussed and the different diagnostic strategies applied for each machine will be properly analysed. Then faults uniquely existing in different machines will be discussed.

Figure 2.3 *Remote monitoring of motor currents. © 2017–2018 IEEE. Reprinted, with permission, from Reference [6]*

2.1.3.1 The eccentricity fault

Eccentricity is the condition where the rotor is abnormally positioned inside the stator, and as a consequence, the air-gap around the rotor circumference is not symmetrical [11]. There are mainly two types of eccentricity: static and dynamic.

In both eccentricity types, the geometrical centre of the rotor is different than that of the stator. Moreover, in the case of the static eccentricity, the centre of rotation is fixed in space and coincides with the rotor geometrical centre [Figure 2.4(a)]. However, in the case of the dynamic eccentricity, the centre of rotation does not have a fixed location but constantly changing in space over time [Figure 2.4(b)]. A combination between the two above-mentioned conditions is called mixed eccentricity. It is important to note that some inherent mixed eccentricity always exists even in new electrical machines [12]. The maximum permitted level of inherent eccentricity is 10%, although in most cases, manufacturers put an effort to keep it significantly less than that.

The static eccentricity is a fault usually introduced during the manufacturing stages of the motor. It can be caused by the ovality of the stator or by the misplacement of the rotor in the stator. On the other hand, the dynamic eccentricity is usually related to bearing failures or bent motor shaft. If not detected at an early stage, it will evolve and lead to the rubbing between rotor and stator, which will cause irreparable damage and possible deformation of the electrical machine iron core (Figure 2.5). Usually, deformation of an electrical machine's iron core leads to long service period and is expensive.

The eccentricity causes an asymmetry of the machine's air-gap geometry, which directly influences the air-gap permeance and will cause an asymmetry in the machine's rotating magnetic field. That asymmetry is expressed via enhanced higher harmonics in the machine's electromagnetic and mechanical operating variables and characteristics.

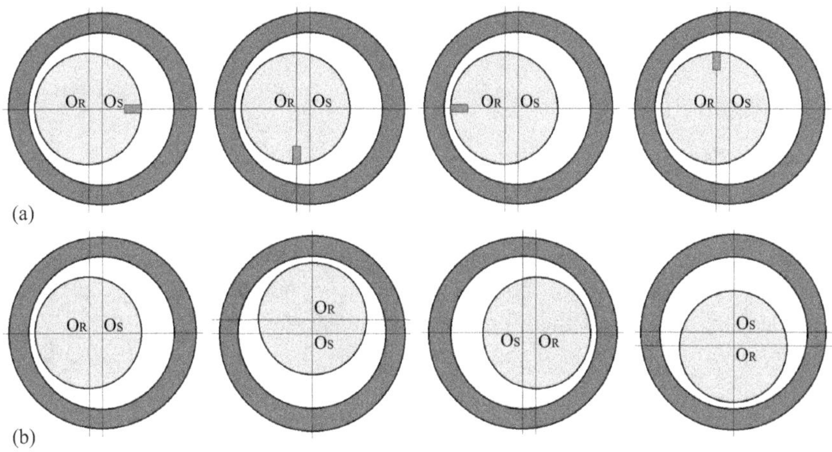

Figure 2.4 *Four different instances of the rotor rotation for (a) static and (b) dynamic eccentricity*

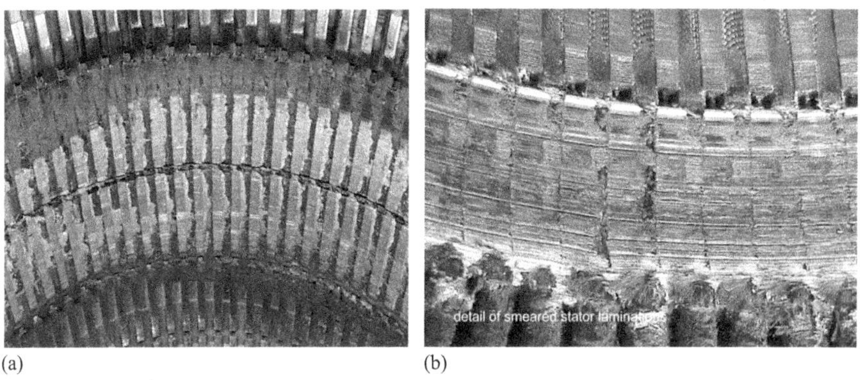

Figure 2.5 *Major mechanical damage to the stator due to rotor rubbing. Not repairable unless (a) the core is dismantled and repaired and (b) the core is restacked or replaced [13] (with permission from EASA)*

Induction motors

For induction motors, the static or dynamic eccentricity fault can be detected through the monitoring of signatures in the stator current located at frequencies [14]:

$$f_{ecc\_IM} = \left[(kR \pm n_d)\left(\frac{1-s}{p}\right) \pm n\right] f_s \tag{2.1}$$

where R is the rotor slot number, k is the integer, s is the slip, p is the pole pairs, n is the stator harmonic ranks, f_s is the supply frequency and n_d is an integer which is zero for static eccentricity and non-zero for dynamic eccentricity.

Furthermore, the following formula [11] has also been proposed to detect the mixed eccentricity fault in the low-frequency area of the stator current:

$$f_{ecc\_IM2} = f_s \pm f_r \tag{2.2}$$

where $f_r = [(1-s)/p]f_s$, that is the mechanical rotation frequency.

It is important to note that formula (2.1) does not always offer reliable results for the cases of only-dynamic or only-static eccentricity diagnosis [15]. Previous works have pointed out that diagnosis is possible only for certain combinations between the rotor slot and the pole pair numbers [16,17]. To be more specific, if $n_d = 0, k = 1$ then formula (2.1) transforms into:

$$f_{PSH} = \left[R\left(\frac{1-s}{p}\right) \pm n\right]f_s \tag{2.3}$$

Equation (2.3) corresponds to the location of the principle slot harmonics (PSH) in the stator current frequency spectrum. Induction motors with a rotor slot number multiple of the pole pair number produce such harmonics in the line current. For those motors, the only-static and only-dynamic eccentricity faults cannot be detected because the fault signatures are located at the same frequencies as the normally existing PSH.

However, for non-PSH induction motors, if the rotor slot number is even, it is possible to monitor the only-static and only-dynamic eccentricities at low or no-load operation. Formula (2.1) is very reliable for non-PSH induction motors with odd rotor slot numbers [15].

Figure 2.6 illustrates the application of MCSA to detect a mixed eccentricity faulty condition in a four-pole cage induction motor with 28 rotor bars.

Permanent magnet machines

When there is eccentricity in permanent magnet (PM) machines, a portion of the stator is closer to the PM of the rotor, thus generating a net attraction force acting on the rotor [19]. As a result, an unbalanced magnetic force is generated between the rotor and the stator. Unbalanced forces may be the most troubling problem when brought by rotor eccentricity. That is because they may cause significant levels of vibration and noise which accelerates the motor degradation [20].

Previous studies have shown that eccentricity affects interior PM (IPM) motors differently than surface mounted PM (SPM) motors [21]. It has been shown that, the eccentricity distorts the air-gap magnetic flux density more in the case of the IPM motors. In the same work, authors came to the conclusion that the magnetic unbalanced forces increase linearly due to relatively small eccentricity ratio in SPM motors. However for the IPM motors, they increase significantly and non-linearly due to severe magnetic saturation with the eccentricity level.

Furthermore, the eccentricity effects are different between symmetrical and asymmetrical PM machines [20]. Rotor eccentricity has a minor impact on asymmetric motors in terms of the magnitude of the radial force. Low detent

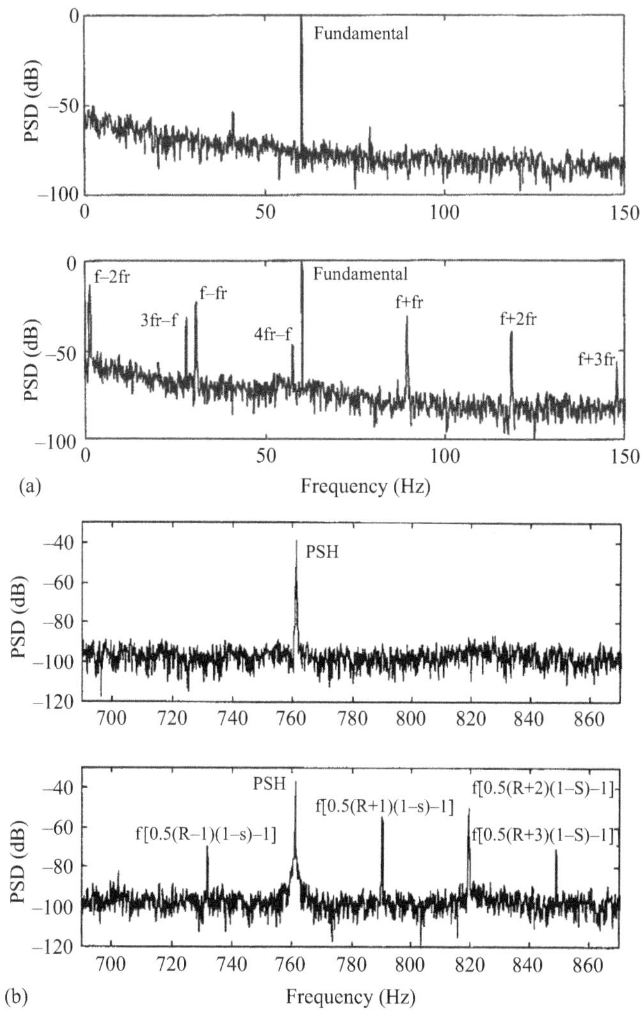

Figure 2.6 Simulated, normalised spectra of the line current of a four-pole induction machine with 28 rotor slots under load (a) around the fundamental and (b) around the PSH. Upper: healthy. Lower: with mixed eccentricity (41.37% SE, 20.69% DE). © 2017–2018 IEEE. Reprinted, with permission, from Reference [18]

torque, the primary reason for using asymmetric motors, is magnified if rotor eccentricity is involved, so the importance of manufacturing precision cannot be overstressed.

One more important finding was reported in [22]. The results reveal that the static eccentricity and the uneven magnetisation cause exactly the same harmonic index of the unbalanced magnetic force. However, in dynamic eccentricity conditions, the amplitude of the DC component has been reported to rise significantly.

Regarding the diagnosis of eccentricity, formula (2.4) was initially proposed for static eccentricity diagnosis via the stator current harmonic index [23]. Further investigation led to the conclusion that the formula is also reliable for the cases of dynamic and mixed eccentricity [24,25]. Results from the application of (2.4) can be seen in Figure 2.7:

$$f_{ecc\_PM} = \left(1 \pm \frac{(2k-1)}{p}\right) f_s \tag{2.4}$$

where k is the integer, p is the pole pairs and f_s is the supply frequency.

In a later work, the following formula was also proposed, the concept of which is very similar to that of (2.2). That means examination of the sidebands around the

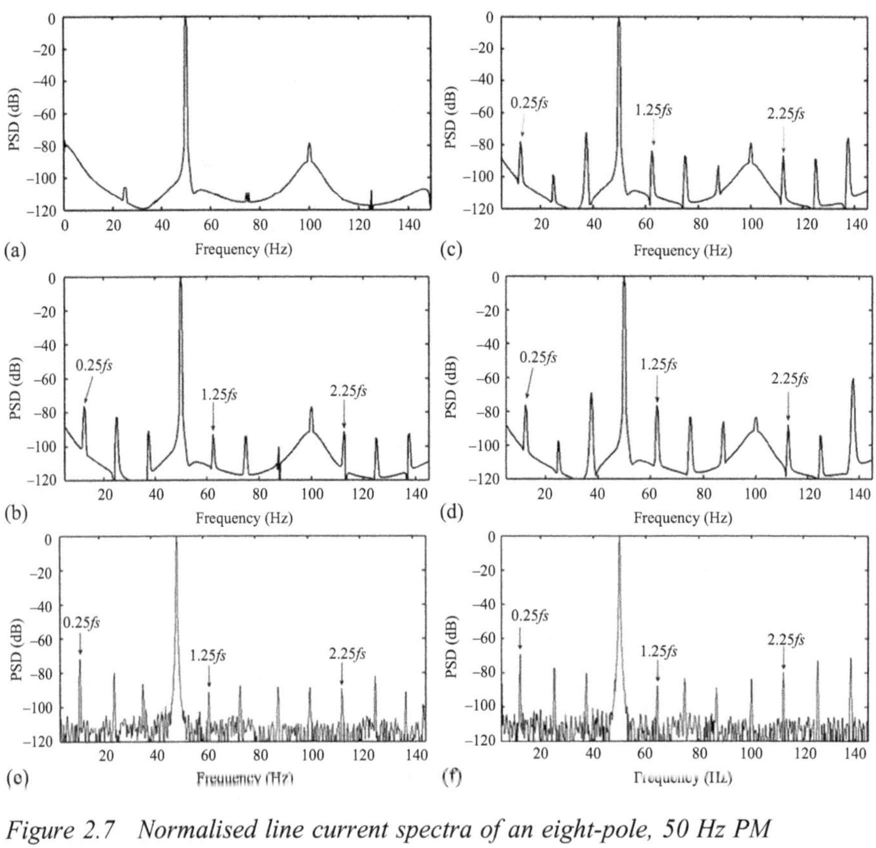

Figure 2.7 Normalised line current spectra of an eight-pole, 50 Hz PM synchronous motor where: (a) healthy, (b) with 30% SE, (c) with 40% DE, (d) with 30% SE and 40% DE, (e) experimental result for 50% DE and (f) experimental result for 50% SE. © 2017–2018 IET. Reprinted, with permission, from Reference [28]

fundamental stator current harmonic located at ± the mechanical rotation frequency [26]:

$$f_{ecc\_PM2} = \left(1 \pm \frac{1}{p}\right) f_s \tag{2.5}$$

A different later work though pointed out that (2.5) describes harmonics produced not only by eccentricity but also demagnetisation and rotor and load imbalances [27], which does not allow reliable discrimination and identification of the exact fault condition.

2.1.3.2 The stator inter-turn fault

Stator electrical faults can be either short or open-circuit failures. Short-circuit faulty conditions include phase-to-phase and phase-to-ground short-circuits. However, in both open and short-circuit conditions, the machine will suffer from high currents and will be severely damaged, thus many protection methods exist the aim of which is to trip the protection relays and disconnect the machine immediately when the fault appears. So, the above-mentioned conditions are not to be discussed any further in this section as they are not related to early fault stages where the prompt diagnosis is meaningful. Instead, this section will focus on the stator inter-turn faults.

Stator inter-turn faults happen between stator winding wires of the same phase because of degradation of the insulation materials [29,30]. The stator inter-turn fault is considered an early fault condition which will evolve into an actual short-circuit and lead to machine damage. However, the term "early" can be misleading because it has been shown that this fault evolves very fast (in seconds [31]) into higher severity levels leading to an actual short-circuit condition and triggering the protection relays to disconnect the machine. Different degradation mechanisms might affect the end-winding portion or the slot portion [32]. In addition, it should be noted that the time progression and extent of the damage depends on the location of the fault and the original number of shorted turns [33].

In order to understand the motor winding configuration in the case of an inter-turn fault, Figure 2.8 is presented, and one can see two neighbouring stator slots where the winding turns go and then the two neighbouring stator slots where the same winding turns return. Each slot's conductors/wires are divided in three groups. In the first and last slots, there are k turns, one single turn and j turns. In the inner slots, there are m turns, a single turn and n turns. It is supposed that the short-circuit happens between the points A and B. Then it is evident that, from left to right, the four slots contain $k + 1$, n, $n + 1$ and k wires, respectively, which belong to the healthy part of the winding. The other wires, which mean j, $1 + m$, m and $j + 1$, form a closed loop where the short-circuit current will develop. It is to be noted that when modelling this condition, a small resistance should be added between the points A and B to account for the contact resistance between the shorted turns. Typically, the contact resistance value is between 0.1 and 1 Ω. Figure 2.9 illustrates a real case of an inter-turn fault which happened to an industrial induction motor.

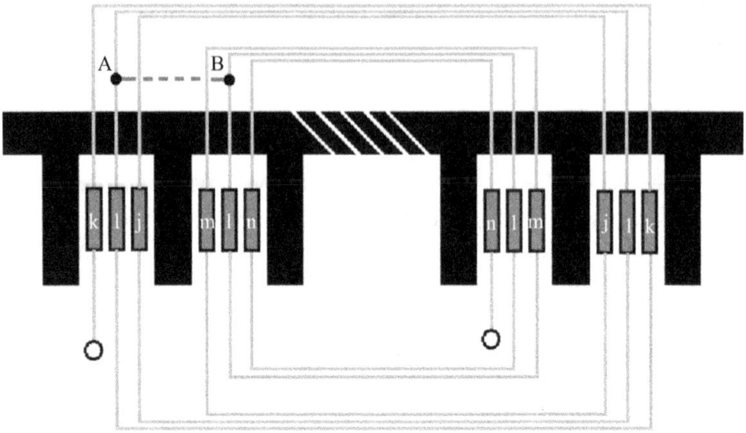

Figure 2.8 Winding distribution in the case of an inter-turn short-circuit

Figure 2.9 Burned out turns of an induction motor's stator winding due to an inter-turn fault (courtesy of Mr M. Thumpy)

Induction machines

When there is an inter-turn short-circuit fault, the resistance of the faulted phase drops by a portion which depends on the fault level severity or in other words the resistive part of the phase which now belongs to the shorted loop. This means that when the machine is supplied by a symmetrical, three-phase, voltage source, the faulty phase will draw more current than the other two. As a result, there is an imbalance between the three phase currents which in principle means an asymmetrical rotating magnetic field. The negative sequence current interacts with the fundamental slip frequency current in the rotor to produce torque pulsation at double the supply frequency [34]. The consequent speed ripple induced harmonic index back to the stator with frequency three times the fundamental one. Due to the above, the stator current will experience an increase of the third harmonic [34–37]. This is clearly shown in Figure 2.10. Saturation plays an important role as it can enhance the amplitude increase of the third harmonic in the stator current [37].

However, the third harmonic increase is also associated with other imbalances [38] such as asymmetrical three-phase voltage supply, inherent asymmetry between the three phase windings, high resistance connections which lead to unbalanced phase currents, etc. Moreover, it was shown that the increase of the third harmonic

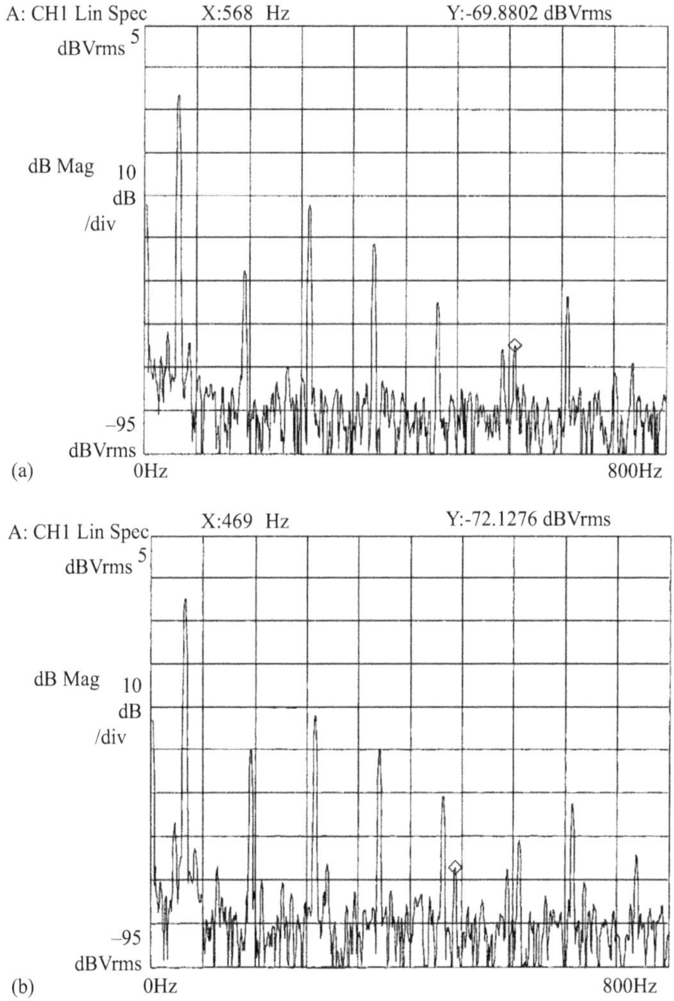

Figure 2.10 Spectral content of the line current of a: (a) healthy motor and (b) motor with stator inter-turn fault operating under $s = 0.028$. © 2017–2018 IEEE. Reprinted, with permission, from Reference [34]

in the stator current does not have a monotonic relation to the fault level severity [39]. So, the monitoring of the third harmonic is not deemed as very reliable for very low severity levels of inter-turn short-circuits as it may lead to a false negative alarm.

Practically, when there is a stator inter-turn fault, the created asymmetrical rotating magnetic field will induce asymmetrical currents in the rotor. As a result, rotor slot related harmonics will rise. So, past works [40] have proposed the following formula for the detection of stator inter-turn faults:

$$f_{sc\_IM} = \left[kR\left(\frac{1-s}{p}\right) \pm 2n_{sa} \pm n\right]f_s, k \in \mathbb{N} \tag{2.6}$$

where R is the rotor slot number, k is the integer, s is the slip, p is the pole pairs, n is the stator harmonic ranks, f_s is the supply frequency and n_{sa} is the rank of the saturation harmonics.

An interesting method able to detect the fault existence as well as the faulty phase with low computational time is the Park's vector approach (PVA) [41]. The method relies on the analysis of the Park's vector components i_d, i_q as follows:

$$i_d = \frac{\sqrt{2}}{\sqrt{3}}i_a - \frac{1}{\sqrt{6}}i_b - \frac{1}{\sqrt{6}}i_c \tag{2.7}$$

$$i_q = \frac{1}{\sqrt{2}}i_b - \frac{1}{\sqrt{2}}i_c \tag{2.8}$$

Under ideal conditions, the three phase currents i_n (where $n = a, b, c$) lead to a Park's vector with the following components:

$$i_d = \frac{\sqrt{6}}{2}i_M \sin \omega t \tag{2.9}$$

$$i_q = \frac{\sqrt{6}}{2}i_M \sin\left(\omega t - \frac{\pi}{2}\right) \tag{2.10}$$

The corresponding representation is a circular locus centred at the origin of the coordinates. Under abnormal conditions, (2.9) and (2.10) are no longer valid, and consequently the observed picture differs from the reference pattern. This can be seen in Figure 2.11 where i_d is on the x axis and i_q on the y axis, respectively. The displacement of the locus reveals the faulty phase.

However, the PVA on its own cannot offer an easy measure of determining the fault level severity with accuracy. That is because the determination of the locus angle shift is influenced by other parameters as well. This is why, the method evolved into the extended PVA (EPVA) which relies on the monitoring of the frequency spectra of the Park's vector modulus [42]. It was found that, the stator inter-turn fault gives rise to harmonics located at twice the supply frequency in the Park's vector modulus spectra (Figure 2.12).

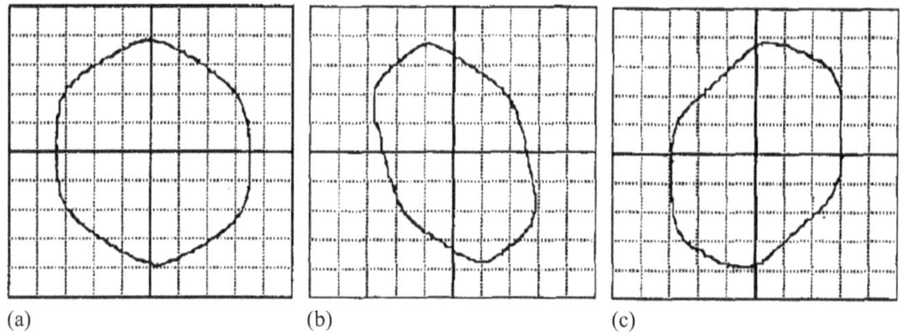

Figure 2.11 Experimentally derived Park's vector pattern for (a) healthy motor, (b) motor with 18 shorted turns in Phase A and (c) motor with 18 shorted turns in Phase B. © 2017–2018 IEEE. Reprinted, with permission, from Reference [41]

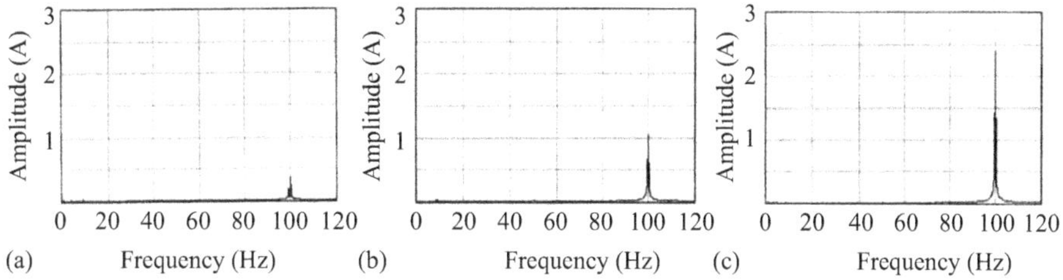

Figure 2.12 EPVA signature corresponding to: (a) healthy motor, (b) motor with 12 shorted turns and (c) motor with 36 shorted turns. © 2017–2018 IEEE. Reprinted, with permission, from Reference [42]

Permanent magnet machines

PM machines are widely used in various special applications where high power density and efficiency is required, while restrictions on the size of the motor apply, such as electric vehicle propulsion, aerospace applications, etc. Due to their critical role associated with safety, prompt and reliable diagnosis is required. This makes the inter-turn short-circuits failure quite a timely and challenging issue because it is the type of fault that progresses fast leading to undesired motor breakdown. It was shown that the inter-turn fault may cause excessive heat that is proportional to the square of the circulating current in the shorted turns [43].

This type of the PM machine plays an important role for the impact of the inter-turn short-circuit faults. Past contributions have shown that the short-circuit current is lower in IPM than SPM machines [44]. This is crucial for the evolution time of the fault. The short-circuit current is relatively high and will generate heat. This will lead not only to extension of the inter-turn fault including more and more turns until the motor breakdown bus also demagnetisation of the PMs [45] (Figure 2.13). However, the required motor toque is the same which leads to higher

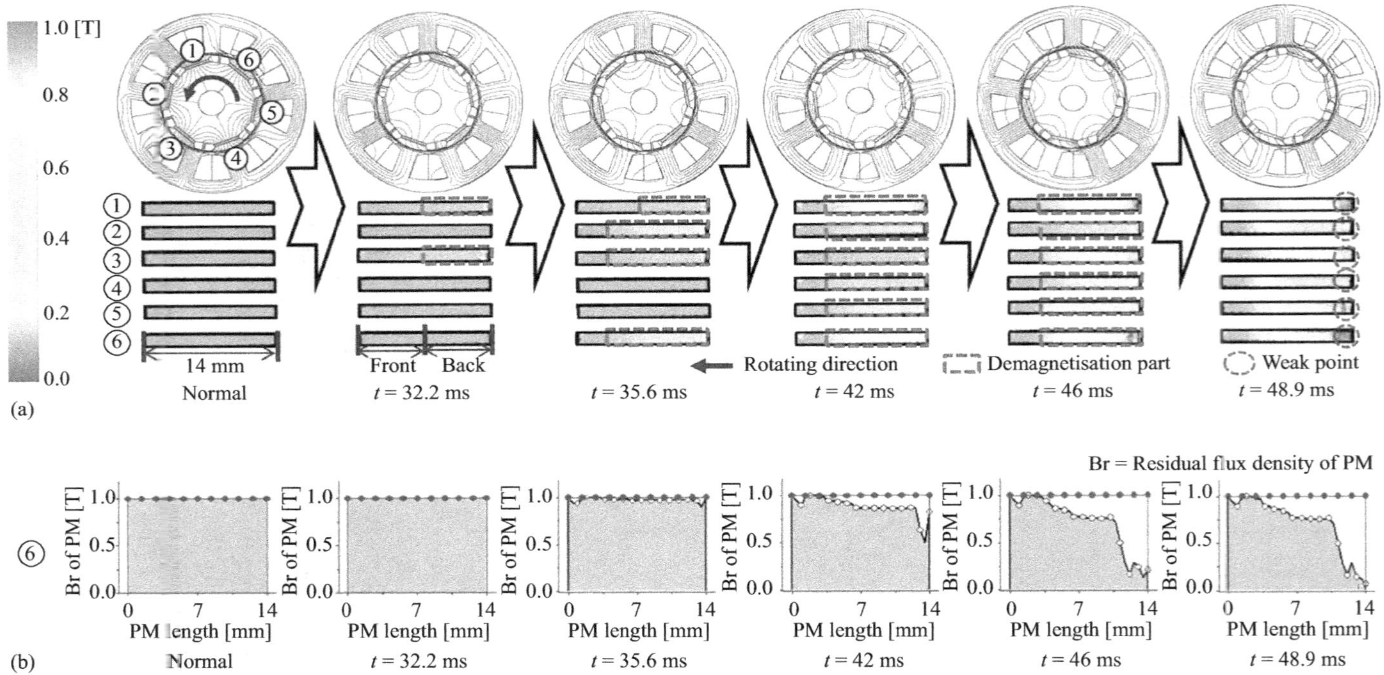

Figure 2.13 (a) Irreversible demagnetisation progress of six PMs while the BLDC machine operates under inter-turn short-circuit fault. (b) Residual flux density of the sixth PM over time. © 2017–2018 IEEE. Reprinted, with permission, from Reference [45]

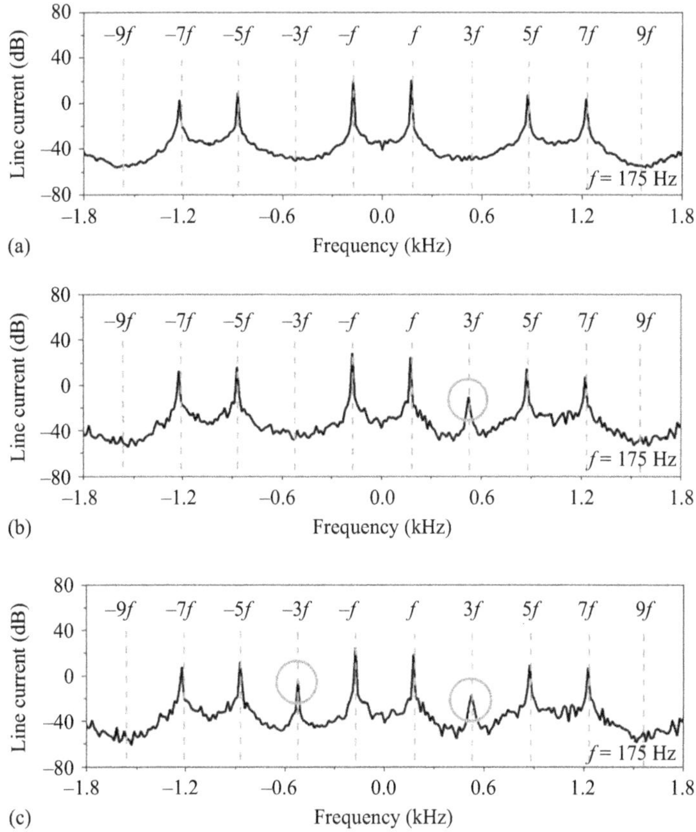

Figure 2.14 Line current frequency spectra of a PM motor at a rated speed and full load where: (a) healthy motor under balanced supply, (b) healthy motor under imbalanced supply and (c) motor with inter-turn fault under balanced supply. © 2017–2018 IEEE. Reprinted, with permission, from Reference [46]

current to serve the load need. Higher current will lead to more heating accelerating both demagnetisation and windings insulation degradation.

Similar to induction motors, the third line current harmonic as well as the other odd triplet harmonics are expected to rise with the inter-turn fault due to the three phase currents asymmetry, while absent in healthy motors [46]. This is shown in Figure 2.14. At the same time, it can be seen that the negative third harmonic frequency increases only in the case of inter-turn fault.

Furthermore, formula (2.11) has been proposed [44] for the detection of inter-turn faults in PM machines through the spectral analysis of the stator current at steady state. It is evident that for $k = 1$, the formula depicts the fault related

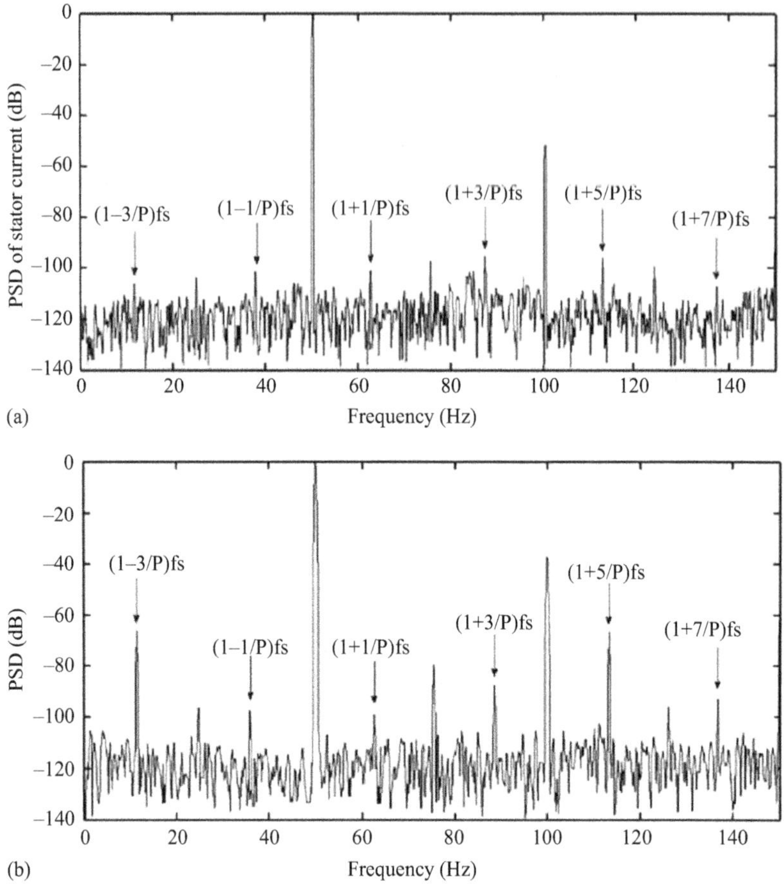

Figure 2.15 Normalised line-current spectra of full-load PMSM in (a) healthy and (b) motor with 1 short-circuited turn. © 2017–2018 IEEE. Reprinted, with permission, from Reference [44]

harmonics around the fundamental stator current harmonic. An example from the application of the formula in a real PM motor is shown Figure 2.15:

$$f_{itsc\_PM} = \left(k \pm \frac{2m+1}{p}\right) f_s \quad , k, m \in \mathbb{N} \tag{2.11}$$

2.1.3.3 Broken rotor bars or end-rings

There are two types of rotor squirrel cages of induction motors; fabricated and cast. Usually, low voltage induction motor rotors are cast aluminium, and high voltage ones are fabricated from copper. Medium voltage induction motor rotors can be of both types. Usually aluminium rotors have skewed bars, while in copper rotors, the bars are usually parallel to the shaft. Another difference between them is that in aluminium rotors, there is no insulation between the bars and the rotor iron core. On the other hand, in copper rotors the bars are firstly insulated and then placed inside the slots of the iron core. Those differences play an important role in the area of diagnostics as it will be discussed below.

A crack or breakage in an aluminium cage usually originates from improper casting which allows air bubbles inside the cage. This phenomenon is known as porosity [47]. These air-bubbles result in local high-resistance areas that cause hotspots and make the cage prone to local breakage [48].

On the other hand, copper rotors bars usually break due to thermal and mechanical stresses. First, the thermal stress will cause thermal expansion of the bars which might disconnect from the end-ring. Second, the mechanical stresses such as vibrations and frequent start-ups may lead to the same result [49]. However, the co-existence of both is probably the reason while the one mechanism enhances the catastrophic effects of the other.

When there is a broken rotor bar, the adjacent bars are overcharged, thus expected to break next [50,51]. This is shown in Figure 2.16. This is the usual case in aluminium rotors; however, multiple cases of non-adjacent broken rotor bars have been reported in large industrial induction motors [52]. Broken rotor bars do not normally result in an immediate failure of the motor. If the fault goes unnoticed and enough bars break then there is a chance that the motor will not be able to develop enough starting torque to accelerate from stall. However, in the past, some catastrophic failures have been reported like the one shown in Figure 2.17 where the rotor bars bent and severely damaged the stator winding [53].

When there is a broken rotor bar fault, two counter rotating magnetic fields are created with slip frequencies sf_s and $-sf_s$. The first one does not interact with the stator, while the second one induces components of frequency $-2sf_s$ back to the stator windings. As a result, the broken rotor bar fault can be identified in the stator current spectrum via the existence of harmonics at $(1-2s)f_s$ [55]. Furthermore, due to the speed ripple effect, a second broken bar fault harmonic appears to the right of the fundamental frequency at frequency: $(1+2s)f_s$ [56]. This interaction between the mechanical and electromagnetic quantities continues, and as a result multiple fault-related signatures are created at equal frequency distances $2sf_s$ from one another. As a result, the following formula has been proposed for the identification of the broken rotor bar fault around the fundamental stator current frequency:

$$f_{bb} = (1 \pm 2ks)f_s, \quad k \in \mathbb{N} \tag{2.12}$$

The stator current is rich in harmonics due to the saturation and other interacting phenomena. The result is that odd multiples of the supply frequency exist in the stator current. Those higher harmonics create additional magnetic fields inside the induction motor which are rotating with higher speeds due to their higher frequencies. Those magnetic fields also interact with the broken rotor bar fault, and as

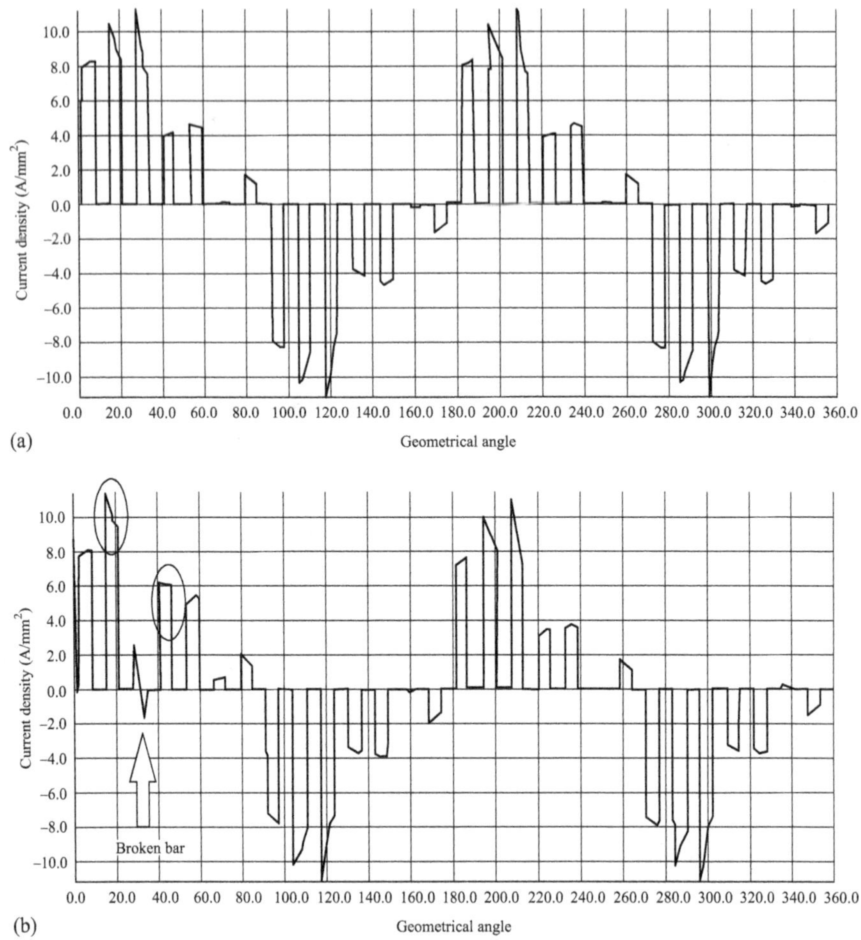

Figure 2.16 Spatial distribution of the amplitude of the current density of the rotor bars along the rotor circumference for (a) healthy machine and (b) machine model with one broken bar. © 2017–2018 IEEE. Reprinted, with permission, from Reference [54]

a result more signatures are created. The following formula has been proposed to include the broken bar fault sidebands at higher harmonics [57]:

$$f_{bb2} = \left[\frac{k}{p}(1-s) \pm s\right] f_s, \quad \frac{k}{p} \in \mathbb{N} \tag{2.13}$$

Figure 2.18 illustrates the application of (2.12) and (2.13) in order to detect a broken rotor bar fault using the stator current frequency spectra. It is evident that the fault creates specific fault-related harmonic sidebands around the odd multiples of supply frequency.

Figure 2.17 *Forced outage of 3.3-kV, 450-kW gasoline transfer pump induction motor due to rotor bar damage. (a) Rotor bar detachment from end ring and damage in rotor core due to arcing. (b) Damage in stator end winding due to protrusion of rotor bar. © 2017–2018 IEEE. Reprinted, with permission, from Reference [53]*

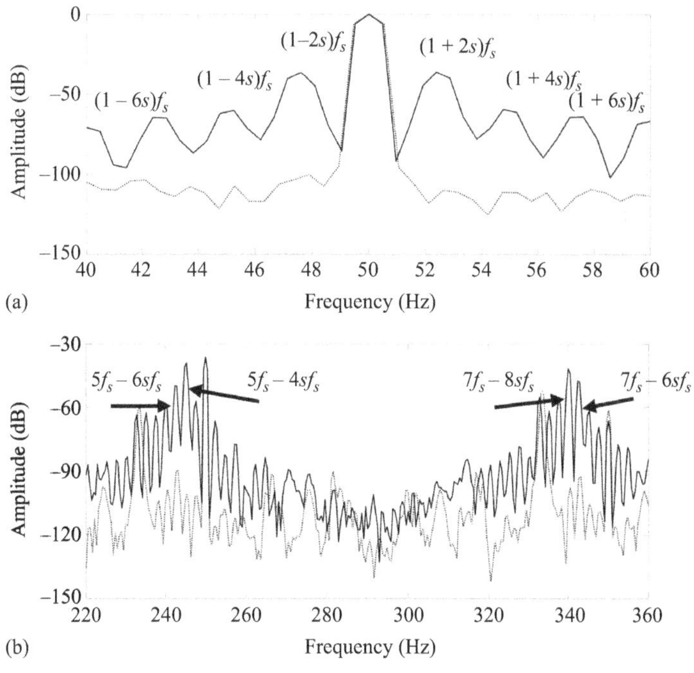

Figure 2.18 *Comparative frequency spectra of the healthy induction motor (dashed) and one with a broken bar fault (solid) around (a) the fundamental stator current harmonic and (b) the fifth (250 Hz) and seventh (350 Hz) stator current harmonics (FEM result)*

Lately, a lot of work has been focused on broken rotor bar fault detection. One of the reasons is that some phenomena exist, which can produce broken rotor bar fault harmonics in the stator current spectra of healthy motors. The misdiagnosis may lead to false positive alarms which may result in high costs for inspection and service without need. Such cases are the following:

- Mechanical load oscillations [58]
- Magnetic anisotropy of the rotor iron core [59]
- Axial cooling rotor air ducts (Figure 2.19) [60]
- Fan blades number in pumping applications [61]

Furthermore, earlier in this section, it was mentioned that in large induction motors, cases have been reported with multiple broken rotor bars (Figure 2.20). Moreover, in some cases, the rotor bars might break in non-adjacent positions. It has been reported that, the stator current analysis is not capable to provide information regarding the health of the rotor cage via the proposed frequency component $(1 - 2s)f_s$ if the broken bars are located electrically $\pi/2$ rad away with respect to each other [51,52].

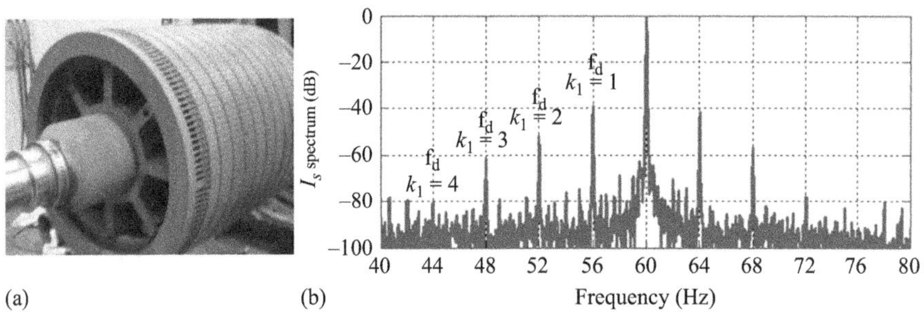

Figure 2.19 (a) A rotor with axial cooling air ducts. (b) Example of the stator current spectrum of a healthy motor (kidney holes) with four magnetic poles and equal axial air ducts operating under rated load conditions. © 2017–2018 IEEE. Reprinted, with permission, from Reference [60]

Figure 2.20 Rotor of a 5-MW, 6-kV cage motor with multiple broken bars. © 2017–2018 IEEE. Reprinted, with permission, from Reference [52]

Mainly due to such unreliability of the traditional diagnostic approach of the fast Fourier transform application at steady state, to detect rotor-related faults, a new trend has been developed. Special focus is now given to the monitoring and analysis of the stator current during start-up. The proposed methods have defined a new area called transient MCSA [62]. In this family, many methods have been proposed so far: the short-time Fourier transform [63], the MUSIC [64] and the Wavelet transform [62,65]. The main idea behind the application of such methods is that the broken bar fault harmonics are slip dependent. As a result, in a time-frequency decomposition, the trajectory of those harmonics frequency will vary versus time. This is not the case for the stator-related harmonics. This allows for the detection of the fault during transients. Another advantage of the stator current monitoring at start-up is that the rotor current gets its maximum, and as a result, rotor electrical faults are magnified during start-up with respect to the steady state. Aiming to illustrate the application of such methods for the detection of broken rotor bars, Figure 2.21(a) shows the application of the analytical wavelet transform (AWT), whereas Figure 2.21(b) shows the application of the discrete wavelet transform (DWT).

2.1.3.4 Demagnetisation of permanent magnets

Demagnetisation is the condition at which the PMs lose partially or fully their ability to magnetise. Motor applications involving PM motors require high power and torque density which cannot be usually supported by electrical machine types applying only electromagnets. If the PMs are demagnetised, the magnetic flux density drops, and with it the output torque and mechanical power. So, the PM machine loses its main selection purpose. It is to be noted that the demagnetisation effect could be irreversible, depending on the operating conditions. As a result, the reliable detection of this fault is crucial for the safe and normal operation of the PM machine.

Before discussing the diagnosis of the demagnetisation any further, it is important to explain the phenomenon and its dependence on temperature. The normal curve of a PM material is shown in Figure 2.22(a), following the line described by the points B_r, a, a', a'', H_c. The curve can be assumed to be linear for high magnetic flux density values followed by a sharp drop until the point of coercive magnetic field strength H_c. The magnet operates at the intersection of the demagnetisation curve and the load characteristic line. Normally, the intersection point exists on the linear part of the curve (around the point a'). However, an increase in the load will drive the operation point below the knee of the curve (around the point a''). If that happens, the PM will not be able to fully recover its former remanence magnetic flux density when the demagnetising effect disappears. Instead it will be characterised by a new B_r' which is less than the original B_r.

A similar approach is followed to explain the irreversible demagnetisation due to the temperature increase. The impact of the temperature on the demagnetisation curve of the PM is shown in Figure 2.22(b). An increase in temperature leads to a new curve characterised by less coercive magnetic field strength as well as less remanence magnetic flux density. As a result, the same load line, which for

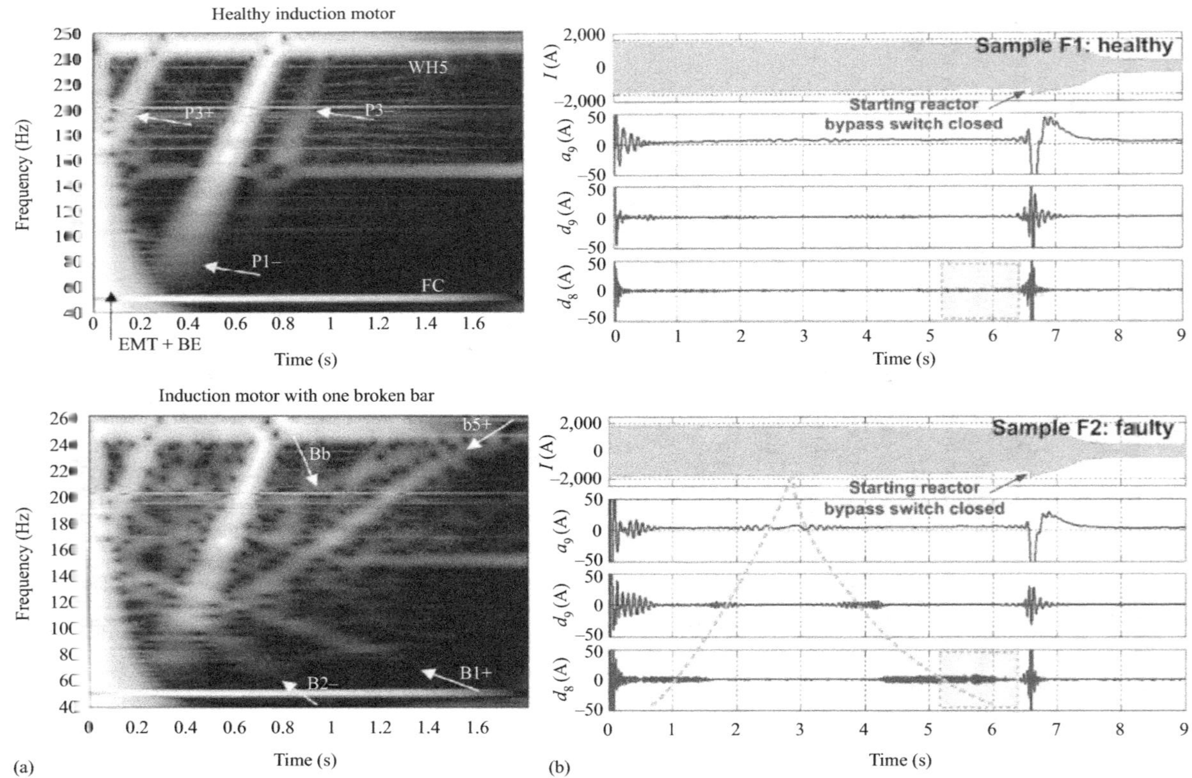

Figure 2.21 *Application of the: (a) AWT [62] and (b) DWT [65] to detect a broken rotor bar during an induction motor's start-up. © 2017–2018 IEEE. Reprinted, with permission, from References [62,65]*

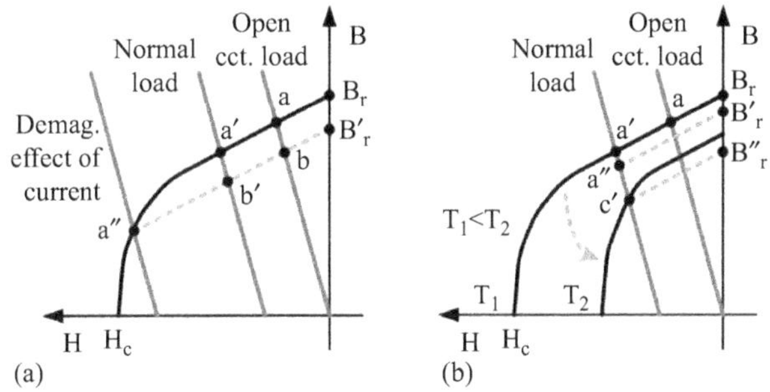

Figure 2.22 PM operating point (demagnetisation curve and load line). Irreversible demagnetisation due to (a) external demagnetising MMF. (b) Operation at high temperature (SmCo- or NdFeB-based magnets).

temperature T_1, intersected the demagnetisation curve at the linear part, now intersects the curve for $T_2 > T_1$ at point c' which is below the knee of the shifted demagnetisation curve. The result is that when the demagnetising effect disappears, the new remanence magnetic flux density will be B_r'' which is lower than the original B_r. Now if the temperature increases back to T_1, the remanence magnetic flux density will increase to B_r' with respect to B_r''; however, it will be less than the original one B_r.

For applications where the load requirements are fixed, it is evident that demagnetisation will evolve into increased fault levels. Due to demagnetisation, the produced torque capability of the motor for a given current will decrease. So, in order to serve the fixed load/torque requirements, the stator winding is forced to draw more current, leading to increased Joule losses and elevation of temperature, thus accelerating the demagnetisation of the PMs and leading to faster degradation of the windings insulation materials [67].

It is to be noted that demagnetisation in electrical machines might be uniform or partial. Uniform demagnetisation is more difficult to detect since its distorting effect on the magnetic field distribution is minimum [68]. However, partial demagnetisation leads to a strong asymmetry of the motor's magnetic field which will unavoidably lead to unbalanced magnetic pull (UMP) [69]. UMP will lead to torque oscillations, vibrations and noise and could cause some level of eccentricity [70]. A second level of fault evolution concerns the bearings which are overstressed and degrade faster.

The PM demagnetisation effect gives rise to stator current frequencies located at

$$f_{dm} = \left(1 \pm \frac{k}{p}\right) f_s, \quad k \in \mathbb{N} \tag{2.14}$$

The effect of partial demagnetisation on the magnetic flux density distribution and stator current frequency spectra is presented in Figure 2.23 for a six-pole PM synchronous machine (PMSM) operating at 6,000 rpm.

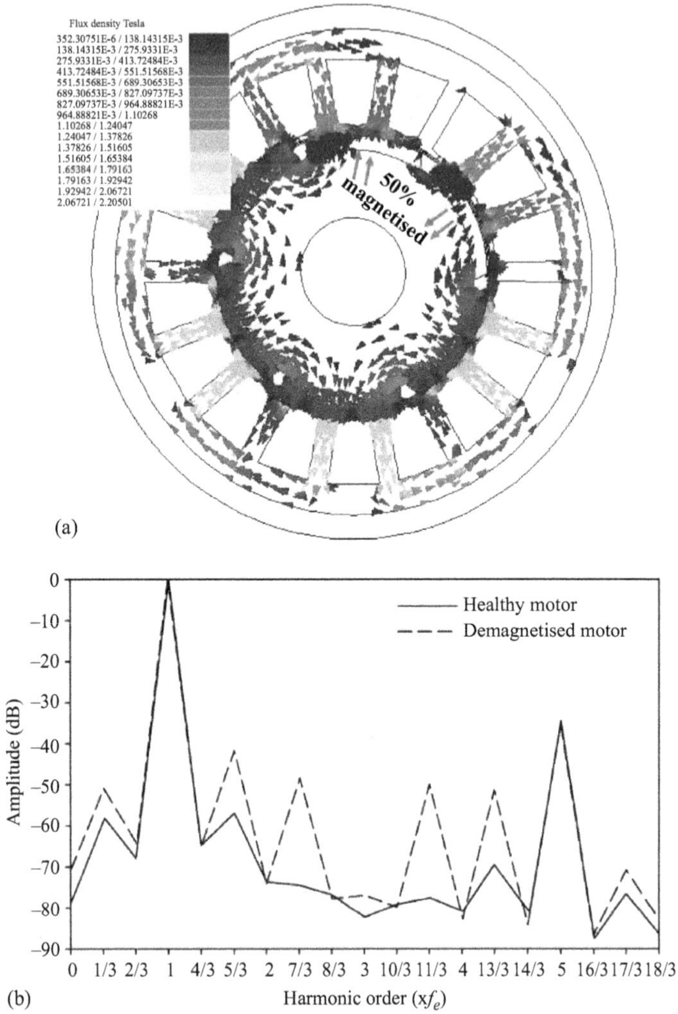

Figure 2.23 (a) Spatial distribution of the magnetic flux density in a partially demagnetised PMSM. (b) Simulated stator current harmonics in a healthy and a partially demagnetised PMSM when running at 6,000 rpm. © 2017–2018 IEEE. Reprinted, with permission, from Reference [71]

2.1.3.5 Bearing faults

Bearings are the electrical machine components which guarantee and secure the appropriate rotor positioning with respect to the stator while allowing rotation. There are two types of bearings, namely rolling elements or ball and sleeve

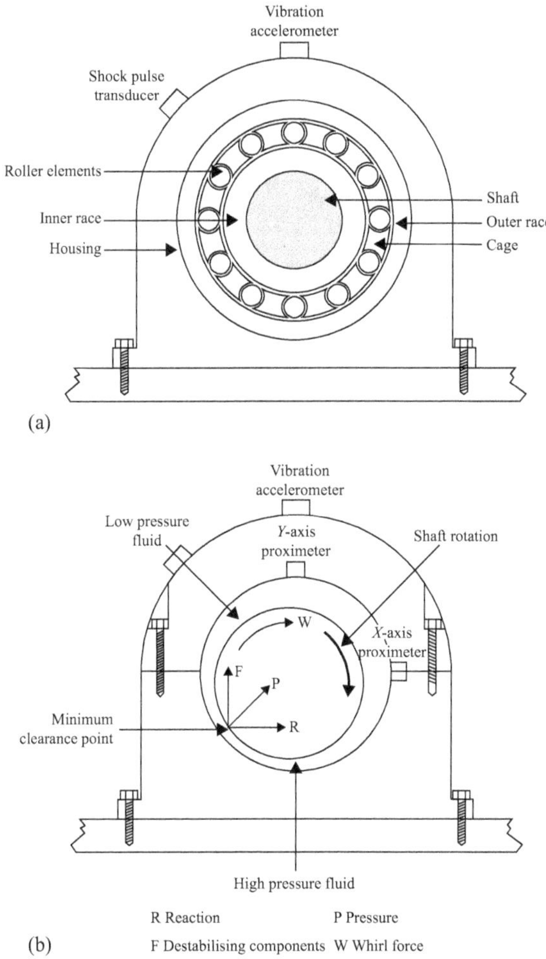

Figure 2.24 Bearing types and components: (a) rolling element bearing and (b) forces acting upon a shaft in a sleeve bearing. © 2017–2018 IET. Reprinted, with permission, from Reference [2]

bearings (Figure 2.24). It was shown earlier in Figure 2.1 that bearing failures are the main fault in low and medium voltage machines, while being of significantly lesser importance in high voltage machines. This is due to the fact that high voltage machines utilise sleeve bearings, while low and medium voltage machines utilise rolling element bearings [3]. For this reason, this section will be focused on the rolling element bearings fault detection.

It has been shown that, the location of the fault or in other words the faulty component of the bearing produces a unique vibrating harmonic response. More specifically, the following formulas have been proposed to detect the origin of the bearing fault [3,72]:

$$\text{Outer race defect} \quad f_o = \frac{N}{2} f_r \left(1 - \frac{D_b}{D_c} \cos \beta \right) \tag{2.15}$$

$$\text{Inner race defect} \quad f_i = \frac{N}{2} f_r \left(1 + \frac{D_b}{D_c} \cos \beta \right) \tag{2.16}$$

$$\text{Ball defect} \quad f_b = \frac{D_c}{D_b} f_r \left[1 - \left(\frac{D_b}{D_c} \cos \beta \right)^2 \right] \tag{2.17}$$

where N is the number of balls, D_b is the ball diameter, D_c is the bearing pitch diameter and β is the contact angle of the balls on the races.

As a result, it is possible to monitor each individual defect via the stator current frequency spectrum by applying the following formula [73]:

$$f_{reb} = f_s \pm m f_c, \quad m \in \mathbb{N} \tag{2.18}$$

where f_c corresponds to the appropriate vibration frequency described by (2.15)–(2.17).

An application of the above formulas can be seen in Figure 2.25 for inner and outer race faults. However, past experience has shown that the use of the above characteristic frequencies is not reliable when trying to detect general bearing faults, such as contamination or degradation. It is to be noted that bearing failures create some level of eccentricity, so there are numerous cases where the detection of eccentricity has led to the detection of bearing failures indirectly.

2.1.4 Alternative diagnostic methods

It is to be expected that although the analysis of the stator current for diagnostic purposes is indeed a powerful tool, it still has weaknesses in some applications or specific problems. This is the reason for the existing rich literature where many researchers propose alternative signals or methods to overcome the MCSA drawbacks. Some of the most frequently met will be discussed in this section.

2.1.4.1 Electromagnetic/mechanical torque monitoring

The torque monitoring has drawn a lot of interest over the years, whether it is the electromagnetic or the mechanical one. The electromagnetic is difficult to measure directly, this is why methods exist to estimate it from current, voltage and/or flux measurements [75,76]. The mechanical torque can be measured with a torque transducer in the lab, however, has limited application in a real industrial environment. However, the signatures existing in the electromagnetic torque are practically the same in the mechanical torque, so multiple cases exist where researchers do not differentiate between the two.

The motor's torque comes as a result of Lenz law, so it is in some way the end effect of the electromechanical energy conversion. Any fault or imbalance causing

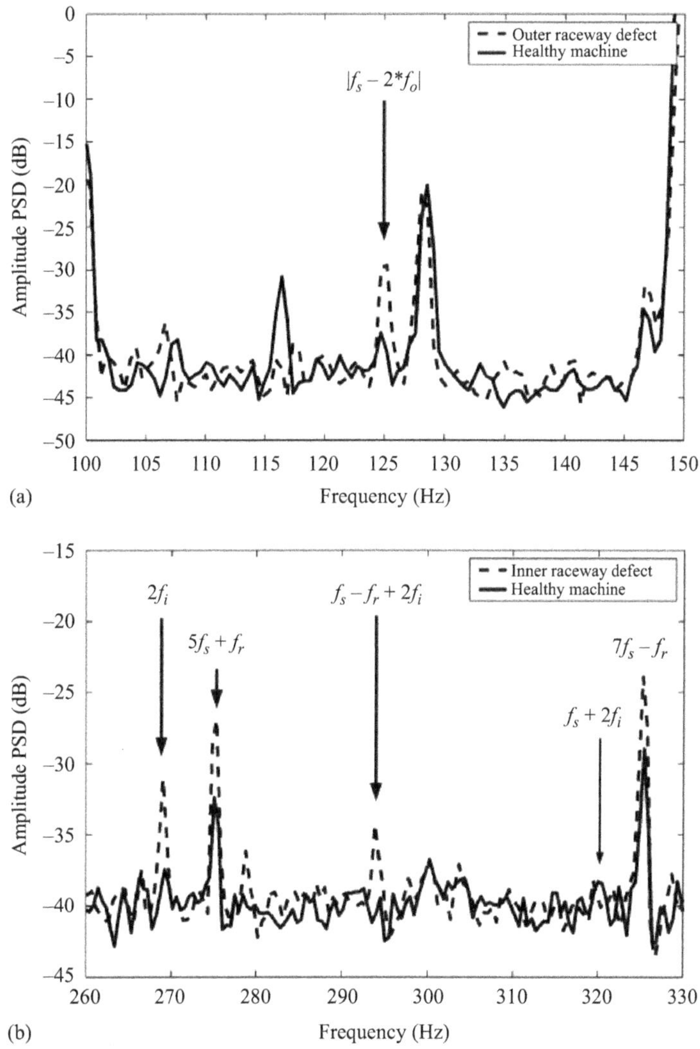

Figure 2.25 Application of MCSA to detect bearing faults where (a) outer raceway defect detection of loaded induction motor and (b) inner raceway defect detection of unloaded induction motor. © 2017–2018 IEEE. Reprinted, with permission, from Reference [74]

an asymmetry in the magnetic field will express itself as torque oscillations. So, the torque is just one signal originating from the synthesis of multiple electrical ones. This is why the torque is considered a diagnostically valuable tool.

To enhance understanding, all faults described earlier produce specific side-band harmonics to the fundamental in the stator current. The exact same sidebands

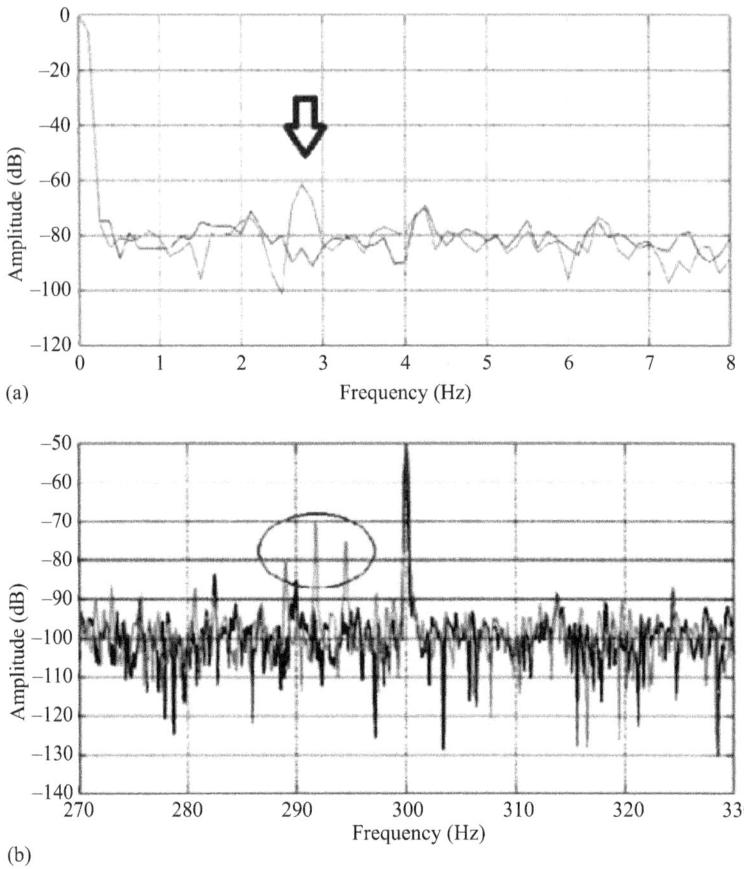

Figure 2.26 Comparative experimental spectra of the motors' (healthy and induction motor with a broken rotor bar) torque for s = 0.027 (a) at the low-frequency range and (b) at the frequency range close to 300 Hz. © 2017–2018 IEEE. Reprinted, with permission, from Reference [54]

(although of different amplitude) exist in the torque spectra around the DC component. Similarly, higher harmonics also exist. Figure 2.26 illustrates the diagnosis of a broken rotor bar fault located at $2sf_s$ and at $(6 - 2ks)f_s$ in the mechanical torque spectra of a four-pole, 400 V, 4 kW, 50 Hz cage induction motor.

2.1.4.2 Magnetic flux monitoring

The main idea of the diagnostic strategies is that a fault will cause an asymmetry in the magnetic field. As a result, many works [35,77–80] have proposed the direct monitoring of the magnetic flux using flux sensors. It is possible to monitor either the radial or axial magnetic flux. The MCSA signatures exist also in the radial flux spectra for radial flux electrical machines. However, the flux sensor is independent from the motor geometry and more specifically does not depend on the number of

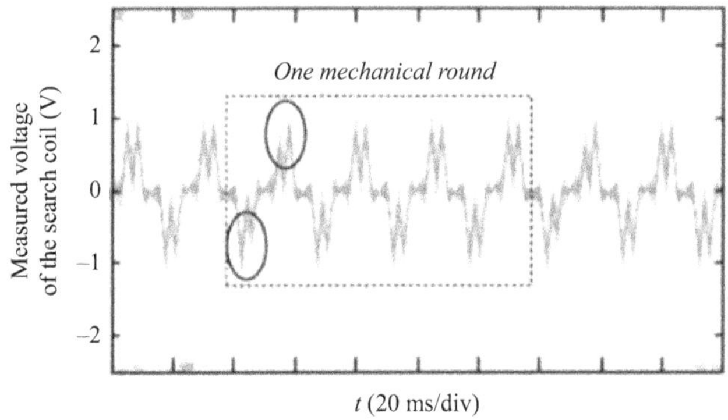

Figure 2.27 Measured voltage of the search coil from the prototype machine with partial demagnetisation

poles like the stator phase winding. As a result, the radial flux spectrum is richer in harmonic index than the stator current.

The application of a search coil to detect partial demagnetisation of a PM machine is shown in Figure 2.27 [81]. The distortion of the flux waveform is easily noticed. However, in most real cases, the fault needs to be diagnosed at incipient stages where it cannot be noticed by the flux waveform. As a result, spectral analysis is applied. Figure 2.28 illustrates the spectra of the radial flux derivative in an induction motor under healthy condition as well as under broken bar fault [82].

2.1.4.3 Single-phase rotation test

The single-phase rotation test (SPRT) is a test applied offline and aiming to detect rotor defects [83]. The rotor is manually rotated under fixed speed while only one of the stator phases is supplied, thus producing a pulsating magnetic field. The aim is to monitor the phase current and voltage, thus being able to calculate the phase impedance. Variations of the impedance during rotation are strong indicators of rotor asymmetries. The test belongs to the non-intrusive methods, so it does not require the motor disassembly. The setup is shown in Figure 2.29(a). Furthermore, the application of the test to detect different levels and types of eccentricity in an induction machine is shown in Figure 2.29(b).

2.1.5 Fault prognosis of electrical machines

There are two end goals in the field of electrical machine prognosis. The first one is to model the degradation mechanisms with accuracy. The second one is to estimate the component or device RUL [84,85]. Degradation is the irreversible process where a material, component or device loses its properties. Usually, the degradation has a strong multi-scientific character, that means many different mechanisms act together to produce the end effect which is the end of the subject's life [86,87]. In electrical machines, the area of fault prognosis is meaningful and thus has met significant interest and progress in the case of windings insulation materials. This is

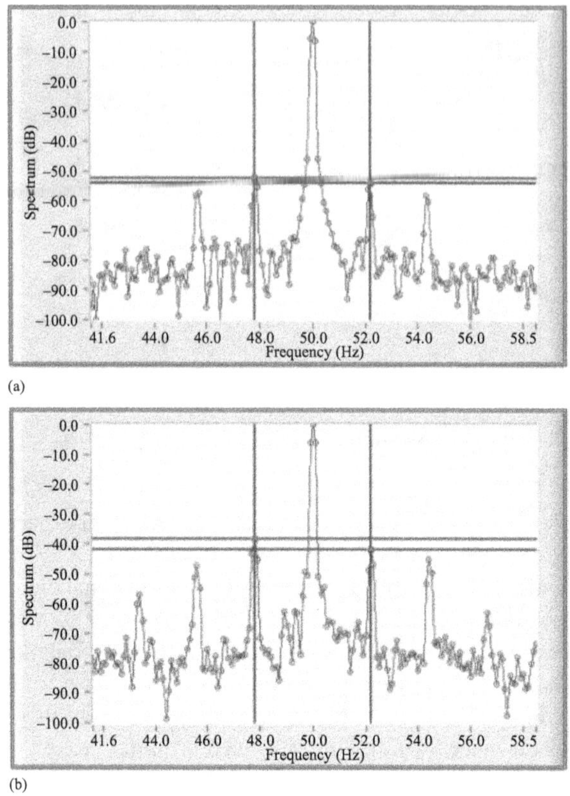

Figure 2.28 Spectra of stray flux derivative at rated load of (a) healthy induction machine and (b) induction machine with a broken rotor bar (four poles, 18.5 kW). © 2017–2018 IEEE. Reprinted, with permission, from Reference [82]

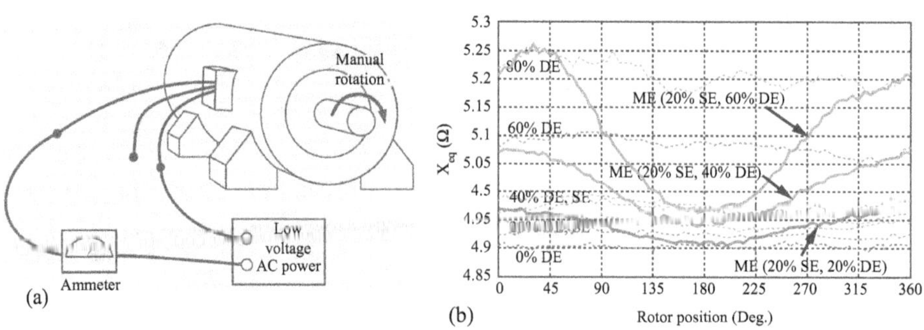

Figure 2.29 (a) The setup of the SPRT and (b) SPRT experimental results of an induction machine suffering from mixed eccentricity. © 2017–2018 IEEE. Reprinted, with permission, from Reference [83]

because the insulating materials are subjected to chemical reactions which change their characteristics permanently.

One of the most reliable surveys on root causes of insulation concerning a large number of electrical machines (in this case large hydro-generators) was published by CIGRE [88]. According to this study in over 1,199 generators' population, 56% of total faults were insulation related. An investigation over the factors leading to those failures has led to the following statistics shown in Figure 2.30 below. The internal partial discharges (PD) and the winding contamination impacts are quite similar, while combined they form about half of the total root causes of insulation damage. Of interest is the fact that the ageing is found to be the main cause of insulation damage.

In literature, it is custom to describe the various ageing factors with the term TEAM, an acronym for thermal, electrical, ambient and mechanical stresses [89]. The role of each one on the degradation of the insulation materials will be discussed below.

2.1.5.1 Thermal stress

First, the thermal ageing is one of the most critical degradation factors of electrical machine winding insulation. Thermal ageing depends on the operating temperature, which is directly related to the electrical machine losses. The losses in electrical machines are of different mechanisms and can be Joule losses due to the rotor and stator conductors' currents, iron losses, which can be further divided into eddy current and hysteresis losses and finally the stray losses. The temperature increase inside the electrical machine leads to unavoidable chemical reactions which transform the insulation material's chemical composition and lead to changes in the material's dielectric properties [90–92]. This will progressively lead to decrease of the insulation resistance [93] and electrical current flow between neighbouring turns creating either inter-turn short-circuits or heavier short-circuits. Furthermore, the different thermal expansion mechanisms of the conductors and the insulation leads to the delamination effect [94]. Delamination may be further enhanced by normally existing vibrations.

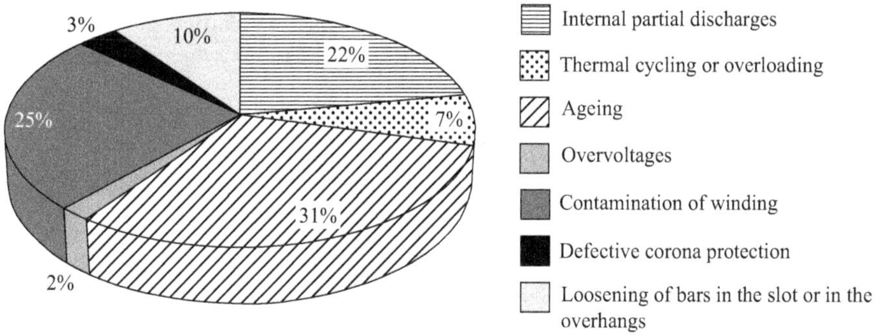

Figure 2.30 Distribution of root causes of insulation damage in large hydro generators

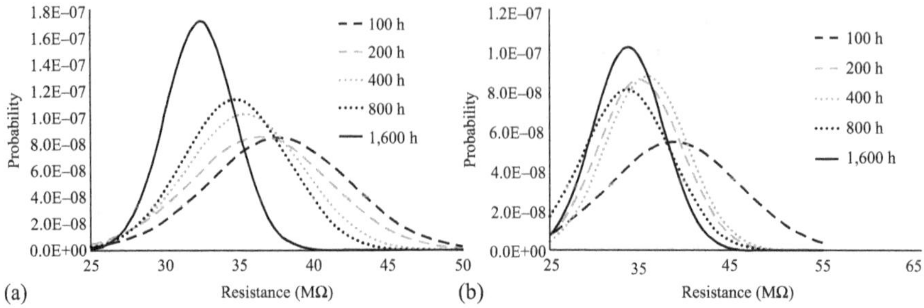

Figure 2.31 Normal distributions of the early breakdown resistance of thin film insulation material and for different ageing periods under (a) 200°C and (b) 230°C thermal stress. © 2017–2018 IEEE. Reprinted, with permission, from Reference [93]

The estimation of the insulation material remaining life as a function of temperature still relies on the well-known Arrhenius equation [95] in many applications:

$$t = K \cdot e^{(A/T)} \tag{2.19}$$

where t is the life of the insulation material, K is a material-dependent proportionality constant, A is the activation energy required to break chemical bonds in the insulation material and T is the temperature in Kelvin. A rule of thumb based on the Arrhenius equation is that the life expectancy of the insulation material decreases by 50% when the temperature increases 10K over its rated value.

However, recent works have questioned the Arrhenius equation for insulation degradation predictions. It seems that the chemical reactions due to the temperature increase differ during the different degradation stages of the electrical machines' insulation materials. This is illustrated in Figure 2.31, where the normal distribution of the early breakdown resistance of windings' thin film insulation material is subjected to fixed thermal stress [93].

This field is open to extensive research lately due to significance of insulation prognosis for motors used in electric vehicle applications where reliability and safety are primary concerns.

2.1.5.2 Electrical stress

When the electrical stress is the only acceleration variable, then the life expectancy of the stator bars may be determined approximately from [96]:

$$\frac{t_t}{t_s} = \left(\frac{E_s}{E_t}\right)^n \tag{2.20}$$

where t_t is the testing time or the time to failure, E_s is the service voltage stress, E_t is the breakdown or test voltage stress and t_s is the service time.

Electrical ageing can be mainly caused by two different physical mechanisms. The first mechanism is related to PD and the second to any fast transient voltage changes. It has been reported in [97,98] that although electrical breakdown is causing the final failure of the electrical insulation, electrical stress is not the dominating factor. It is rather believed that the ageing mechanism is mainly due to thermal degradation of the binder resin, mechanical stress caused by vibration and switching pulses and stress caused by the different thermal expansion coefficients of the different materials [99].

First, it is necessary to relate the PD activity with some serious ageing factors of the insulation material. Such factors were mentioned in [100], and they concern mica–resin coils. The ageing factors are ageing of the resign because of depolymerisation and carbonisation, moisture absorption, insulation voids, cracks and delamination. It is also important to mention the impact of humidity although its full impact mechanism has not been fully revealed [101,102].

According to [103], a common breakdown mechanism was found: local PD form in the voids. Those PD grow and eventually form a bridge between the local voltage difference, causing the breakdown.

The second mechanism usually occurs in electrical machines which are fed by an inverter and where there are fast voltage changes due to the great switching frequencies [104,105].

2.1.5.3 Mechanical stress

According to [106], the main causes for insulation failure due to mechanical stresses are coil movement and strikes from the rotor. Strikes from the rotor can happen during starting or due to eccentricity caused by bearing faults or bent shaft. However, in [94], the thermomechanical effects are also considered. Thermomechanical effects originate from the different expansion and contraction rates of different materials with the change of temperature. The insulation materials are characterised by lower thermal expansion coefficient compared to copper. Also, its temperature is lower than that of copper. As a result, a shear stress is developed due to the different expansion rates. The shear stress is believed to be more severe when the windings are cold [89]. A combination of the above leads to loosened stator bars in the slots resulting in a vertical vibration [107]. This phenomenon results to abrasion and erosion of the insulation material and consequently causes increased PD which accelerate the ageing process.

2.1.5.4 Discussion

The ageing mechanism of the winding insulation materials is complex and involves many different mechanisms which sometimes influence each other. For example, thermal stress can deteriorate the insulation material, and at the same time it causes the thermomechanical effect which changes the materials dimensions. Furthermore, the resistance of the conducting materials is dependent on the temperature, and as a consequence there will be a change on the flowing current which will influence the electrical stress level as well as the ohmic losses of the conductor itself which in turn will change the temperature level. This is just a short logical sequence of events to show how much complex it can be to discriminate and study each stress type independently. This is the reason why a lot of researchers propose the use of accelerated multi-stress ageing tests when they need to predict the remaining life of insulation materials [108]. Despite that, a lot of work has been accomplished by researchers who try to isolate the stress impact. This is generally an on-going research field where a lot of work is yet to be accomplished.

2.2 Fault diagnostic techniques applied to voltage source inverter-fed drives

2.2.1 Introduction

The use of two-level voltage source inverters (VSIs) in three-phase variable speed drive applications has become a standard. Although this technology has reached a good maturation level, since these devices are still quite complex and taking into account that they often operate under high stressful conditions [109], VSIs are very susceptible to suffer critical failures that may negatively affect the entire drive system operation.

Over the last decades, several statistical studies have been published regarding the fault distribution in these power electronic converters [110–114]. As depicted in Figure 2.32, it can be concluded that the majority of faults occur in power devices, capacitors and gate drivers.

Moreover, by aggregating the fault distributions related to power semiconductors and their corresponding control circuits (power devices and gate drivers), it becomes evident that these components are the most susceptible ones, reaching about 50% of the total failure rate in industrial power converters.

Therefore, and with the aim to reduce the negative impact of these faults as well as to improve the system reliability, during the last decades, the development of fault diagnostic methods for these fault types has gained a lot of interest.

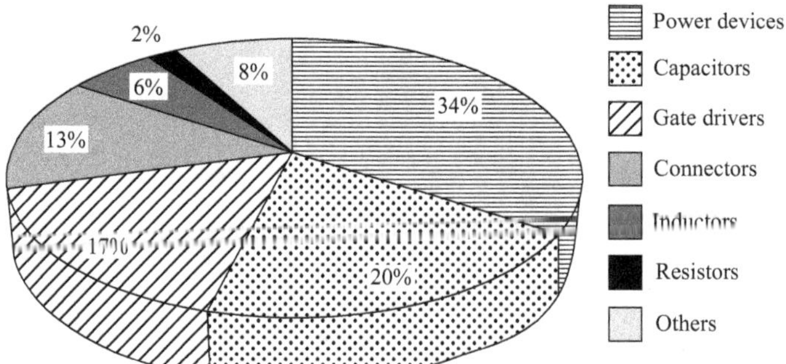

Figure 2.32 Fault distribution in industrial power electronic converters

2.2.2 Fault diagnostic approaches

As shown before, the aggregate failure rate related to inverter power switches represents more than one-half of the total failures in industrial power converters. In general, inverter power device failures can be broadly classified as open-circuit faults and short-circuit faults. Considering the typical structure of a two-level VSI used in variable speed drives and shown in Figure 2.33, a short-circuit fault in these elements, in the majority of the situations, causes an overcurrent peak due to DC bus shoot-through fault, which can be very destructive.

As a result, short-circuit protection is a standard feature that it is integrated in today's industrial converters, allowing to shut down the drive immediately and avoid further damages. On the other hand, open-circuit faults do not necessarily cause the system shutdown and can remain undetected for an extended period of time. This may lead to overstress on the remaining power devices, resulting in secondary faults in the converter or in the remaining drive components, leading to the total system shutdown and high repairing costs. Considering this, the research and development of suitable diagnostic methods for VSIs power switch open-circuit faults has attracted lot of attention over the last decades. These methods can be typically classified as current-based approaches and voltage-based approaches. Some reviews and surveys can be found in the literature addressing these two groups of fault diagnostic methods for VSIs power switch open-circuit faults [115–119]. Nevertheless, each of these fault diagnosis approaches is addressed in more detail in the next subsections.

2.2.2.1 Current-based approaches

Today's standard power converters based on VSIs are equipped with DC bus voltage sensor and output current sensors, required for control and protection purposes. The availability of these default current signals is the obvious choice for fault diagnostic purposes, and therefore, current-based methods have been widely proposed in the literature. These current signals are already available and can be easily used to extract information about the faulty devices.

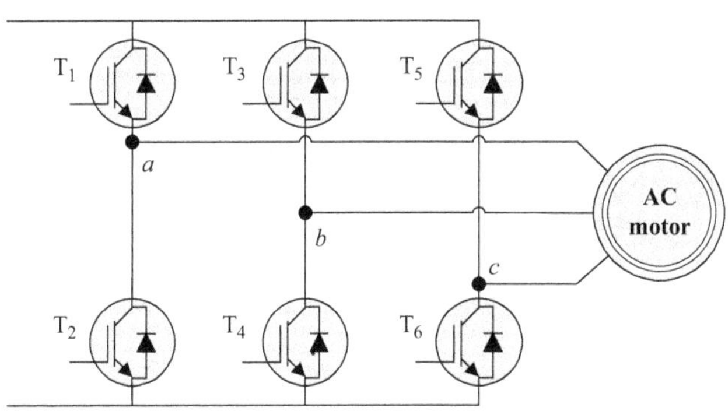

Figure 2.33 Structure of a typical two-level VSI feeding an AC motor

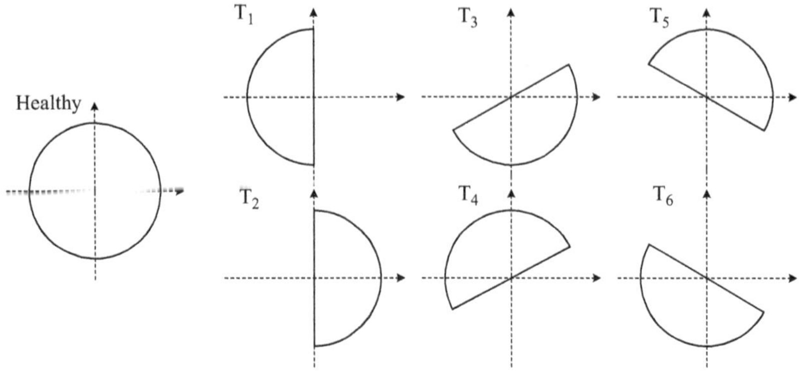

Figure 2.34 Typical Park's vector shapes for the healthy case and different VSI single power switch open-circuit faults

Park's vector approach

One of the first methods to effectively diagnose open-circuit faults in power converters was the PVA [120–125]. With this technique, the three VSI output currents are acquired and the Park's transformation already defined by (2.7) and (2.8) is used to plot the *dq* characteristic graphs (Figure 2.34).

Under normal operating conditions, the resulting pattern is similar to a circle as demonstrated by (2.9) and (2.10). Under a single power switch open-circuit fault, part of the circle is missing, and the current pattern becomes biased towards the direction of faulty switch.

Basically, with this technique, the damaged power switches can be detected and localised through the analysis of patterns. However, this approach requires very complex pattern recognition algorithms which are not suitable for integration into the drive controller.

Also related to pattern analysis, the current space vector trajectory diameter was also used for the localisation of the faulty switch [126,127]. Its slope provides information that allows for the faulty leg diagnosis, while the current waveform missing half-cycle is used to locate the faulty power device. Despite being relatively simple, this technique has serious drawbacks related to large detection time, tuning issues and poor performance under low current values.

The currents Park's vector phase and its polarity can also be used to effectively diagnose power converter open-circuit faults [128,129]. The Park's vector phase θ can be directly calculated from the corresponding *dq* current components:

$$\theta = \tan^{-1}(i_q/i_d) \tag{2.21}$$

For detection purposes, the derivative of the current Park's vector phase d is defined by

$$d = \left| \frac{d}{dt} |\theta| \right| \tag{2.22}$$

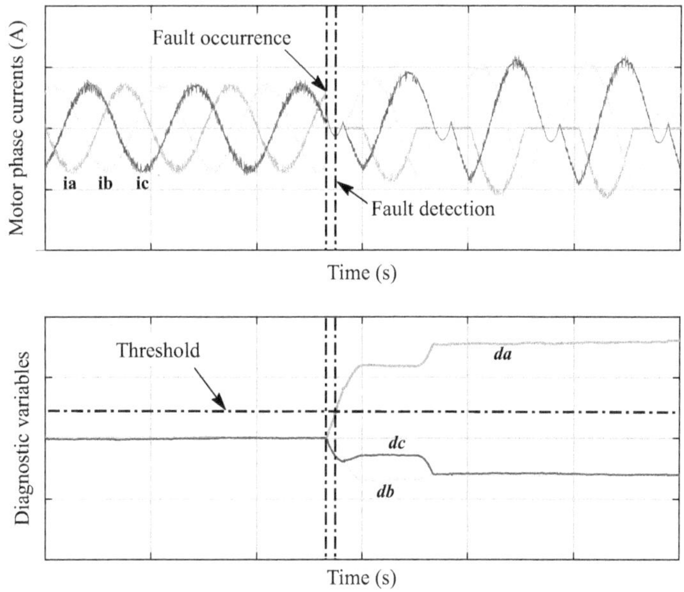

Figure 2.35 Typical motor phase current waveforms and diagnostic variables (da, db and dc) based on average values calculation under a VSI open-circuit fault

The detection principle relies on the comparison of d with a defined threshold value. In order to locate the faulty devices, the algorithm relies on the currents polarity by calculating the corresponding average values. The obtained results show multiple fault diagnosis capabilities and a very robust behaviour to the issue of false positives.

Average values approaches

When an open-circuit power switch occurs, the motor phase currents i_n, where ($n = a, b, c$) present a distinct waveform and their corresponding average values will no longer be zero. Therefore, this characteristic can be exploited to develop fault diagnostic algorithms where the diagnostic variables are compared with threshold values (Figure 2.35). The general fault diagnostic structure of this approach is represented in Figure 2.36.

Based on the average currents calculation, an improved version of the PVA was developed [130]. The use of the instantaneous current values was replaced by the calculation of the corresponding average values in order to determine the resulting Park's vector. The vector absolute value was used to detect the fault, whereas its corresponding phase provides information to identify the faulty VSI power switch. Despite some additional improvements [131,132], this technique still presents some limitations related to the tuning effort due to the fact of being load dependent, together with some false alarms issues.

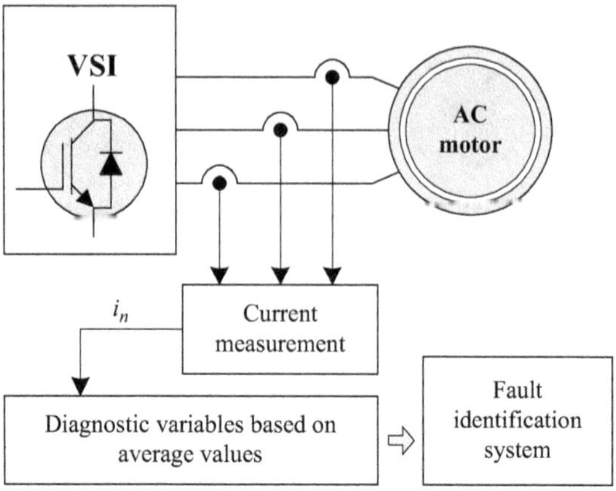

Figure 2.36 General structure of current-based fault diagnostic approaches based on average values calculations

A mean to mitigate the generation of false positives resulting from the misinterpretation of large transient variations of the currents average values, was the development of normalised diagnostic variables as defined below [133,134]:

$$r_n = \langle \omega_n \rangle \frac{\langle i_n \rangle}{\langle |i_n| \rangle} \tag{2.23}$$

The three-phase diagnostic variables r_n are normalised by dividing the currents average values $\langle i_n \rangle$ by the corresponding currents average absolute values $|\langle i_n \rangle|$. The auxiliary variables $\langle \omega_n \rangle$ are used in order to improve the algorithm's performance.

A distinct real-time open-circuit fault diagnostic method for VSI fed motor drives was proposed in [135]. The inverter output currents i_n are normalised using the corresponding Park's vector modulus:

$$i_{nN} = \frac{i_n}{\sqrt{i_d^2 + i_q^2}} \tag{2.24}$$

The diagnostic variables e_n are then obtained by

$$e_n = \xi - \langle i_{nN} \rangle \tag{2.25}$$

where ξ is a constant. These variables allow to detect the faulty phase, while the average values provide information about the faulty device. Despite being relatively simple and of easy integration into the control system, this technique shows a very high immunity against the issue of false alarms, fast detection time, and it does not depend on the motor operating conditions.

Reference currents errors

In vector controlled drives, the reference current signals current i_n^* are easily available in the control system. Therefore, they can be used for diagnostic purposes as shown in Figure 2.37.

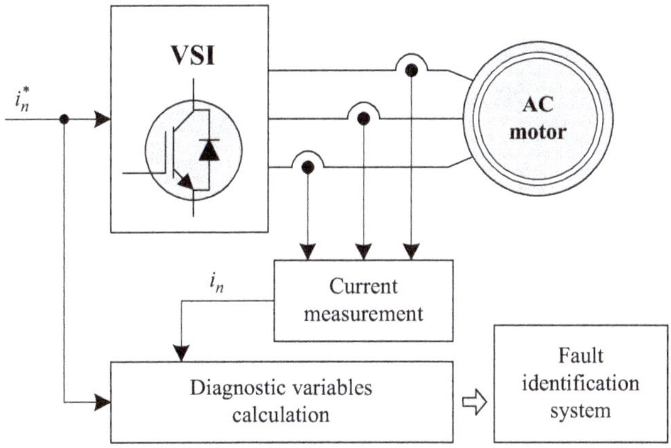

Figure 2.37 General structure of current-based fault diagnostic approaches based on reference current errors

The three-phase reference current signals i_n^* can be compared with the corresponding measured values i_n. The average values of the obtained residuals are then normalised using the average absolute values of the motor phase currents $\langle |i_n| \rangle$ [136]:

$$d_n = \frac{\langle i_n^* - i_n \rangle}{\langle |i_n| \rangle} \tag{2.26}$$

The obtained results demonstrate that this approach allows to achieve detection times as fast as 5% of the fundamental period, being independent of the load and speed values, and with great robustness against transients.

A very similar approach has also proposed where the Park's vector modulus of the reference current signals is used as normalising quantity, instead of the average absolute values of the motor phase currents [137]. The obtained results also demonstrate that similar performance levels can be achieved.

Other similar technique using the *dq* reference signals can be used instead of the three-phase reference current signals [138]. Accordingly, each reference signal is compared to the equivalent measured feedback signal, resulting in an error. Threshold values are employed to determine if there is fault or not. In order to improve the algorithm's reliability, the proposed method also has a counter. If a fault is detected, then the counter starts incrementing until it reaches a jitter-free count. If the state is not stable, the counter resets to its initial value. The faulty devices are then identified by using localisation stages defined according to the current vector rotating angle.

Sustained near-zero current values

As shown in Figure 2.35, when a single power switch open-circuit fault occurs, the faulty device will no longer be able to switch on, leading to near-zero current values in the affected phase during approximately half-cycle of the fundamental period. This typical characteristic is clearly shown in Figure 2.38.

Consequently, this fault signature can be exploited for the development and implementation of fault diagnostic methods (Figure 2.39).

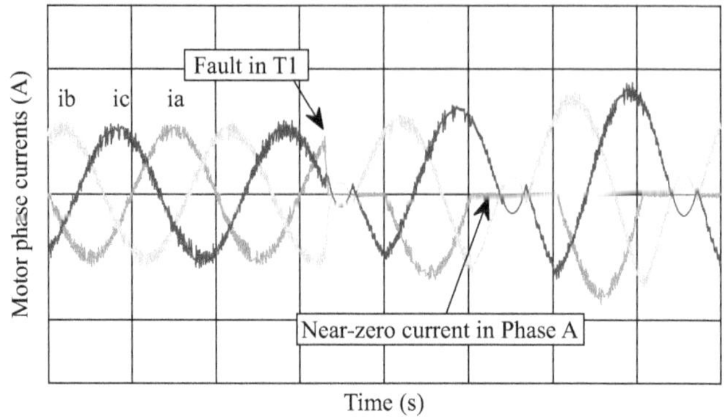

Figure 2.38 Phase A near-zero current after an inverter open-circuit fault in T1

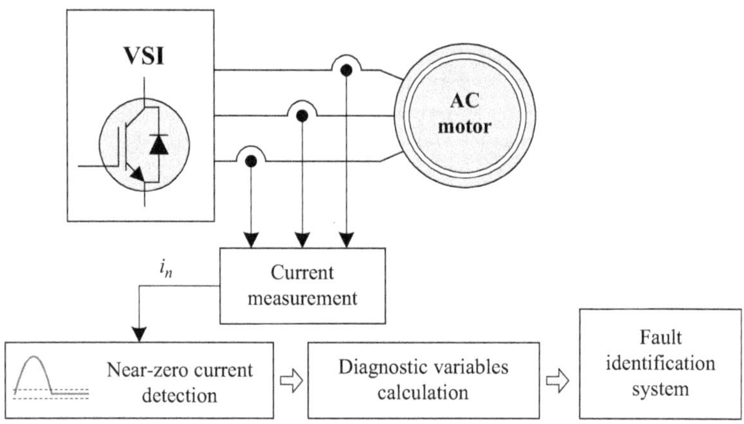

Figure 2.39 General structure of fault diagnostic approaches based on sustained near-zero current values

A fault diagnostic approach that relies on the use of near-zero current values was proposed in [139]. The fault detection is achieved by evaluating the nearly zero current values and also by comparing if the relative magnitude of the second-order harmonic is larger than a defined threshold value. Using this combined analysis, it is possible to detect and localise multiple semiconductor open-circuit faults.

In [140], the VSI symmetry under normal and faulty operating conditions is described based on the concept of allelic points and phase functions. The residuals between any two-phase functions are used for diagnostic purposes. Under normal operation, all the residuals are closed to zero. In the case of an open-circuit fault, the residuals will be larger than predefined thresholds, since the near-zero current samples per fundamental period will increase. Moreover, its proportions in a period will be different when different fault classes occur, allowing for multiple fault diagnosis.

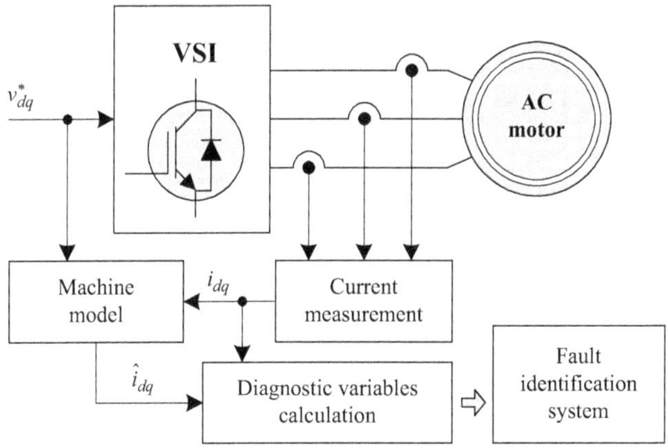

Figure 2.40 General structure of current-based fault diagnostic approaches based on observer techniques

Observer-based diagnosis schemes

Observer-based diagnostic techniques can also be used by implementing a *dq* machine model to estimate the motor phase currents, as depicted in Figure 2.40.

The *dq* reference voltage signals v_{dq}^* together with the measured *dq* motor phase currents i_{dq} are used by an observer relying on the machine model. The obtained estimated motor phase currents $\hat{i}_{dq}$ together with the measured currents are then used to compute the diagnostic variables.

Taking this into account, Luenberger observers have been widely used for VSI fault diagnosis. The approach proposed in [141] uses robust residuals constructed by combining two nonlinear Luenberger observers. The algorithm proves to be load independent, does not require additional sensors and allows for multiple fault diagnosis. A different approach for multiple open-circuit fault diagnosis in VSIs was proposed [142], and later improved in [143] with added current sensors diagnostic capabilities. The machine's mathematical model was used for the implementation of the Luenberger observer. The current form factors of both actual and estimated currents are calculated, being the last ones used as adaptive thresholds. These quantities are compared and the obtained residuals are used to detect the faulty phase, whereas the measured currents average values are used for fault localisation. The obtained results proved that the use of an adaptive threshold enables very high immunity against false diagnostics under strong current transients and fast speed variations.

Model reference adaptive system (MRAS) techniques can also be effectively used for open-circuit fault diagnosis [144]. The MRAS is built based on the electric motor dynamic model, having as inputs the measured currents and the reference voltage signals. The MRAS output currents are then compared with measured currents, where the resulting error is fed into a voltage distortion observer. When it reaches a defined threshold, the fault is detected.

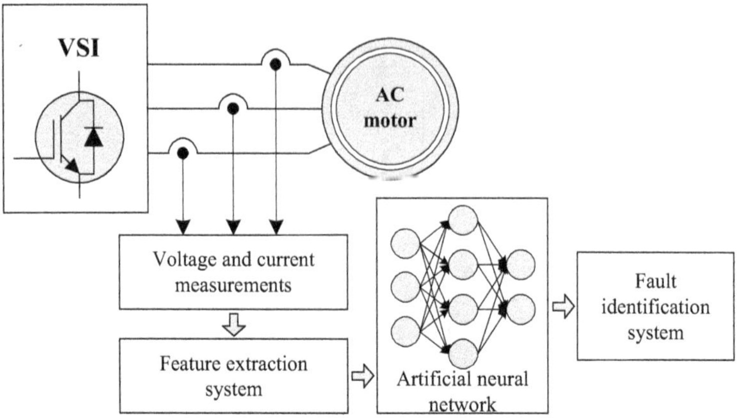

Figure 2.41 General structure of current-based fault diagnostic approaches based on ANN techniques

In order to avoid the influence of load and controller effects, it was proposed in [145] to use a mixed logical dynamic model of the motor to estimate the currents. Again, the error between the estimated and actual currents is used for fault detection purposes. For fault localisation, the vector plane is divided into six sectors, and it is evaluated in which the error lies. The use of thresholds and sectoring improve the immunity to parameter variations and transients to avoid false alarms.

Artificial neural network techniques

The development of artificial neural network (ANN) techniques to the diagnostic of inverter open-circuit faults was also effectively validated. These approaches rely on the measured VSI output currents and voltages that are fed into a feature extraction routine. Then, a fault table is built based on the extracted features, and the results are used to train an ANN. The final diagnosis information is finally processed by a fault identification system (Figure 2.41).

The overall structure of these techniques is very similar, and the main differences are related to the various feature extraction techniques that can be used and the number of nodes/layers of the neural network. Typically, the most common feature extraction approaches are based on the Clark transformation [146], DWT [147], principal component analysis [148] or 3D current state space [149]. Despite these, more advanced techniques allow for an intelligent diagnostic without detailed background knowledge of the drive system; they are relatively complex and require high tuning effort to implement and train the neural networks.

2.2.2.2 Voltage-based approaches

Although current-based fault diagnostic approaches are the most common ones since they do not rely on extra sensors and are relatively simple to implement, a lot of research has been made focusing on the development of voltage-based methods. Typically, these techniques require some additional hardware but present some

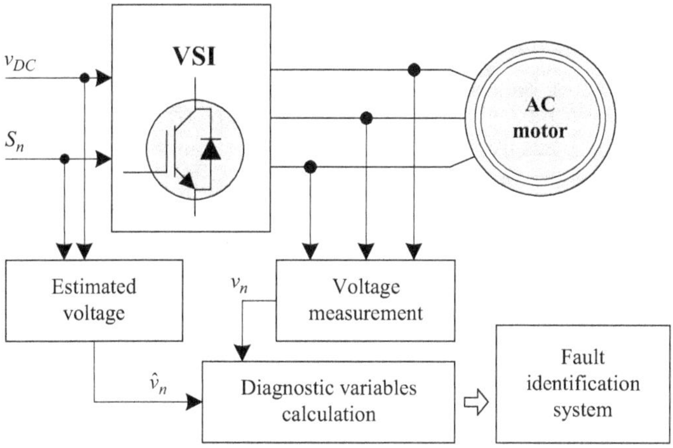

Figure 2.42 General structure of voltage-based fault diagnostic approaches based on the use of additional voltage sensors

excellent features such as an inherent independence of load operating conditions and fast detection times.

Techniques requiring additional voltage sensors

In a similar way to current-based approaches where the actual signals, required for control purposes, are also used to extract information on VSI faults, additional voltage sensors can be implemented with the aim to use the inverter output voltage signals for the development of fault diagnostic algorithms (Figure 2.42).

Regarding this, and depending on the way that output voltages are measured, these methods are based on line-to-line voltages [150–155] or rely on phase voltage signals [156–160]. The typical implementation of these fault diagnostic approaches is generally based on the comparison of the actual voltages v_n, obtained by the sensor measurements, with the estimated voltages $\hat{v}_n$, calculated using the gate command signals S_n and the DC bus voltage v_{DC} or using mathematical models. The resulting residuals can be then processed in order to detect and localise the faulty devices. Other approaches, based on ANN techniques, may use different signature extraction features to obtain data to feed and train the neural networks. The diagnostic methods based on additional voltage sensors usually present fast detection times, allowing a faulty VSI device to be detected in less than a PWM carrier period. Further than the additional costs of using extra sensors, these approaches also require some tuning due to operating voltage, current and temperature levels and their effects on electronics propagation times.

Simple hardware techniques

Voltage-based fault diagnostic techniques can also be implemented through the indirect sensing of inverter terminal voltages [161–165]. For this particular situation, simple analogue circuits comprising passive components, basic electronic integrated circuits and optocouplers (for galvanic insulation) are used to get data regarding the VSI output voltages. Then, the diagnostic strategy follows the similar approach as the ones based on additional voltage sensors, where the sensed voltage information (which may not be exactly the real inverter output voltage) together with the DC bus measurement that is typically available, is compared with the estimated output voltages, calculated based on the gate command signals provided by the control system or through a mathematical model. Naturally, by using simpler hardware, a good diagnostic performance can be achieved with lower costs, comparing to the use of dedicated voltage sensors. Nevertheless, these approaches have the same disadvantages related to the use of additional voltage sensors, meaning that extra tuning effort is required to adjust the time delay values according to electronics propagation times.

A more dedicated circuit can be integrated with the device triggering circuit, allowing to achieve open-circuit and short-circuit fault detections in less than 10 μs [166]. However, the analogue circuit requires some voltage measurements, which increases the system complexity, and the technique cannot be applied to all power devices since it strongly depends on the power device characteristics.

Available control variables

With the aim to avoid the use of additional voltage sensors or any kind of extra hardware, other voltage-based approaches have been developed relying on the use of voltage variables that can be obtained indirectly from the control system as depicted in Figure 2.43.

In closed-loop vector controlled drives, the reference voltage signals v_n^* available in the voltage loop, can be compared with the estimated machine voltages $\hat{v}_n$, calculated using a flux observer [167,168]. The machine parameters may change according to the operating conditions, and therefore, the diagnostic performance and robustness of these approaches can be negatively affected.

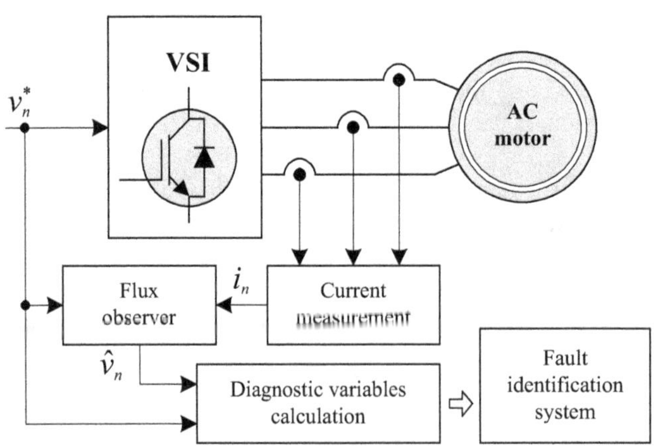

Figure 2.43 General structure of voltage-based fault diagnostic approaches based on the available control variables

Since the inverter output current signals are available in standard VSI fed drives, other approaches were developed taking into account the implementation of a current-based flux observer that can be used to estimate the motor terminal voltages [169]. As a result, this approach can be applied to a great variety of closed-loop control systems, such as the ones based on direct torque control (DTC) and vector control.

More specific voltage-based diagnostic methods can be developed according to the PWM strategy used. Considering this, a dedicated fault diagnostic approach for vector controlled drives based on space vector PWM was developed [170,171]. The diagnostic principle takes into account the monitoring of the voltage vector in the complex plane, and depending on the vector speed and faulty device, the reference voltage vector is forced in one characteristic sector during a much longer time-period than in the case of some other ones.

In opposition to the previous diagnostic methods based on additional sensors or hardware, due to the fact that these techniques only rely on voltage signals available from the control system, they are simpler to implement. However, as main disadvantage, it is pointed out the longer detection times that became similar to the ones related to current-based approaches.

2.3 Fault-tolerant techniques applied to VSI-fed drives

2.3.1 Introduction

The susceptibility of VSI drives to suffer critical failures can be extremely undesirable, especially for critical applications, such as in the aerospace industry, traction systems and pumping/cooling applications in nuclear power plants. For these critical situations, unplanned stoppages due to power converter failure can result in very large costs or in even more catastrophic consequences. Hence, the development of fault-tolerant remedial strategies applied to variable speed AC drive systems has become a very important issue over the last decades.

The fault-tolerant remedial strategies that can be applied to two-level VSI-fed drives can be generally classified into two distinct power converter topologies: the non-redundant topologies and the redundant ones.

Regarding the non-redundant VSI configurations, in general they only rely on the use of supplementary and cheaper devices. Under post-fault operating conditions, the aim is to create a different path for load current flow without any additional redundant hardware, maintaining the drive operation at the maximum possible performance conditions. For the case of redundant fault-tolerant power converter topologies, typically, extra inverter legs similar to the ones already used under normal operating conditions are added (at least one) and used as back-up units. After fault detection and isolation, this extra inverter leg can be used, maintaining the drive normal operation.

Taking this into account, it becomes clear that the first group of topologies comprise lower costs than the second one and may be considered acceptable in some applications, requiring also a slightly modification of the control strategy.

On the other hand, redundant topologies are more expensive but present the advantage of integrally saving the drive system operation.

2.3.2 Non-redundant topologies

As far as non-redundant fault-tolerant strategies is concerned, the most common topologies are depicted in Figure 2.44.

For the first configuration, the inverter faulty phase is connected to the DC link capacitors midpoint through triacs [Figure 2.44(a)]. The fault-tolerant converter is based on the traditional three-phase VSI, and each motor phase is connected to the DC bus with three extra triacs. On the other side, for the second fault-tolerant topology, the machine neutral point is connected to the DC link capacitors midpoint [Figure 2.44(b)]. For this case, the fault-tolerant converter is based on the conventional three-phase six-switch VSI with just one additional triac that connects the motor neutral point to the DC bus capacitors midpoint.

2.3.2.1 Phase connection to capacitors midpoint

Regarding the topology that consists on the faulty phase connection to the capacitors midpoint, a lot of research work has been done on this topic [172–180]. Under normal operating conditions all triacs are off and a total of eight voltage

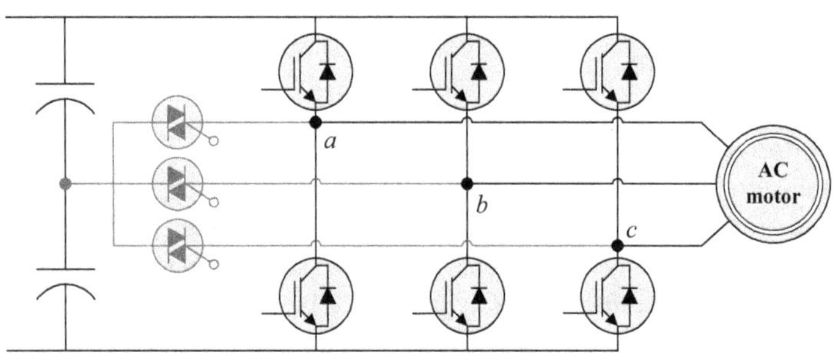

(a)

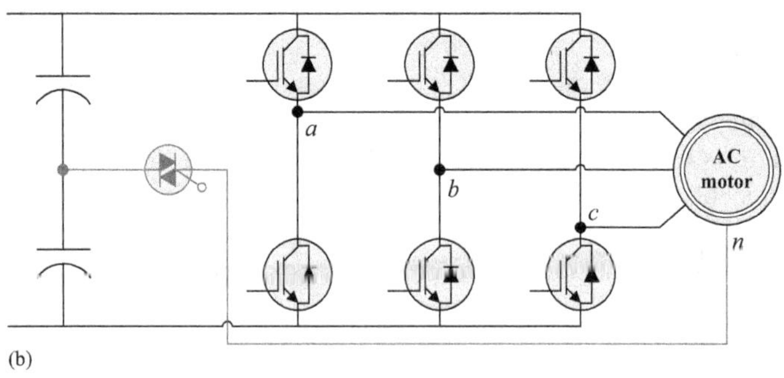

(b)

Figure 2.44 Non-redundant two-level power converter topologies: (a) phase connection to capacitors midpoint; (b) neutral connection to capacitors midpoint

vectors, six active vectors (V_1–V_6) and two zero vectors (V_7 and V_8) are generated by the inverter to control the machine [Figure 2.45(a)].

Under post-fault operating conditions, one of the motor phases is directly connected to the DC bus and therefore, its corresponding voltage is imposed, while the other phases are supplied by the remaining healthy inverter legs. As a result, the motor is fed by an asymmetrical power converter topology (four-switch three-phase power converter), that will control the motor by generating four active non-balanced voltage space vectors, as shown in Figure 2.45(b).

Comparing the post-fault operation with the healthy case, it becomes clear that the maximum voltage phasor amplitude r becomes limited to one-half of the original value. Consequently, and despite this post-fault reconfiguration allows for rated torque operation, in order to maintain acceptable torque ripple/pulsation, the operating speed must be limited to 50% of the motor rated speed value.

Depending on the control scheme used to control the motor speed, other actions may be required in order to allow the proper drive operation. For the case of a vector control strategy employing hysteresis current controllers, due to the direct control of each motor phase current, no additional software changes are required.

For the case where a different modulation strategy such as space vector modulation (SVM) is used, and since under post-fault operating conditions it is not possible to generate zero voltage vectors, the modulation scheme must be adapted. For this specific situation, and with the aim to compensate for the generation of zero vectors, the remaining time can be compensated by using two vectors with opposite direction, applied in a flyback mode for the same amount of time. Consequently, the flux linkage vector trajectory travels back and forward for the same period of time, resulting in the generation of virtual zero vector [181–188]. Also related to this modulation approach, other issues may arise due to the voltage unbalance across the DC bus capacitors. Under these circumstances, the generated voltage vectors become even more distorted, which may lead to a large DC link

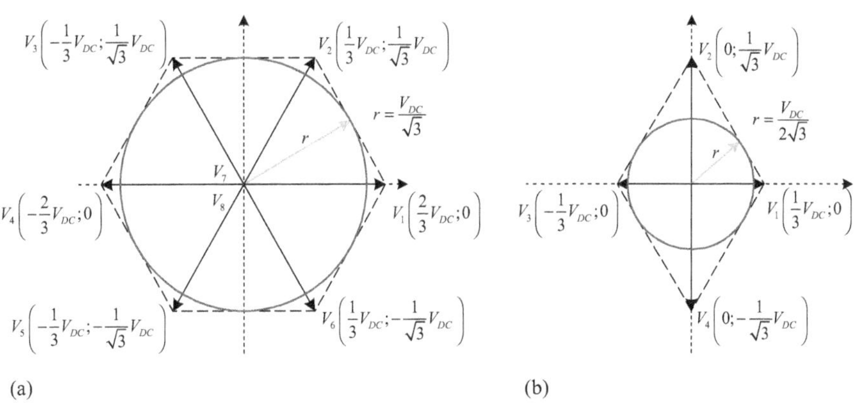

Figure 2.45 Voltage space vector representation under: (a) normal operating conditions and (b) Phase A connection to the capacitors midpoint

voltage ripple. This situation has been also investigated and dedicated modulation strategies have been proposed in order to compensate for this problem [189–193].

Regarding DTC variable speed drives, the control strategy must be also changed in order to optimise the drive overall performance. As explained before, since under post-fault operating conditions the voltage vectors generation is different, the typical switching table derived for the normal DTC implementation must be modified by adapting the voltage vectors selection for each new sector [194–198].

2.3.2.2 Neutral connection to capacitors midpoint

For this non-redundant topology, and comparing with the previous one that consists on the faulty phase connection to the capacitors midpoint, it becomes clear that it is much more simpler and less expensive since only one triac is required [173,174,199–203]. However, it must be taken into account that in order to use this topology, the machine windings must be connected in star configuration with the neutral point available for external connections.

In this topology, one of the motor phases is in open-circuit and as a result, its phase voltage is floating. Hence, and further than the incapability of generating zero voltage vectors, the generated vectors do not present a well-defined amplitude and phase. For this case, the generated voltage vectors are depicted in Figure 2.46.

The vertical vectors are uniformly distributed along the points BC and EF while the horizontal vectors present an uneven distribution along the real axis. Thus, in order to determine the voltage space vector limit, it is required to calculate the horizontal vectors average coordinates (points A and D). From the resulting complex representation, it can be concluded that the voltage vector amplitude limit is 3/4 of the one obtained under normal operating conditions [204]. Consequently, in order to maintain proper drive performance with low torque pulsation, under post-fault operating conditions, the drive speed must be limited to 75% of the rated motor value.

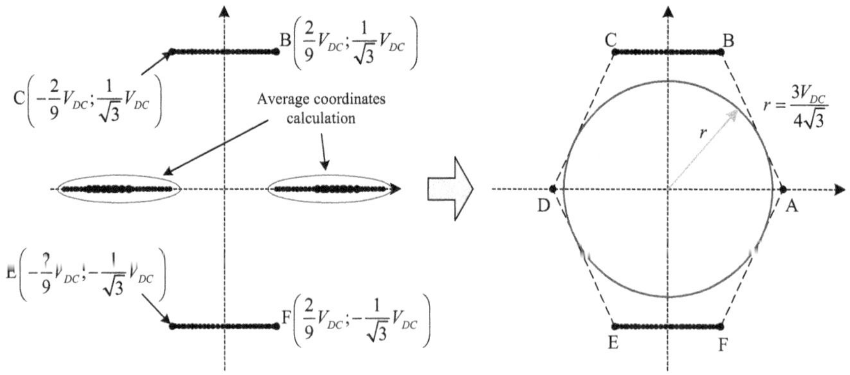

Figure 2.46 Voltage space vector limits for the neutral connection to capacitors midpoint

In terms of other control software modifications and with the aim to reach the same magneto-motive force obtained under normal operating conditions, in this topology, the motor phase currents must increase by a factor of $\sqrt{3}$ with an imposed phase-shift of 60° between the healthy phases. This control adaptation is mandatory for vector controlled drives using hysteresis current controllers. Hence, if rated torque operation is mandatory, both the machine and the VSI must be oversized by an equivalent factor, leading to an increasing of the overall system cost. On the contrary, it can also be guaranteed that under post-fault operating conditions the available torque must be limited to $1/\sqrt{3}$ of the machine rated value. Additionally, more power losses are generated due to the larger current values.

As mentioned previously, this reconfiguration does not allow the converter to generate zero voltage vectors. Therefore, the same post-fault control optimisations used for the phase connection to the capacitors midpoint topology, can also be applied to SVM or DTC techniques.

2.3.3 Redundant topologies

Concerning the redundant fault-tolerant topologies that can be applied to two-level inverter-fed variable speed AC drives, there are also two common approaches as shown in Figure 2.47.

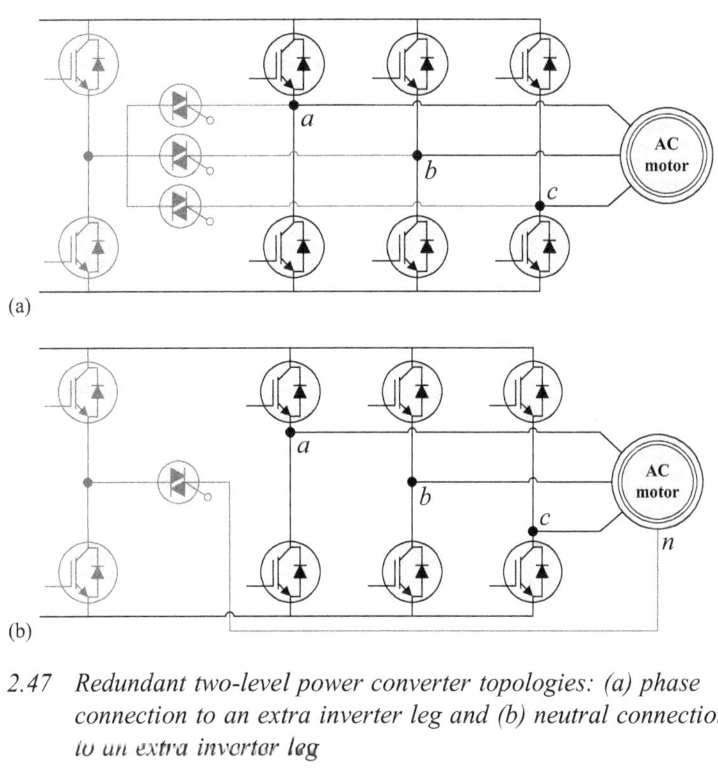

Figure 2.47 Redundant two-level power converter topologies: (a) phase connection to an extra inverter leg and (b) neutral connection to an extra inverter leg

For the first topology, the converter faulty phase is connected to an extra inverter leg through the use of a triac, one for each phase [Figure 2.47(a)]. Similarly, the second topology comprises the machine neutral point connection to an inverter additional leg [Figure 2.47(b)].

A different group of less common redundant topologies based on two-level VSI series configurations will be also addressed.

2.3.3.1 Phase connection to an extra inverter leg

Considering the first redundant topology depicted in Figure 2.47(a) and assuming a three-phase machine, the fault-tolerant power converter comprises a four-phase VSI with three additional triacs that are used to connect each motor phase to the inverter extra leg [205–210]. As a result, after detecting and localising the inverter faulty phase, the corresponding motor phase is connected to a redundant inverter leg, with the same specifications of the original converter.

Naturally, this fault-tolerant converter topology has the great advantage of enabling the post-fault operating conditions to be the same as the healthy operating mode. As far as post-fault modifications are concerned, further than this hardware reconfiguration, it is only needed to redirect the gate command signals of the affected phase to the extra inverter leg. The original control scheme used for normal operating conditions is also valid, and therefore, there is no need for other software reconfigurations.

Despite these great benefits, this topology is also much more expensive. In addition to the three triacs, this approach requires more power semiconductors, including two more extra power switches for the fourth leg with the same ratings of the conventional inverter.

2.3.3.2 Neutral connection to an extra inverter leg

A simpler approach consists of the connection of the machine neutral point to an extra inverter leg by a single triac [211–218] as shown in Figure 2.47(b). Comparing to the previous redundant fault-tolerant converter, instead of using three triacs, only one triac is required, reducing the system complexity and cost.

However, under post-fault operating conditions, this converter topology does not permit the drive to maintain the same performance level as for the healthy case.

Despite rated speed operation is still possible, when the machine is supplied by the neutral connection and the two remaining healthy phases, in order to produce the same magneto-motive force magnitude as for healthy operation, the motor phase currents must increase by a factor of $\sqrt{3}$ with a phase-shift of 60°. Accordingly, under post-fault operating conditions, the available torque must be limited to $1/\sqrt{3}$ of the machine rated value. If rated motor operation is required, other option is to oversize both the machine and the converter by a factor of $\sqrt{3}$.

With respect to the post-fault software reconfigurations, further than the need to redirect the gate command signals of the affected phase to the extra inverter leg, there is no need for other software reconfigurations.

2.3.3.3 Series VSI topologies

If all the machine six winding connections are available, it is possible to connect two VSIs in a series configuration [219–221], as shown in Figure 2.48.

As it can be observed, there are two distinct ways to connect the VSIs. If only one DC power supply is available, both VSIs are connected to the same power supply as depicted in Figure 2.48(a). Under these conditions, each motor phase winding is fed by an H-bridge inverter. The other topology consists on supplying each VSI by independent DC electrical sources [Figure 2.48(b)].

Since these configurations use twice the number of active power semiconductors in comparison with the conventional two-level VSI (12 instead of 6), these topologies have an intrinsic fault-tolerant capability.

When an inverter failure occurs, the versatility of these power converters allows for the creation of additional paths for the current flow and permit the drive continuous operation. For example, if a VSI single power switch open-circuit occurs, the remaining controllable power switches allow for the creation of a neutral point, enabling the machine continuous operation. Naturally, this approach implies the reduction of total voltage vector magnitude applied to the motor, limiting therefore the maximum post-fault operating speed. Another option could be to simply force the current in the faulty phase to zero and control the remaining power switches to generate new and optimised currents for the two healthy phases, compensating for the electromagnetic torque pulsations.

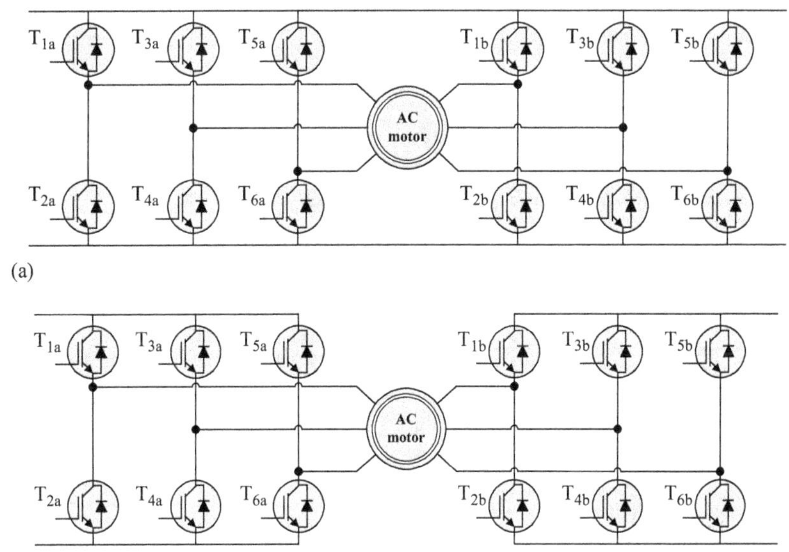

Figure 2.48 Series fault-tolerant converter topologies: (a) same DC supply; (b) two independent DC sources

Another advantage of this topology is related to the occurrence of short-circuits since after guaranteeing the pulse inhibition to the complementary device, equivalent post-fault control options can be adopted in order to allow the continuous drive operation.

Acknowledgement

This work was supported by the European Regional Development Fund (ERDF) through the Operational Programme for Competitiveness and Internationalization (COMPETE 2020), under Project POCI-01-0145-FEDER-029494, and by National Funds through the FCT – Portuguese Foundation for Science and Technology, under Projects PTDC/EEI-EEE/29494/2017 and UID/EEA/04131/2013.

References

[1] S. Nandi, H. A. Toliyat and X. Li, "Condition monitoring and fault diagnosis of electrical motors—a review", *IEEE Transactions on Energy Conversion*, Vol. 20, No. 4, pp. 719–729, 2005.

[2] P. J. Tavner, "Review of condition monitoring of rotating electrical machines", *IET Electric Power Applications*, Vol. 2, No. 4, pp. 215–247, 2008.

[3] P. Zhang, Y. Du, T. G. Habetler and B. Lu, "A survey of condition monitoring and protection methods for medium-voltage induction motors", *IEEE Transactions on Industry Applications*, Vol. 47, No. 1, pp. 34–46, 2011.

[4] W. T. Thomson and I. D. Stewart, "Online current monitoring for fault diagnosis in inverter fed induction motors", *Third International Conference on Power Electronics and Variable-Speed Drives*, pp. 432–435, 1988.

[5] G. B. Kliman, R. A. Koegl, J. Stein, R. D. Endicott and M. W. Madden, "Noninvasive detection of broken rotor bars in operating induction motors", *IEEE Transactions on Energy Conversion*, Vol. 3, No. 4, pp. 873–879, 1988.

[6] J. A. Antonino-Daviu, A. Quijano-Lopez, V. Fuster-Roig and C. Nevot, "Case stories of induction motors fault diagnosis based on current analysis", *PCIC Europe*, pp. 1–9, 2016.

[7] A. Yazidi, H. Henao, G. A. Capolino, M. Artioli, F. Filippetti and D. Casadei, "Flux signature analysis: an alternative method for the fault diagnosis of induction machines", *IEEE Russia Power Tech*, pp. 1–6, 2005.

[8] J. S. Hsu, "Monitoring of defects in induction motors through air-gap torque observation", *IEEE Transactions on Industry Applications*, Vol. 31, No. 5, pp. 1016–1021, 1995.

[9] S. F. Legowski, A. H. M. Sadrul Ula and A. M. Trzynadlowski, "Instantaneous stator power as a medium for the signature analysis of induction motors", *Conference Record of the IEEE Industry Applications Conference*, Vol. 1, pp. 619–624, 1995.

[10] J. Milimonfared, H. M. Kelk, S. Nandi, A. D. Minassians and H. A. Toliyat, "A novel approach for broken-rotor-bar detection in cage induction

motors", *IEEE Transactions on Industrial Applications*, Vol. 35, No. 5, pp. 1000–1006, 1999.

[11] S. Nandi, R. M. Bharadwaj, H. A. Toliyat and A. G. Parlos, "Performance analysis of a three-phase induction motor under mixed eccentricity condition", *International Conference on Power Electronic Drives and Energy Systems for Industrial Growth*, 1998.

[12] D. G. Dorrell, W. T. Thomson and S. Roach, "Analysis of airgap flux, current, and vibration signals as a function of the combination of static and dynamic airgap eccentricity in 3-phase induction motors", *IEEE Transactions on Industry Applications*, Vol. 33, No. 1, pp. 24–34, 1997.

[13] "The effect of repair/rewinding on motor efficiency", *EASA/AEMT Rewind Study and Good Practice Guide to Maintain Motor Efficiency*, 2003.

[14] P. Vas, "Parameter Estimation, Condition Monitoring, and Diagnosis of Electrical Machines", Oxford, U.K.: Clarendon Press, 1993.

[15] S. Nandi, S. Ahmed and H. A. Toliyat, "Detection of rotor slot and other eccentricity related harmonics in a three phase induction motor with different rotor cages", *IEEE Transactions on Energy Conversion*, Vol. 16, No. 3, pp. 253–260, 2001.

[16] S. Nandi, T. C. Ilamparithi, S. B. Lee and D. Hyun, "Detection of eccentricity faults in induction machines based on nameplate parameters", *IEEE Transactions on Industrial Electronics*, Vol. 58, No. 5, pp. 1673–1683, 2011.

[17] X. Li, Q. Wu and S. Nandi, "Performance analysis of a three-phase induction machine with inclined static eccentricity", *IEEE Transactions on Industry Applications*, Vol. 43, No. 2, pp. 531–541, 2007.

[18] S. Nandi, R. M. Bharadwaj and H. A. Toliyat, "Performance analysis of a three-phase induction motor under mixed eccentricity condition", *IEEE Transactions on Energy Conversion*, Vol. 17, No. 3, pp. 392–399, 2002.

[19] T. Yoon, "Magnetically induced vibration in a permanent-magnet brushless DC motor with symmetric pole-slot configuration", *IEEE Transactions on Magnetics*, Vol. 41, No. 6, pp. 2173–2179, 2005.

[20] J. P. Wang and D. K. Lieu, "Magnetic lumped parameter modeling of rotor eccentricity in brushless permanent-magnet motors", *IEEE Transactions on Magnetics*, Vol. 35, No. 5, pp. 4226–4231, 1999.

[21] K. T. Kim, K. S. Kim, S. M. Hwang, T. J. Kim and Y. H. Jung, "Comparison of magnetic forces for IPM and SPM motor with rotor eccentricity", *IEEE Transactions on Magnetics*, Vol. 37, No. 5, pp. 3448–3451, 2001.

[22] C. I. Lee and G. H. Jang, "Experimental measurement and simulated verification of the unbalanced magnetic force in brushless DC motors", *IEEE Transactions on Magnetics*, Vol. 44, No. 11, pp. 4377–4380, 2008.

[23] B. M. Ebrahimi, J. Faiz, M. Javan-Roshtkhari and A. Z. Nejhad, "Static eccentricity fault diagnosis in permanent magnet synchronous motor using time stepping finite element method", *IEEE Transactions on Magnetics*, Vol. 44, No. 11, pp. 4297–4300, 2008.

[24] B. M. Ebrahimi, J. Faiz and M. J. Roshtkhari, "Static-, dynamic-, and mixed-eccentricity fault diagnoses in permanent-magnet synchronous motors",

IEEE Transactions on Industrial Electronics, Vol. 56, No. 11, pp. 4727–4739, 2009.

[25] B. M. Ebrahimi, J. Faiz and B. N. Araabi, "Pattern identification for eccentricity fault diagnosis in permanent magnet synchronous motors using stator current monitoring", *IET Electric Power Applications*, Vol. 4, No. 6, pp. 418–430, 2010.

[26] B. M. Ebrahimi and J. Faiz, "Configuration impacts on eccentricity fault detection in permanent magnet synchronous motors", *IEEE Transactions on Magnetics*, Vol. 48, No. 2, pp. 903–906, 2012.

[27] J. Hong, S. Park, D. Hyun, *et al.*, "Detection and classification of rotor demagnetization and eccentricity faults for PM synchronous motors", *IEEE Transactions on Industry Applications*, Vol. 48, No. 3, pp. 923–932, 2012.

[28] B. M. Ebrahimi and J. Faiz, "Diagnosis and performance analysis of three-phase permanent magnet synchronous motors with static, dynamic and mixed eccentricity", *IET Electric Power Applications*, Vol. 4, No. 1, pp. 53–66, 2010.

[29] M. Sumislawska, K. N. Gyftakis, D. F. Kavanagh, M. McCulloch, K. J. Burnham and D. A. Howey, "The impact of thermal degradation on properties of electrical machine winding insulation material", *IEEE Transactions on Industry Applications*, Vol. 52, No. 4, pp. 2951–2960, 2016.

[30] C. Zoeller, M. A. Vogelsberger, T. M. Wolbank and H. Ertl, "Impact of SiC semiconductor switching transition speed on insulation health state monitoring of traction machines", *IET Power Electronics*, Vol. 9, No. 15, pp. 2769–2775, 2016.

[31] R. M. Tallam, T. G. Habetler and R. G. Harley, "Experimental testing of a neural-network-based turn-fault detection scheme for induction machines under accelerated insulation failure conditions", *4th IEEE SDEMPED*, 2003, pp. 58–62.

[32] A. Siddique, G. S. Yadava and B. Singh, "A review of stator fault monitoring techniques of induction motors", *IEEE Transactions on Energy Conversion*, Vol. 20, No. 1, pp. 106–114, 2005.

[33] K. N. Gyftakis and A. J. M. Cardoso, "A new space vector approach to detect stator faults in induction motors", *IEEE WEMDCD 2017*, Nottingham, UK, 2017.

[34] G. Joksimovic and J. Penman, "The detection of interturn short-circuits in the stator windings of operating motors", *IEEE Transactions on Industrial Electronics*, Vol. 47, No. 5, pp. 1078–1084, 2000.

[35] H. Henao, C. Demian and G. A. Capolino, "A frequency-domain detection of stator winding faults in induction machines using an external flux sensor", *IEEE Transactions on Industry Applications*, Vol. 39, No. 5, pp. 1272–1279, 2003.

[36] S. M. A. Cruz and A. J. M. Cardoso, "Diagnosis of stator inter-turn short-circuits in DTC induction motor drives", *IEEE Transactions on Industry Applications*, Vol. 40, No. 5, pp. 1349–1360, 2004.

[37] S. Nandi, "A detailed model of induction machines with saturation extendable for fault analysis", *IEEE Transactions on Industry Applications*, Vol. 40, No. 5, pp. 1302–1309, 2004.

[38] H. A. Toliyat, S. Nandi, S. Choi and H. Meshgin-Kelk, "Electric Machines Modelling, Condition Monitoring and Fault Diagnosis", Boca Raton, FL: CRC Press, Taylor and Francis Group, pp. 118, 2013.

[39] K. N. Gyftakis, M. Drif and A. J. M. Cardoso, "Thorough investigation of the third current harmonic in delta-connected induction motors suffering from a stator inter-turn fault", *IEEE SDEMPED 2015*, Guarda, Portugal, 2015.

[40] A. Stavrou, H. Sedding and J. Penman, "Current monitoring for detecting inter-turn short-circuits in induction motors", *IEEE Transactions on Energy Conversion*, Vol. 16, No. 1, pp. 32–37, 2001.

[41] A. J. M. Cardoso, S. M. A. Cruz and D. S. B. Fonseca, "Inter-turn stator winding fault diagnosis in three-phase induction motors by Park's vector approach", *IEEE Transactions on Energy Conversion*, Vol. 14, No. 3, pp. 595–598, 1999.

[42] S. M. A. Cruz and A. J. M. Cardoso, "Stator winding fault diagnosis in three-phase synchronous and asynchronous motors, by the extended Park's vector approach", *IEEE Transactions on Industry Applications*, Vol. 37, No. 5, pp. 1227–1233, 2001.

[43] J. K. Park, C. L. Jeong, S. T. Lee and J. Hur, "Early detection technique for stator winding inter-turn fault in BLDC motor using input impedance", *IEEE Transactions on Industry Applications*, Vol. 51, No. 1, pp. 240–247, 2015.

[44] B. M. Ebrahimi and J. Faiz, "Feature extraction for short-circuit fault detection in permanent-magnet synchronous motors using stator-current monitoring", *IEEE Transactions on Power Electronics*, Vol. 25, No. 10, pp. 2673–2682, 2010.

[45] Y. S. Lee, K. T. Kim and J. Hur, "Finite-element analysis of the demagnetization of IPM-type BLDC motor with stator turn fault", *IEEE Transactions on Magnetics*, Vol. 50, No. 2, pp. 1–4, 2014.

[46] S. T. Lee and J. Hur, "Detection technique for stator inter-turn faults in BLDC motors based on third-harmonic components of line currents", *IEEE Transactions on Industry Applications*, Vol. 53, No. 1, pp. 143–150, 2017.

[47] M. Jeong, J. Yun, Y. Park, S. B. Lee and K. N. Gyftakis, "Off-line flux injection test probe for screening defective rotors in squirrel cage induction machines", *IEEE SDEMPED 2017*, Tinos, Greece, 2017.

[48] C. D. Pitis, "Thermo-mechanical stresses of the squirrel cage rotors in adverse load conditions", *IEEE International Symposium on Electrical Insulation*, ISEI 2008.

[49] M. F. Cabanas, J. L. Ruiz Gonzalez, J. L. B. Sampayo, *et al.*, "Analysis of the fatigue causes on the rotor bars of squirrel cage asynchronous motors: experimental analysis and modelling of medium voltage motors", *Proc. SDEMPED*, pp. 247–252, 2003.

[50] G. Y. Sizov, A. Sayed-Ahmed, C. C. Yeh and N. A. O. Demerdash, "Analysis and diagnostics of adjacent and nonadjacent broken-rotor-bar faults in squirrel-cage induction machines", *IEEE Transactions on Industrial Electronics*, Vol. 56, No. 11, pp. 4627–4641, 2009.

[51] T. J. Sobczyk and W. Maciolek, "Does the component $(1-2s)f0$ in stator current is sufficient for detection of rotor cage faults?", *Int. Symp. Diagnostic Elect. Mach., Power Electron. Drives (SDEMPED 2005)*, Vienna, Austria, 2005.

[52] M. Riera-Guasp, M. F. Cabanas, J. A. Antonino-Daviu, M. Pineda-Sanchez and C. H. R. Garcia, "Influence of nonconsecutive bar breakages in motor current signature analysis for the diagnosis of rotor faults in induction motors", *IEEE Transactions on Energy Conversion*, Vol. 25, pp. 80–89, 2010.

[53] J. A. Antonino-Daviu, J. Pons-Llinares and S. B. Lee, "Advanced rotor fault diagnosis for medium-voltage induction motors via continuous transforms", *IEEE Transactions on Industry Applications*, Vol. 52, No. 5, pp. 4503–4509, 2016.

[54] K. N. Gyftakis, D. V. Spyropoulos, J. Kappatou and E. D. Mitronikas, "A novel approach for broken bar fault diagnosis in induction motors through torque monitoring", *IEEE Transactions on Energy Conversion*, Vol. 28, No. 2, pp. 267–277, 2013.

[55] G. B. Kliman, W. J. Premerlani, B. Yazici, R. A. Koegl and J. Mazereeuw, "Sensorless, online motor diagnostics", *IEEE Computer Applications in Power*, Vol. 10, No. 2, pp. 39–43, 1997.

[56] F. Filippetti, G. Franceschini, C. Tassoni and P. Vas, "AI techniques in induction machines diagnosis including the speed ripple effect", *IEEE Transactions on Industry Applications*, Vol. 34, No. 1, pp. 98–108, 1998.

[57] W. Deleroi, "Broken bar in squirrel cage rotor of an induction motor, Part 1: Description by superimposed fault currents", *Archiv für Elektrotechnik*, Vol. 67, pp. 91–99, 1984.

[58] M. Drif and A. J. M. Cardoso, "Discriminating the simultaneous occurrence of three-phase induction motor rotor faults and mechanical load oscillations by the instantaneous active and reactive power media signature analyses", *IEEE Transactions on Industrial Electronics*, Vol. 59, No. 3, pp. 1630–1639, 2012.

[59] S. Shin, J. Kim, S. B. Lee, C. Lim and E. J. Wiedenbrug, "Evaluation of the influence of rotor magnetic anisotropy on condition monitoring of two-pole induction motors", *IEEE Transactions on Industry Applications*, Vol. 51, No. 4, pp. 2896–2904, 2015.

[60] S. Lee, J. Hong, S. B. Lee, E. J. Wiedenbrug, M. Teska and H. Kim, "Evaluation of the influence of rotor axial air ducts on condition monitoring of induction motors", *IEEE Transactions on Industry Applications*, Vol. 49, No. 5, pp. 2024–2033, 2013.

[61] Y. Park, M. Jeong, S. B. Lee, J. A. Antonino-Daviu and M. Teska, "Influence of blade pass frequency vibrations on MCSA-based rotor fault

detection of induction motors", *IEEE Transactions on Industry Applications*, Vol. 53, No. 3, pp. 2049–2058, 2017.

[62] J. Pons-Llinares, J. A. Antonino-Daviu, M. Riera-Guasp, M. Pineda-Sanchez and V. Climente-Alarcon, "Induction motor diagnosis based on a transient current analytic wavelet transform via frequency B-splines", *IEEE Transactions on Industrial Electronics*, Vol. 58, No. 5, pp. 1530–1544, 2011.

[63] G. Georgoulas, P. Karvelis, C. D. Stylios, I. P. Tsoumas, J. A. Antonino-Daviu and V. Climente-Alarcon, "Automatizing the broken bar detection process via short time Fourier transform and two-dimensional piecewise aggregate approximation representation", *IEEE Energy Conversion Congress and Exposition (ECCE)*, pp. 3104–3110, 2014.

[64] D. Morinigo-Sotelo, R. J. Romero-Troncoso, P. A. Panagiotou, J. A. Antonino-Daviu and K. N. Gyftakis, "Reliable detection of broken rotor bars in induction motors via MUSIC and ZSC methods", *IEEE Transactions on Industry Applications*, Vol. 54, No. 2, pp. 1224–1234, 2018.

[65] C. Yang, T. J. Kang, D. Hyun, S. B. Lee, J. A. Antonino-Daviu and J. Pons-Llinares, "Reliable detection of induction motor rotor faults under the rotor axial air duct influence", *IEEE Transactions on Industry Applications*, Vol. 50, No. 4, pp. 2493–2502, 2014.

[66] J. Hong, D. Hyun, S. B. Lee, J. Y. Yoo and K. W. Lee, "Automated monitoring of magnet quality for permanent-magnet synchronous motors at standstill", *IEEE Transactions on Industry Applications*, Vol. 46, No. 4, 2010.

[67] K. C. Kim, S. B. Lim, D. H. Koo and J. Lee, "The shape design of permanent magnet for permanent magnet synchronous motor considering demagnetization", *IEEE Transactions on Magnetics*, Vol. 42, No. 10, pp. 3485–3487, 2006.

[68] J. Hong, S. Park, D. Hyun, *et al.*, "Detection and classification of rotor demagnetization and eccentricity faults for PM synchronous motors", *IEEE Transactions on Industry Applications*, Vol. 48, No. 3, pp. 923–932, 2012.

[69] J. Y. Song, K. J. Kang, C. H. Kang and G. H. Jang, "Cogging torque and unbalanced magnetic pull due to simultaneous existence of dynamic and static eccentricities and uneven magnetization in permanent magnet motors", *IEEE Transactions on Magnetics*, Vol. 53, No. 3, pp. 1–9, 2017.

[70] A. J. Pina Ortega and L. Xu, "Investigation of effects of asymmetries on the performance of permanent magnet synchronous machines", *IEEE Transactions on Energy Conversion*, Vol. 32, No. 3, pp. 1002–1011, 2017.

[71] J. R. R. Ruiz, J. A. Rosero, A. G. Espinosa and L. Romeral, "Detection of demagnetization faults in permanent-magnet synchronous motors under nonstationary conditions", *IEEE Transactions on Magnetics*, Vol. 45, No. 7, pp. 2961–2969, 2009.

[72] T. A. Harris, "Rolling Bearing Analysis", 2nd edition, New York: Wiley, 1984.

[73] R. R. Schoen, T. G. Habetler, F. Kamran and R. G. Bartfield, "Motor bearing damage detection using stator current monitoring", *IEEE Transactions on Industry Applications*, Vol. 31, No. 6, pp. 1274–1279, 1995.
[74] M Blodt, P. Granjon, B. Rainos and G. Rostaing, "Models for bearing damage detection in induction motors using stator current monitoring", *IEEE Transactions on Industrial Electronics*, Vol. 55, No. 4, pp. 1813–1822, 2008.
[75] J. S. Hsu, "Monitoring of defects in induction motors through air-gap torque observation", *IEEE Transactions on Industry Applications*, Vol. 31, No. 5, pp. 1016–1021, 1995.
[76] K. N. Gyftakis, S. B. Lee, J. C. Kappatou and J. A. Antonino-Daviu, "Identification of the broken bar fault in induction motors with rotor air ducts through the torque spectrum", *Int. Conf. Electrical Machines, ICEM '14*, Berlin, Germany, pp. 1608–1614, 2014.
[77] J. Penman, M. N. Dey, A. J. Tait and W. E. Bryan, "Condition monitoring of electrical drives", *Proc. IEEE*, Vol. 133, Pt. B, pp. 142–148, 1986.
[78] J. Antonino-Daviu, A. Quijano-López, V. Climente-Alarcon and H. Razik, "Evaluation of the detectability of rotor faults and eccentricities in induction motors via transient analysis of the stray flux", *IEEE Energy Conversion Congress and Exposition (ECCE)*, pp. 3559–3564, 2017.
[79] A. Ceban, R. Pusca and R. Romary, "Study of rotor faults in induction motors using external magnetic field analysis", *IEEE Transactions on Industrial Electronics*, Vol. 59, No. 5, pp. 2082–2093, 2012.
[80] H. Henao, G. A. Capolino and C. Martis "On the stray flux analysis for the detection of the three-phase induction machine faults", *IEEE-IAS'03*, S. Lake City, 2003.
[81] Z. Liu, J. Huang and B. Li, "Diagnosing and distinguishing rotor eccentricity from partial demagnetisation of interior PMSM based on fluctuation of high-frequency d-axis inductance and rotor flux", *IET Electric Power Applications*, Vol. 11, pp. 1265–1275, 2017.
[82] A. Yazidi, H. Henao and G. -A. Capolino, "Broken rotor bars fault detection in squirrel cage induction machines", *IEEE International Conference on Electric Machines and Drives*, pp. 741–747, 2005.
[83] D. Hyun, S. Lee, J. Hong, S. B. Lee and S. Nandi, "Detection of airgap eccentricity for induction motors using the single-phase rotation test", *IEEE Transactions on Energy Conversion*, Vol. 27, No. 3, pp. 689–696, 2012.
[84] S. R. Movahed, S. H. Oraee Mirzamani, A. Rajabi and H. Daneshvar, "Estimation of insulation life of inverter-fed induction motors", *2010 1st Power Electronic & Drive Systems & Technologies Conference (PEDSTC)*, pp. 335–339, 2010.
[85] G. C. Stone, H. G. Sedding, B. A. Lloyd and B. K. Gupta, "The ability of diagnostic tests to estimate the remaining life of stator insulation", *IEEE Transactions on Energy Conversion*, Vol. 3, No. 4, pp. 833–841, 1988.

[86] N. Lahoud, J. Faucher, D. Malec and P. Maussion, "Electrical aging of the insulation of low-voltage machines: model definition and test with the design of experiments", *IEEE Transactions on Industrial Electronics*, Vol. 60, No. 9, pp. 4147–4155, 2013.

[87] F. Salameh, A. Picot, M. Chabert, E. Leconte, A. Ruiz-Gazen and P. Maussion, "Variable importance assessment in lifespan models of insulation materials: a comparative study", *2015 IEEE 10th SDEMPED*, Guarda, Portugal, 2015.

[88] CIGRE Study Committee SC11, EG11.02, "Hydrogenerator Failures-Results of the Survey", 2003.

[89] G. C. Stone, E. A. Boulter, I. Culbert and H. Dhirani, "Electrical insulation for rotating machines—design, evaluation, aging, testing and repair", *IEEE Press Series on Power Engineering*, 2004.

[90] M. Sumislawska, K. N. Gyftakis, D. F. Kavanagh, M. McCulloch, K. J. Burnham and D. A. Howey, "The impact of thermal degradation on properties of electrical machine winding insulation material", *IEEE Transactions on Industry Applications*, Vol. 52, No. 4, pp. 2951–2960, 2016.

[91] M. Farahani, H. Borsi, E. Gockenbach, K. Schafer and M. Kaufhold, "Behavior of machine insulation systems subjected to accelerated thermal ageing test", *IEEE Transactions on Dielectrics and Electrical Insulation*, Vol. 17, No. 5, pp. 1364–1372, 2010.

[92] H. Torkaman and F. Karimi, "Influence of ambient and test conditions on insulation resistance/polarization index in HV electrical machines—a survey", *IEEE Transactions on Dielectrics and Electrical Insulation*, Vol. 22, No. 1, pp. 241–250, 2015.

[93] K. N. Gyftakis, P. A. Panagiotou, N. Lophitis, D. A. Howey and M. D. McCulloch, "Breakdown resistance analysis of traction motor winding insulation under thermal ageing", *IEEE ECCE 2017*, Cincinnati, OH, USA, 2017.

[94] M. Istad, M. Runde and A. Nysveen, "A review of results from thermal cycling tests of hydrogenerator stator windings", *IEEE Transactions on Energy Conversion*, Vol. 26, No. 3, pp. 890–903, 2011.

[95] 1972 Annual Report Conference on Electrical Insulation and Dielectric Phenomena, National Academy of Sciences, Washington, DC, 1973.

[96] R. Morin, J. P. Novak, R. Bartnikas and R. Ross, "Analysis of in-service aged stator bars", *IEEE Transactions on Energy Conversion*, Vol. 10, No. 4, pp. 645–654, 1995.

[97] A. Wichmann, P. Grunewald and J. Weidner, "Betriebliche Einflussgrossen auf die elektrische Lebensdauer von Hochspannungsisolierungen in Turbogeneratoren", *ETG Fachberichte*, Vol. 16, pp. 44–48, 1985.

[98] K. Kimura and Y. Kaneda, "Breakdown voltage and observation on mica insulation systems after thermal aging and mechanical fatigue", *Proc. 7th Int. Symp. High. Volt. Eng.*, Dresden, pp. 151–154, 1991.

[99] R. Brutsch, M. Tari, K. Frohlich, T. Weilers and R. Vogelsang, "Insulation failure mechanisms of power generators", *IEEE Electrical Insulation Magazine*, Vol. 24, No. 4, pp. 17–25, 2008.

[100] K. Kadotani, T. Miyashita, F. Aki and K. Matsunobu, "An approach for insulation diagnosis of mica-resin coils", *IEEE Transactions on Power Apparatus and Systems*, Vol. PAS-100, No. 9, pp. 4136–4141, 1981.

[101] M. Fenger and G. C. Stone, "Investigations into the effect of humidity on stator winding partial discharges", *IEEE Transactions on Dielectrics and Electrical Insulation*, Vol. 12, No. 2, pp. 341–346, 2005.

[102] M. A. R. M. Fernando, W. M. L. B. Naranpanawa and R. M. H. M. Rathnayake, "Condition assessment of stator insulation during drying, wetting and electrical ageing", *IEEE Transactions on Dielectrics and Electrical Insulation*, Vol. 20, No. 6, pp. 2081–2090, 2013.

[103] T. Tsukui, M. Takamura and Y. Kako, "Correlations between non-destructive and destructive tests on high-voltage coil insulations for rotating machines", *IEEE Transactions on Electrical Insulation*, Vol. EI-15, No. 2, pp. 118–127, 1980.

[104] W. Yin, "Failure mechanism of winding insulations in inverter-fed motors", *IEEE Electrical Insulation Magazine*, Vol. 13, pp. 18–23, 1997.

[105] B. Bose, "Power electronics and motor drives recent progress and perspective", *IEEE Transactions on Industrial Electronics*, Vol. 56, No. 2, pp. 581–588, 2009.

[106] S. Grubic, Jose M. Aller, B. Lu and T. G. Habetler, "A survey on testing and monitoring methods for stator insulation systems of low-voltage induction machines focusing on turn insulation problems", *IEEE Transactions on Industrial Electronics*, Vol. 55, No. 12, pp. 4127–4136, 2008.

[107] R. Morin and R. Bartnikas, "Multistress aging of stator bars in a three-phase model stator under load cycling conditions", *IEEE Transactions on Energy Conversion*, Vol. 27, No. 2, pp. 374–381, 2012.

[108] R. Bartnikas and R. Morin, "Multi-stress aging of stator bars with electrical, thermal, and mechanical stresses as simultaneous acceleration factors", *IEEE Transactions on Energy Conversion*, Vol. 19, No. 4, pp. 702–714, 2004.

[109] V. Smet, F. Forest, J. Huselstein, *et al.*, "Ageing and failure modes of IGBT modules in high-temperature power cycling", *IEEE Transactions on Industrial Electronics*, Vol. 58, No. 10, pp. 4931–4941, 2011.

[110] H. B. A. Sethom and M. A. Ghedamsi, "Intermittent misfiring default detection and localization on a PWM inverter using wavelet decomposition", *Journal of Electrical Systems*, Vol. 4, No. 2, pp. 222–234, 2008.

[111] S. Yang, A. Bryant, P. Mawby, D. Xiang, L. Ran and P. Tavner, "An industry-based survey of reliability in power electronic converters", *IEEE Transaction on Industry Applications*, Vol. 47, No. 3, pp. 1441–1451, 2011.

[112] S. Yang, A. Bryant, P. Mawby, D. Xiang, L. Ran and P. Tavner, "An industry-based survey of reliability in power electronic converters", *IEEE Energy Conversion Congress and Exposition*, pp. 3151–3157, 2009.

[113] S. Yang, D. Xiang, A. Bryant, P. Mawby, L. Ran and P. Tavner, "Condition monitoring for device reliability in power electronic converters—a

review", *IEEE Transactions on Power Electronics*, Vol. 25, No. 11, pp. 2734–2752, 2010.

[114] F. W. Fuchs, "Some diagnosis methods for voltage source inverters in variable speed drives with induction machines—a survey", *29th Annual Conference of the IEEE Industrial Electronics Society*, Vol. 2, pp. 1378–1385, Nov. 2003.

[115] K. Rothenhagen and F. W. Fuchs, "Performance of diagnosis methods for IGBT open-circuit faults in three phase voltage source inverters for ac variable speed drives", *European Conference on Power Electronics and Applications*, 10 pp., 2005.

[116] B. Lu and S. K. Sharma, "A survey of IGBT fault diagnostic methods for three-phase power inverters", *International Conference on Condition Monitoring and Diagnosis*, pp. 756–763, 2008.

[117] B. Lu and S. K. Sharma, "A literature review of IGBT fault diagnostic and protection methods for power inverters", *IAS Annual Meeting Industry Applications Conference*, 8 pp., 2008.

[118] B. Lu and S. K. Sharma, "A literature review of IGBT fault diagnostic and protection methods for power inverters", *IEEE Transactions on Industry Applications*, Vol. 45, No. 5, pp. 1770–1777, 2009.

[119] A. Tommaso, F. Genduso, R. Miceli and G. R. Galluzzo, "A review of multiple faults diagnosis methods in voltage source inverters", *International Conference on Renewable Energy Research and Applications*, Palermo, Italy, pp. 1376–1381, 2015.

[120] A. M. S. Mendes, A. J. M. Cardoso and E. S. Saraiva, "Voltage source inverter fault diagnosis in variable speed AC drives, by Park's vector approach", *Seventh International Conference on Power Electronics and Variable Speed Drives*, Conf. Publ. No. 456, pp. 538–543, 1998.

[121] P. Gilreath and B. N. Singh, "A new centroid based fault detection method for 3-phase inverter-fed induction motors", *IEEE Power Electronics Specialists Conference*, pp. 2664–2669, 2005.

[122] F. Zidani, D. Diallo and M. Benbouzid and R. Nait-Said, "A fuzzy-based approach for the diagnosis of fault modes in a voltage-fed PWM inverter induction motor drive", *IEEE Transactions on Industrial Electronics*, Vol. 55, No. 2, pp. 586–593, 2008.

[123] D. Foito, J. F. Martins, V. F. Pires and J. Maia, "An eigenvalue/eigenvector 3D current reference method for detection and fault diagnosis in a voltage source inverter", *Annual Conference of IEEE Industrial Electronics*, pp. 190–194, 2009.

[124] T. Orlowska-Kowalska and P. Sobanski, "Simple diagnostic technique of a single IGBT open-circuit faults for a SVM-VSI vector controlled induction motor drive", *Bulletin of the Polish Academy of Sciences Technical Sciences*, Vol. 63, No. 1, pp. 281–288, 2015.

[125] N. Raja, J. Mathewb, J. Ga and S. Georgea, "Open-transistor fault detection and diagnosis based on current trajectory in a two-level voltage source inverter", *Procedia Technology*, Vol. 25, pp. 669–675, 2016.

[126] R. Peuget, S. Courtine and J. P. Rognon, "Fault detection and isolation on a PWM inverter by knowledge-based model", *IEEE Transactions on Industry Applications*, Vol. 34, No. 6, pp. 1318–1326, 1998.

[127] M. Trabelsi, M. Boussak and M. Gossa, "Multiple IGBTs open-circuit faults diagnosis in voltage source inverter fed induction motor using modified slope method", *XIX International Conference on Electrical Machines*, 6 pp., 2010.

[128] N. Freire, J. Estima and A. J. M. Cardoso, "Open-circuit fault diagnosis in PMSG drives for wind turbine applications", *IEEE Transactions on Industrial Electronics*, Vol. 60, No. 9, pp. 3957–3967, 2013.

[129] I. Jlassi and A. J. M. Cardoso, "A single fault diagnostics approach for power switches, speed sensors and current sensors in regenerative PMSM drives", *IEEE 11th International Symposium on Diagnostics for Electrical Machines, Power Electronics and Drives*, Tinos, Greece, pp. 366–372, 2017.

[130] A. M. S. Mendes and A. J. M. Cardoso, "Voltage source inverter fault diagnosis in variable speed AC drives, by the average current Park's vector approach", *IEEE International Electric Machines and Drives Conference*, Seattle, WA, USA, pp. 704–706, 1999.

[131] J. A. A. Caseiro, A. M. S. Mendes and A. J. M. Cardoso, "Fault diagnosis on a PWM rectifier AC drive system with fault tolerance using the average current Park's vector approach", *IEEE International Electric Machines and Drives Conference*, Miami, FL, USA, pp. 695–701, 2009.

[132] J. A. A. Caseiro, A. M. S. Mendes and A. J. M. Cardoso, "The usage of a PWM rectifier to improve three-phase induction motor drive performance under fault conditions", *European Conference on Power Electronics and Applications*, Barcelona, Spain, 10 pp., 2009.

[133] S. Abramik, W. Sleszynski, J. Niezaniski and H. Piquet, "A diagnostic method for on-line fault detection and localization in VSI-fed AC drives", *European Conference on Power Electronics and Applications*, 8 pp., 2003.

[134] W. Sleszynski, J. Nieznanski and A. Cichowski, "Open-transistor fault diagnostics in voltage-source inverters by analyzing the load currents", *IEEE Transactions on Industrial Electronics*, Vol. 56, No. 11, pp. 4681–4688, 2009.

[135] J. O. Estima and A. J. M. Cardoso, "A new approach for real-time multiple open-circuit fault diagnosis in voltage source inverters", *IEEE Transactions on Industry Applications*, Vol. 47, No. 6, 8 pp., 2011.

[136] J. O. Estima and A. J. M. Cardoso, "A new algorithm for real-time multiple open-circuit fault diagnosis in voltage fed PWM motor drives by the reference current errors", *IEEE Transactions on Industrial Electronics*, Vol. 60, No. 8, pp. 3496–3505, 2013.

[137] L. Yu, Y. Zhang, W. Huang and K. Teffah, "A fast-acting diagnostic algorithm of insulated gate bipolar transistor open-circuit faults for power inverters in electric vehicles", *Energies*, Vol. 10, No. 4, 6 pp., 2017.

[138] J. Zhang, J. Zhao, D. Zhou and C. Huang, "High-performance fault diagnosis in PWM voltage-source inverters for vector-controlled induction motor drives", *IEEE Transactions on Power Electronics*, Vol. 29, No. 11, pp. 6087–6099, 2014.

[139] J. H. Choi, S. Kim, D. S. Yoo and K. H. Kim, "A diagnostic method of simultaneous open-switch faults in inverter-fed linear induction motor drive for reliability enhancement", *IEEE Transactions on Industrial Electronics*, Vol. 62, No. 7, pp. 4065–4077, 2015.

[140] F. Wu and J. Zhao, "A real-time multiple open-circuit fault diagnosis method in voltage-source-inverter fed vector controlled drives", *IEEE Transactions on Power Electronics*, Vol. 31, No. 2, pp. 1425–1437, 2016.

[141] D. U. Campos-Delgado and D. R. Espinoza-Trejo, "An observer-based diagnosis scheme for single and simultaneous open-switch faults in induction motor drives", *IEEE Transactions on Industrial Electronics*, Vol. 58, No. 2, pp. 671–679, 2011.

[142] I. Jlassi, J. O. Estima, S. K. El Khil, N. M. Bellaaj and A. J. M. Cardoso, "Multiple open-circuit faults diagnosis in back-to-back converters of PMSG drives for wind turbine systems", *IEEE Transactions on Power Electronics*, Vol. 30, No. 5, pp. 2689–2702, 2015.

[143] I. Jlassi, J. O. Estima, S. K. El Khil, N. M. Bellaaj and A. J. M. Cardoso, "A robust observer-based method for IGBTs and current sensors fault diagnosis in voltage-source inverters of PMSM drives", *IEEE Transactions on Industry Applications*, Vol. 53, No. 3, pp. 2894–2905, 2017.

[144] S. M. Jung, J. S. Park, H. W. Kim, K. Y. Cho and M. J. Youn, "An MRAS-based diagnosis of open-circuit fault in PWM voltage-source inverters for PM synchronous motor drive systems", *IEEE Transactions on Power Electronics*, Vol. 28, No. 5, pp. 2514–2526, 2013.

[145] Q. T. An, L. Sun and L. Z. Sun, "Current residual vector-based open-switch fault diagnosis of inverters in PMSM drive systems", *IEEE Transactions on Power Electronics*, Vol. 30, No. 5, pp. 2814–2827, 2015.

[146] F. Asghar, M. Talha and S. Kim, "Neural network based fault detection and diagnosis system for three-phase inverter in variable speed drive with induction motor", *Journal of Control Science and Engineering*, Vol. 2016, Article ID 1286318, 12 pp., 2016.

[147] A. Rohan and S. Kim, "Fault detection and diagnosis system for a three-phase inverter using a DWT-based artificial neural network", *International Journal of Fuzzy Logic and Intelligent Systems*, Vol. 16, No. 4, pp. 238–245, 2016.

[148] J. F. Martins, V. F. Pires, C. Lima and A. J. Pires, "Fault detection and diagnosis of grid-connected power inverters using PCA and current mean value", *38th Annual Conference on IEEE Industrial Electronics Society*, Montreal, Canada, pp. 5185–5190, 2012.

[149] F. Asghar, M. Talha and S. Kim, "A Matlab and Simulink based three-phase inverter fault diagnosis method using three-dimensional features", *International Journal of Fuzzy Logic and Intelligent Systems*, Vol. 16, No. 3, pp. 173–180, 2016.

[150] M. Shahbazi, M. Zolghadri, P. Poure and S. Saadate, "Fast detection of open-switch faults with reduced sensor count for a fault-tolerant three-phase converter", *Power Electronics, Drive Systems and Technologies Conference*, Tehran, pp. 546–550, 2011.

[151] M. Trabelsi, M. Boussak and M. Gossa, "PWM-switching pattern-based diagnosis scheme for single and multiple open-switch damages in VSI-fed induction motor drives", *ISA Transactions*, Vol. 51, No. 2, pp. 333–344, 2012.

[152] S. Ghorpade and S. R. Jagtap, "Detection and identification of open-circuit faults in VSI-fed induction motor drive", *IEEE International Conference on Advanced Communications, Control and Computing Technologies*, Ramanathapuram, pp. 306–308, 2014.

[153] C. Shu, C. Ya-Ting, Y. Tian-Jian and W. Xun, "A novel diagnostic technique for open-circuited faults of inverters based on output line-to-line voltage model", *IEEE Transactions on Industrial Electronics*, Vol. 63, No. 7, pp. 4412–4421, 2016.

[154] Z. Li, Z. Bai, H. Ma and Y. Wang, "Fast open-transistor fault diagnosis based on calculated bridge arm pole-to-pole voltages in voltage-source inverters", *IEEE 2nd Annual Southern Power Electronics Conference*, Auckland, 6 pp., 2016.

[155] B. Cai, Y. Zhao, H. Liu and M. Xie, "A data-driven fault diagnosis methodology in three-phase inverters for PMSM drive systems", *IEEE Transactions on Power Electronics*, Vol. 32, No. 7, pp. 5590–5600, 2017.

[156] R. L. A. Ribeiro, C. B. Jacobina, E. R. C. Silva and A. M. N. Lima, "Fault detection of open-switch damage in voltage-fed PWM motor drive systems", *IEEE Transactions on Power Electronics*, Vol. 18, No. 2, pp. 587–593, 2003.

[157] M. Alavi, D. Wang and M. Luo, "Short-circuit fault diagnosis for three-phase inverters based on voltage-space patterns", *IEEE Transactions on Industrial Electronics*, Vol. 61, No. 10, pp. 5558–5569, 2014.

[158] V. Gomathy and S. Selvaperumal, "Fault detection and classification with optimization techniques for three phase single inverter circuit", *Journal of Power Electronics*, Vol. 16, No. 3, 14 pp., 2016.

[159] Z. Li, Y. Wang, H. Ma and L. Hong, "Open-transistor faults diagnosis in voltage-source inverter based on phase voltages with sliding-window counting method", *42nd Annual Conference of the IEEE Industrial Electronics Society*, Florence, pp. 435–440, 2016.

[160] A. Adouni, B. Francois and L. Sbita, "Open-circuit fault detection and diagnosis in pulse-width modulation voltage source inverters based on novel pole voltage approach", *Transactions of the Institute of Measurement and Control*, Vol. 38, No. 7, pp. 795–804, 2016.

[161] Q. T. An, L. Z. Sun, K. Zhao and T. M. Jahns, "Low-cost diagnostic method for open-switch faults in inverters", *Electronics Letters*, Vol. 46, No. 14, pp. 1021–1022, 2010.

[162] Q. T. An, L. Z. Sun, K. Zhao and L. Sun, "Switching function model based fast-diagnostic method of open-switch faults in inverters without sensors", *IEEE Transactions on Power Electronics*, Vol. 26, No. 1, pp. 119–126, 2011.

[163] M. Trabelsi, M. Boussak, P. Mestre and M. Gossa, "Pole voltage based-approach for IGBTs open-circuit fault detection and diagnosis in PWM-VSI-fed induction motor drives", *International Conference on Power Engineering, Energy and Electrical Drives*, 6 pp., 2011.

[164] M. Trabelsi, M. Boussak, P. Mestre and M. Gossa, "An improved diagnosis technique for IGBTs open-circuit fault in PWM-VSI-fed induction motor drive", *IEEE International Symposium on Industrial Electronics*, pp. 2111–2117, 2011.

[165] Y. Wang, Z. Li and H. Ma, "Comparison of two diagnosis methods based on switching voltage pattern for IGBTs open-circuit faults in voltage-source inverters", *IEEE 24th International Symposium on Industrial Electronics*, Buzios, pp. 257–262, 2015.

[166] M. A. Rodríguez-Blanco, A. Claudio-Sánchez, D. Eilliol, *et al.*, "A failure-detection strategy for IGBT based on gate-voltage behavior applied to a motor drive system", *IEEE Transactions on Industrial Electronics*, Vol. 58, No. 5, pp. 1625–1633, 2011.

[167] S.-M. Jung, J.-S. Park, H.-S. Kim, H.-W. Kim and M.-J. Youn, "Simple switch open fault detection method of voltage source inverter", *IEEE Energy Conversion Congress and Exposition*, pp. 3175–3181, 2009.

[168] Y. Wang, Z. Li and L. Lin, "A novel diagnosis method based on flexible error voltage for IGBTs open-circuit faults in voltage-source inverters", *41st Annual Conference of the IEEE Industrial Electronics Society*, Yokohama, pp. 19–24, 2015.

[169] N. M. A. Freire, J. O. Estima and A. J. M. Cardoso, "A voltage-based approach without extra hardware for open-circuit fault diagnosis in closed-loop PWM AC regenerative drives", *IEEE Transactions on Industrial Electronics*, Vol. 61, No. 9, pp. 4960–4970, 2014.

[170] P. Sobanski, "Fuzzy-logic-based approach to voltage inverter fault diagnosis in induction motor drive", *Przegląd Elektrotechniczny*, Vol. 90, No. 6, pp. 149–153, 2014.

[171] S. Piotr and T. Orłowska-Kowalska, "Detection of single and multiple IGBTs open-circuit faults in a field-oriented controlled induction motor drive", *Archives of Electrical Engineering*, Vol. 66, No. 1, pp. 89–104, 2017.

[172] J.-R. Fu and T. A. Lipo, "A strategy to isolate the switching device fault of a current regulated motor drive", *IAS Annual Meeting Industry Applications Conference*, Vol. 2, pp. 1015–1020, 1993.

[173] B. A. Welchko, T. A. Lipo, T. M. Jahns and S. E. Schulz, "Fault tolerant three-phase AC motor drive topologies; a comparison of features, cost, and

limitations", *IEEE Transactions on Power Electronics*, Vol. 19, No. 4, pp. 1108–1116, 2004.

[174] A. M. S. Mendes and A. J. M. Cardoso, "Fault-tolerant operating strategies applied to three-phase induction-motor drives", *IEEE Transactions on Industrial Electronics*, Vol. 53, No. 6, pp. 1807–1817, 2006.

[175] F. Genduso and R. Miceli, "A general mathematical model for non-redundant fault-tolerant inverters", *IEEE International Electric Machines & Drives Conference*, pp. 705–710, 2011.

[176] J. O. Estima and A. J. M. Cardoso, "Fast fault detection, isolation and reconfiguration in fault-tolerant permanent magnet synchronous motor drives", *IEEE Energy Conversion Congress and Exposition*, Raleigh, NC, USA, 2012.

[177] J. O. Estima and A. J. M. Cardoso, "Efficiency evaluation of fault-tolerant operating strategies applied to three-phase permanent magnet synchronous motor drives", *International Conference on Electrical Machines*, Marseille, France, 7 pp., 2012.

[178] A. Stabile, J. O. Estima, C. Boccaletti and A. J. M. Cardoso, "Converter power loss analysis in a fault-tolerant permanent-magnet synchronous motor drive", *IEEE Transactions on Industrial Electronics*, Vol. 62, No. 3, pp. 1984–1996, 2015.

[179] D. Zhou, J. Zhao and Y. Liu, "Independent control scheme for non-redundant two-leg fault-tolerant back-to-back converter-fed induction motor drives", *IEEE Transactions on Industrial Electronics*, Vol. 63, No. 11, pp. 6790–6800, 2016.

[180] K. Ni, Y. Hu, Y. Liu and C. Gan, "Overview on fault-tolerant four-switch three-phase voltage source converters", *Chinese Journal of Electrical Engineering*, Vol. 3, No. 2, pp. 87–101, 2017.

[181] J. Klima, "Analytical investigation of an induction motor fed from four-switch VSI with a new space vector modulation strategy", *IEEE Transactions on Energy Conversion*, Vol. 21, No. 4, pp. 832–838, 2006.

[182] M. B. R. Correa, C. B. Jacobina, E. R. C. Silva and A. M. N. Lima, "A general PWM strategy for four-switch three-phase inverters", *IEEE Transactions on Power Electronics*, Vol. 21, No. 6, pp. 1618–1627, 2006.

[183] P. Q. Dzung, L. M. Phuong, P. Q. Vinh, N. M. Hoang and T. C. Binh, "New space vector control approach for four switch three phase inverter (FSTPI)", *International Conference on Power Electronics and Drive Systems*, pp. 1002–1008, 2007.

[184] B. El Badsi, A. Guermazi and A. Masmoudi, "On the comparison between different space vector PWM strategies implemented in FSTPI-fed induction motor drives", *International Journal for Computation and Mathematics in Electrical and Electronic Engineering*, Vol. 26, No. 1, pp. 127–147, 2007.

[185] M. Monfared, H. Rastegar, H. M. Kojabadi, "Overview of modulation techniques for the four-switch converter topology", *IEEE International Power and Energy Conference*, pp. 803–807, 2008.

[186] Q.-T. An, L. Sun, K. Zhao and T. M. Jahns, "Scalar PWM algorithms for four-switch three-phase inverters", *Electronics Letters*, Vol. 46, No. 13, pp. 900–902, 2010.

[187] H. S. Alavije and M. Akhbari, "Investigation of induction motor drive behavior in low-cost fault tolerant control for electric vehicles", *International Power Engineering and Optimization Conference*, pp. 176–181, 6–7 2011.

[188] S. Bhattacharya, P. Deb, S. K. Biswas and S. K. Chowdhury, "A comprehensive study of modulation strategies for three phase low cost PWM converter", *International Journal of Engineering Science and Technology*, Vol. 3, No. 7, pp. 5772–5777, 2011.

[189] F. Blaabjerg, D. O. Neacsu and J. K. Pedersen, "Adaptive SVM to compensate DC-link voltage ripple for four-switch three-phase voltage-source inverters", *IEEE Transactions on Power Electronics*, Vol. 14, No. 4, pp. 743–752, 1999.

[190] D. Sun, X. Liu, L. Shang and Y. B. Ivonne, "Four-switch three-phase inverter fed DTC system considering DC-link voltage imbalance", *International Conference on Electrical Machines and Systems*, pp. 1068–1072, 2008.

[191] S. Kazemlou and M. R. Zolghadri, "Direct torque control of four-switch three phase inverter fed induction motor using a modified SVM to compensate dc-link voltage imbalance", *International Conference on Electric Power and Energy Conversion Systems*, pp. 1–6, 2009.

[192] H.-H. Lee, P. Q. Dzung, L. D. Khoa and L. M. Phuong, "Development of space vector PWM for four switch three phase inverter fed induction motor with DC-link voltage imbalance", *IEEE International Conference on Industrial Technology*, pp. 1–6, 2009.

[193] H.-H. Lee, P. Q. Dzung, L. D. Khoa and L. M. Phuong, "Optimized adaptive space vector pulse width modulation for four switch three phase inverter under DC-link voltage ripple condition", *IEEE Region 10 Conference*, pp. 1–6, 2009.

[194] D. Sun, Z. He, Y. He and Y. Guan, "Four-switch inverter fed PMSM DTC with SVM approach for fault tolerant operation", *IEEE International Electric Machines & Drives Conference*, Vol. 1, pp. 295–299, 2007.

[195] B. Wang, Y. He and Y. B. Ivonne, "Four switch three phase inverter fed PMSM DTC system with nonlinear perpendicular flux observer and sliding mode control", *International Conference on Electrical Machines and Systems*, pp. 3206–3211, 2008.

[196] Y. B. Ivonne, D. Sun and Y.-K. He, "Study on inverter fault-tolerant operation of PMSM DTC", *Journal of Zhejiang University SCIENCE A*, Vol. 9, No. 2, pp. 156–164, 2008.

[197] S. B. Ozturk, W. C. Alexander and H. A. Toliyat, "Direct torque control of four-switch brushless DC motor with non-sinusoidal back EMF", *IEEE Transactions on Power Electronics*, Vol. 25, No. 2, pp. 263–271, 2010.

[198] K. D. Hoang, Z. Q. Zhu and M. P. Foster, "Influence and compensation of inverter voltage drop in direct torque-controlled four-switch three-phase PM brushless AC drives", *IEEE Transactions on Power Electronics*, Vol. 26, No. 8, pp. 2343–2357, 2011.

[199] T. H. Liu, J. R. Fu and T. A. Lipo, "A strategy for improving reliability of field-oriented controlled induction motor drives", *IEEE Transactions on Industry Applications*, Vol. 29, No. 5, pp. 910–918, 1993.

[200] B. R. Correa, C. B. Jacobina, E. R. C. Silva and A. M. N. Lima, "An induction motor drive system with improved fault tolerance", *IEEE Transactions on Industry Applications*, Vol. 37, No. 3, pp. 873–879, 2001.

[201] O. Wallmark, L. Harnefors and O. Carlson, "Post-fault operation of fault-tolerant inverters for PMSM drives", *European Conference on Power Electronics and Applications*, 10 pp., 2005.

[202] Z. Q. Zhu, K. Utaikaifa, K. Hoang, Y. Liu and D. Howe, "Direct torque control of three-phase PM brushless AC motor with one phase open-circuit fault", *IEEE International Electric Machines and Drives Conference*, pp. 1180–1187, 2009.

[203] B. Mirafzal, "Survey of fault-tolerance techniques for three-phase voltage source inverters", *IEEE Transactions on Industrial Electronics*, Vol. 61, No. 10, pp. 5192–5202, 2014.

[204] J. Estima, "Development and analysis of permanent magnet synchronous motor drives with fully integrated inverter fault-tolerant capabilities", *PhD Thesis*, University of Coimbra, 2012.

[205] R. L. A. Ribeiro, C. B. Jacobina, E. R. C. Silva and A. M. N. Lima, "A fault tolerant induction motor drive system by using a compensation strategy on the PWM-VSI topology", *IEEE 32nd Annual Power Electronics Specialists Conference*, Vol. 2, pp. 1191–1196, 2001.

[206] R. L. A. Ribeiro, C. B. Jacobina, E. R. C. Silva and A. M. N. Lima, "Fault-tolerant voltagefed PWM inverter AC motor drive systems", *IEEE Transactions on Industrial Electronics*, Vol. 51, No. 2, pp. 439–446, 2004.

[207] R. R. Errabelli and P. Mutschler, "A fault tolerant control and power electronic for a permanent magnet synchronous motor drive", *European Conference on Power Electronics and Applications*, 10 pp., 2011.

[208] A. Cordeiro, J. Palma, J. Maia and M. Resende, "Combining mechanical commutators and semiconductors in fast changing redundant inverter topologies", *IEEE International Conference on Computer as a Tool*, pp. 1–4, 2011.

[209] R. Errabelli and P. Mutschler, "Fault tolerant voltage source inverter for permanent magnet drives", *IEEE Transactions on Power Electronics*, Vol. 27, No. 2, 2012.

[210] P. Garg, S. Essakiappan, H. S. Krishnamoorthy and P. N. Enjeti, "A fault-tolerant three-phase adjustable speed drive topology with active common-mode voltage suppression", *IEEE Transactions on Power Electronics*, Vol. 30, No. 5, pp. 2828–2839, 2015.

[211] R. L. A. Ribeiro, C. B. Jacobina, A. M. N. Lima and E. R. C. Silva, "A strategy for improving reliability of motor drive systems using a four-leg three-phase converter", *IEEE Applied Power Electronics Conference and Exposition*, Vol. 1, pp. 385–391, 2001.

[212] N. Bianchi, S. Bolognani, M. Zigliotto, M. Zordan, "Innovative remedial strategies for inverter faults in IPM synchronous motor drives", *IEEE Transactions on Energy Conversion*, Vol. 18, No. 2, pp. 306–314, 2003.

[213] B. A. Welchko, T. M. Jahns and T. A. Lipo, "Fault interrupting methods and topologies for interior PM machine drives", *IEEE Power Electronics Letters*, Vol. 2, No. 4, pp. 139–143, 2004.

[214] O. Wallmark, L. Harnefors and O. Carlson, "Control algorithms for a fault-tolerant PMSM drive", *IEEE Transactions on Industrial Electronics*, Vol. 54, No. 4, pp. 1973–1980, 2007.

[215] L. Lillo, L. Empringham, P. Wheeler, S. Khwan and C. Gerada, "Emulation of faults and remedial control strategies in a multiphase power converter drive used to analyse fault tolerant drive systems for aerospace applications", *European Conference on Power Electronics and Applications*, 6 pp., 2009.

[216] O. Jasim, M. Sumner, C. Gerada and J. Arellano-Padilla, "Development of a new fault-tolerant induction motor control strategy using an enhanced equivalent circuit model", *IET Electric Power Applications*, Vol. 5, No. 8, pp. 618–627, 2011.

[217] F. Meinguet, X. Kestelynx, E. Semailx and J. Gyselinck, "Fault detection, isolation and control reconfiguration of three-phase PMSM drives", *IEEE International Symposium on Industrial Electronics*, pp. 2091–2096, 27–30 2011.

[218] W. Zhang, D. Xu, P. N. Enjeti, H. Li, J. T. Hawke and H. S. Krishnamoorthy, "Survey on fault-tolerant techniques for power electronic converters", *IEEE Transactions on Power Electronics*, Vol. 29, No. 12, pp. 6319–6331, 2014.

[219] M.-A. Shamsi-Nejad, B. Nahid-Mobarakeh, S. Pierfederici and F. Meibody-Tabar, "Control strategies for fault tolerant PM drives using series architecture", *IEEE Vehicle Power and Propulsion Conference*, pp. 1–6, 1–3 2010.

[220] M.-A. Shamsi-Nejad, B. Nahid-Mobarakeh, S. Pierfederici and F. Meibody-Tabar, "Series architecture for fault tolerant PM drives: operating modes with one or two DC voltage source(s)", *IEEE International Conference on Industrial Technology*, pp. 1525–1530, 2010.

[221] J. A. Restrepo, A. Berzoy, A. E. Ginart, J. M. Aller, R. G. Harley and T. G. Habetler, "Switching strategies for fault tolerant operation of single DC-link dual converters", *IEEE Transactions on Power Electronics*, Vol. 27, No. 2, pp. 509–518, 2012.

Chapter 3
Switched reluctance machine drives

Davide S. B. Fonseca[1] *and Natália S. Gameiro*[1,2]

Switched reluctance machine drives present a unique ability for fault-tolerant applications due to a high level of electric and magnetic independence between motor phases. This feature allows the machine to keep running with a high degree of safety and low-performance degradation, when a fault occurs in one or more phases.

A fault in one of the system phases, either a motor winding fault or a power converter switch fault, can be detected and isolated without causing harmful effects in the other phases. Performance degradation is, of course, proportional to the number of faulty phases. Therefore, fault diagnosis is essential to isolate the fault and promote control and/or hardware reconfigurations to achieve a better performance.

In this chapter, the following issues will be addressed:

- Overall characteristics of switched reluctance motor (SRM) drives.
- Comprehensive SRM drives fault analysis.
- Description of suitable diagnostic techniques oriented to machine faults and/or power converter faults.
- Analysis of diagnostic techniques for open- and/or short-circuit fault detection in power switches that provide faulty element identification.
- Post-fault reconfiguration strategies based on control reconfiguration and/or hardware reconfiguration.

3.1 The switched reluctance motor

Whenever a coil is excited, if a mechanical movement promotes the increasing of the inductance, this movement takes place to maximize the coil's inductance. It is based on this principle that reluctance machines, such as a SRM, works.

Variable reluctance stepper motors and SRM share the same working principles. However, they are designed with different goals.

SRM is designed for efficient power conversion. The phases are fed in synchronization with rotor position, requiring shaft position feedback information.

[1]CISE – Electromechatronic Systems Research Centre, Universidade da Beira Interior, Portugal
[2]Polytechnic Institute of Leiria, Portugal

Variable reluctance stepper motors, on the other hand, are designed as torque motors, with a limited speed range, to maintain step-integrity rather than to achieve efficient power conversion. The variable reluctance stepper motor phases are usually fed with a square wave current without rotor position feedback [1].

The SRM presents salient poles on both stator and rotor. The winding is placed on the stator in the form of concentrated coils. Each phase winding comprises two coils, commonly wound on diametrically opposite poles. The coils of a phase can be connected in series or parallel, but they must generate addictive magnetic fluxes. In the present text, the most common situation of a series connection between the coils of a phase will be assumed. This connection naturally ensures a more balanced distribution of the radial forces.

To enable torque production and to obtain a machine with self-starting capability for any rotor position, the SRM design must avoid areas of constant inductance and ensure there is always a stator phase for which a variation of the rotor position implies an increasing of the inductance. For that, the number of poles in the stator (N_S) and the number of poles in the rotor (N_R) must be different (usually $N_S>N_R$) [2,3].

Several combinations of stator and rotor pole numbers have been presented over the years; however, when considering fault-tolerant applications, the following considerations must be taken:

- The number of phases must be as high as possible to assure a smaller impact when a motor phase is disabled.
- The phase operation independency must be assured both in the motor and in the power converter.

However, motors with a high number of phases, such as 5 or 7, will have lower torque density, because torque density is directly related to the ratio between the maximum and minimum inductances. This ratio decreases with the increasing of the number of poles. In addition, the increment of the number of motor phases leads to higher commutation frequencies and higher losses [2].

Thus, the most common SRM drives are the four-phase 8/6 SRM ($N_S=8$ and $N_R=6$) and the three-phase 6/4 SRM ($N_S=6$ and $N_R=4$), both presented in Figure 3.1. However, the three-phase 6/4 SRM has a higher torque ripple, due to significant torque dips, and in case of a phase open-circuit fault, the motor has some positions where the self-starting capability is lost.

During motoring mode, each SRM phase is fed if its inductance increases with the desired rotational movement of the shaft. Figure 3.2 exemplifies how an 8/6 SRM works. It can be easily seen that the motor develops small displacements (steps) for each excited phase.

The value of these small displacements, designated as step angle or stroke angle, is given by

$$\varepsilon = \frac{2\pi}{mN_R} \tag{3.1}$$

where m is the number of stator phases, usually equal to half of the number of stator poles (N_S).

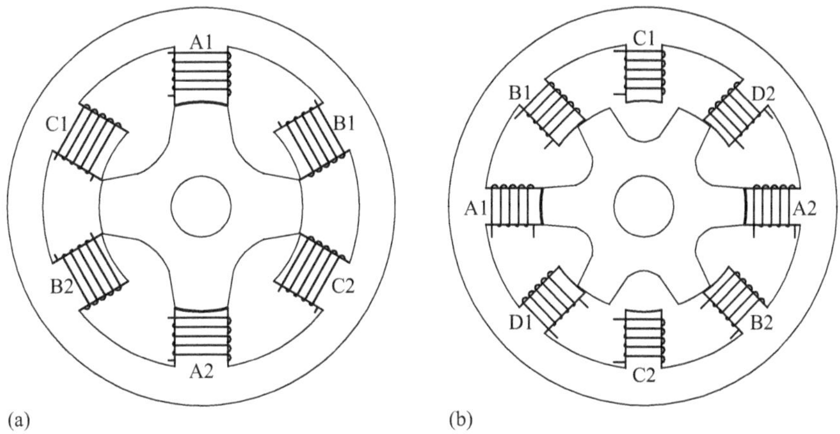

Figure 3.1 Cross section of: (a) 6/4 SRM (b) 8/6 SRM

It can be also concluded that, for the rotor to perform a complete rotation, $m{\cdot}N_R$ steps are required, and each phase is excited N_R times (equally spaced in time in the case of constant speed) developing motoring torque between the unaligned position and the aligned position.

1. The aligned position
   When the rotor is in a position where the symmetry axis of its polar zone is aligned with a phase, as show in Figure 3.3 for phase A, then this phase is in the aligned position.

   If phase A is driven by a current, no torque is created by that current because the air gap in this position takes its minimum value; a position of maximum inductance. However, if there is a small displacement in the rotor, this current already produces a torque that will cause the rotor to recover the position of the initial alignment.

   Because the flux path has a minimum length in the air gap, the current at which saturation begins is very small compared with other rotor positions, particularly the unaligned position. Thus, in the aligned position, magnetic saturation phenomena is very common, as can be seen in Figure 3.4 were a complete set of magnetization curves, for several rotor positions and different magneto-motive forces generated by the phase current, are presented.
2. The unaligned position
   When the rotor is in a position where the symmetry axis of its interpolar zone is aligned with a phase, then this phase is in the unaligned position, as show in Figure 3.5, for phase A.

   If the phase is driven by a current, no torque is developed because the air gap in this position assumes its maximum value, thus the phase inductance presents its minimum value. However, in the case of a small displacement in the rotor, this current will produces a torque that will take the rotor to the next alignment position. Thus, the unaligned position is a position of unstable equilibrium.

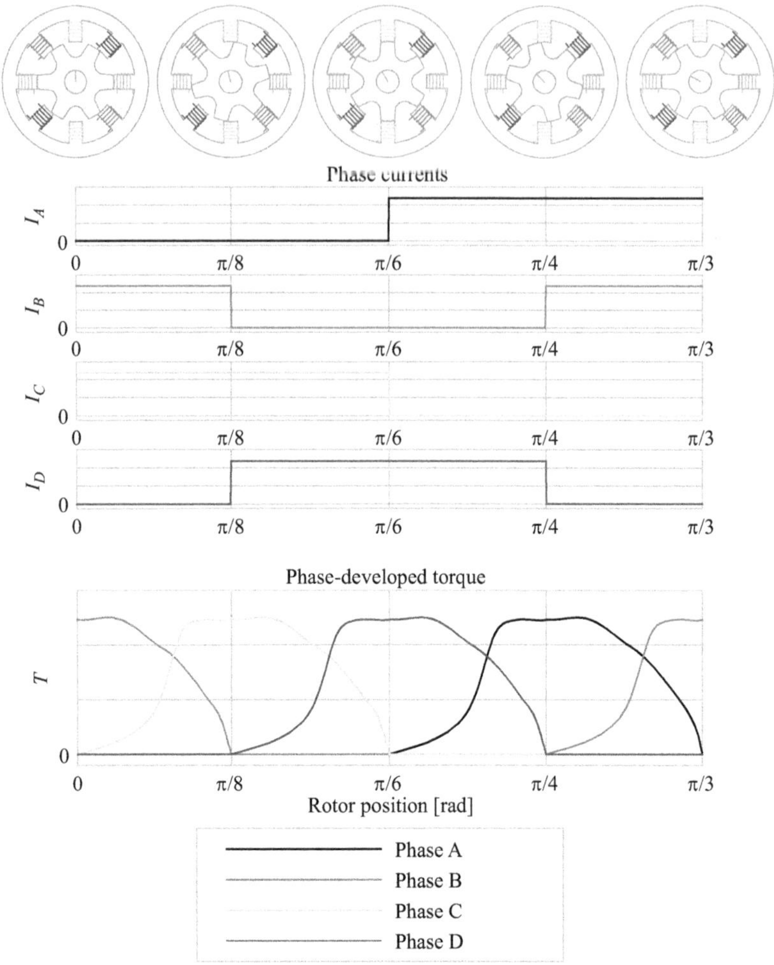

Figure 3.2 Working principle of a four-phase SRM with eight poles in the stator and six poles in the rotor (8/6 SRM)

3. Other positions
   Between the unaligned position and the aligned position, the phase inductance presents a positive rate of change; thus, the phase current produces a motoring torque.

   Beyond the aligned position, the phase inductance presents a negative rate of change; thus, the resulting torque will be negative, i.e. contrary to the rotational movement of the shaft.

   Close to the aligned and unaligned positions, the inductance rate of change is small, allowing a fast change of the phase current. The greatest variation of the phase inductance occurs, obviously, during the overlap of the stator and rotor poles, as presented in Figure 3.6.

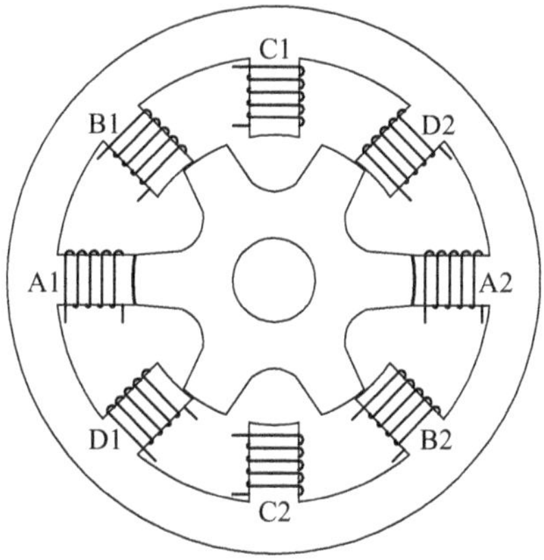

Figure 3.3 Rotor aligned position for phase A

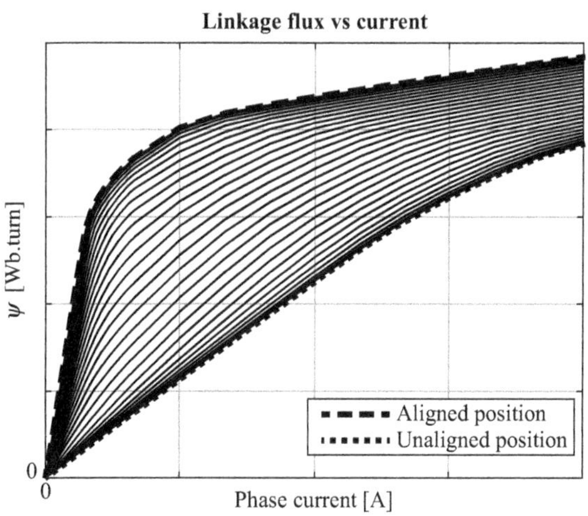

Figure 3.4 SRM magnetization curves

3.1.1 Performance analysis

SRM's torque development is possible because the air gap is not uniform, allowing a variation of the reluctance of the magnetic circuit of the stator phases, and consequentially a variation of the phase inductance.

The most general expression for the instantaneous torque produced by the variable reluctance effect can be derived from D'Alembert's principle, in the following way:

$$T = \frac{dW'(\theta, I)}{d\theta} \tag{3.2}$$

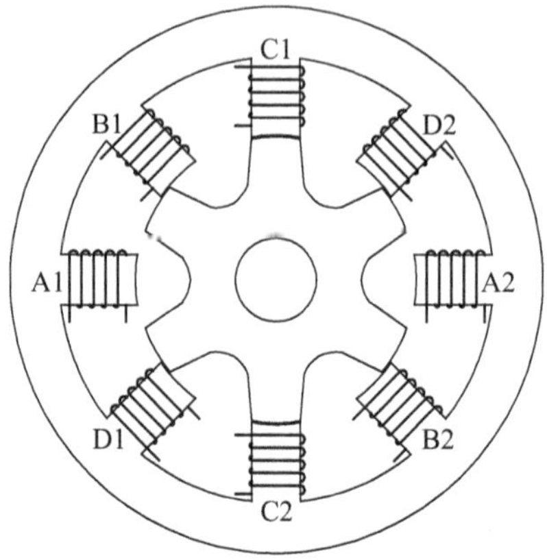

Figure 3.5 Rotor unaligned position for phase A

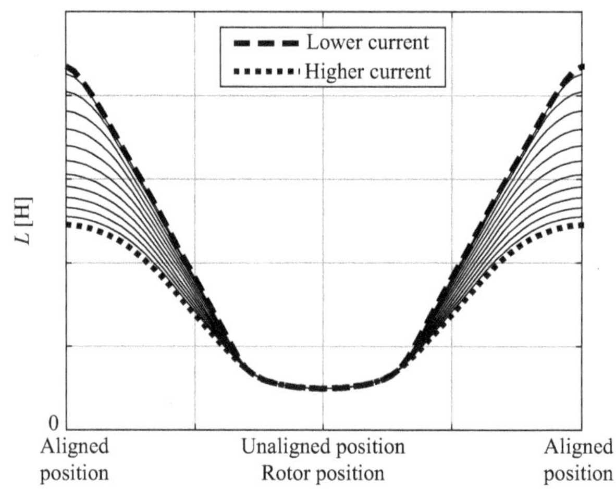

Figure 3.6 Inductance vs rotor position for different current values

where I is the phase current, θ is the rotor angular position and W' is the magnetic co-energy, due to the magnetic flux generated by the phase current, defined as

$$W'(\theta, I) = \int_0^I \psi(\theta, i) di \tag{3.3}$$

where ψ is the phase magnetic linkage flux [2,4].

Since,

$$L = \frac{\psi}{I} \tag{3.4}$$

considering a constant value of the phase current, I, and the motor operating with no magnetic saturation, the torque becomes:

$$T(\theta, I) = \frac{I^2}{2} \frac{dL(\theta)}{d\theta} \tag{3.5}$$

where $L(\theta)$ is the phase inductance at the rotor position θ.

Considering the phase circuit, if the phase is fed by a voltage source, one obtains

$$u = Ri + \frac{d\psi}{dt} \tag{3.6}$$

where u is the instantaneous value of the voltage applied to the phase, i is the instantaneous value of the current and R the phase ohmic resistance. Thus, if the supply voltage, U_S, is constant, one obtains

$$\frac{d\psi(\theta)}{d\theta} = \frac{U_S - Ri(\theta)}{\omega(\theta)} \tag{3.7}$$

where ω is the rotor angular speed. Thus, the linkage flux, for the rotor position θ_1, is given by

$$\psi(\theta_1) = \frac{1}{\omega(\theta)} \int_{\theta_0}^{\theta_1} U_S - Ri(\theta)\, d\theta + \psi(\theta_0) \tag{3.8}$$

For the performance analysis, in normal operating conditions, a complete independency between phases can be considered. The phase is connected to the voltage source when the rotor is positioned in θ_0 (ignition angle), near the unaligned position, and the magnetic flux is established, every step, from zero, i.e. $\psi(\theta_0) = 0$.

At the rotor angular position θ_C, designated as commutation angle, near the aligned position, the phase voltage switch signal and the phase current start to decrease until the extinguishment, flowing through freewheel diodes.

Due to the air-gap geometry and magnetic saturation, the SRM phase linkage flux depends on both the rotor position and phase current resulting in a nonlinear function difficult to interpolate. Several methods have been presented in the literature to model the SRM performance. Since early times, the most usual and accurate way to do the performance analysis of a SRM is by a computational approach model, where the magnetization curves are available in the form of lookup tables [3–5].

Considering a constant value of the phase current and neglecting the phase resistance, the energy delivered by the source to the motor between θ_0 and θ_C is

$$W = \psi(\theta_C, I) \cdot I \tag{3.9}$$

and the mechanical energy developed in the shaft is

$$W_{mt} = \frac{W'(\theta_C, I)}{\theta_C - \theta_0} \cdot \omega \cdot (t_C - t_0) = W'(\theta_C, I) \tag{3.10}$$

At the commutation angle, part of the energy delivered by the source between θ_0 and θ_C is stored in the magnetic field, being given by

$$W_{fc} = W - W_{mt} \tag{3.11}$$

After θ_C, part of this energy is used to produce mechanical energy, developed in the shaft, W_{md}, and part is returned to the source, W_R.

As it can be seen, saturation is desired, because high levels of saturation lead to a better energy ratio, E, proposed by Lawrenson *et al.*, and defined as

$$E = \frac{W_D}{W_D + W_R} \tag{3.12}$$

where W_D is the energy developed in the air gap ($W_D = W_{mt} + W_{md}$) [2].

3.2 Switched reluctance motor operation

Since its appearance in the nineteenth century, the major drawback of SRMs was their need for a controller able to switch ON and OFF the motor phase currents in synchronism with the rotor position.

Power electronics have allowed this drawback to be overcome. However, since SRM drives have to incorporate a power electronics converter, their predisposition for fault tolerance, as well as the post-failure control strategy, is largely dependent on the power electronics converter topology.

The converter topologies, published so far, can be classified according to the following characteristics [6]:

- Number of active switches per phase;
- Strategy to reduce demagnetization time;
- Ability to apply a high positive excitation voltage for building up a higher phase current, which may improve the output power of the motor;
- Utilization mode of the demagnetization energy;
- Ability to reduce switching losses by zero-voltage or zero-current switching.

It is concluded that the classical topology (Figure 3.7), also known as asymmetrical H-bridge, is the power converter applied to SRM drives with the greatest predisposition for fault-tolerant applications [6,7].

Although this converter presents the highest number of switches per phase, without allowing a phase voltage boost to promote a fast increase or decrease of the phase current, the choice of this topology is due to the following main reasons: independence between phases; low cost, due to the almost absence of passive elements (inductances and capacitors); and also because the active devices, although in greater number, are cheaper since their rated voltage is much lower (about half of the value) as compared to converters with a single switch per phase or converters with voltage boost for fast magnetization and demagnetization.

As already mentioned, the SRM magnetic flux is not constant, but must be established from zero, at each step. Since there is no residual magnetization, motor

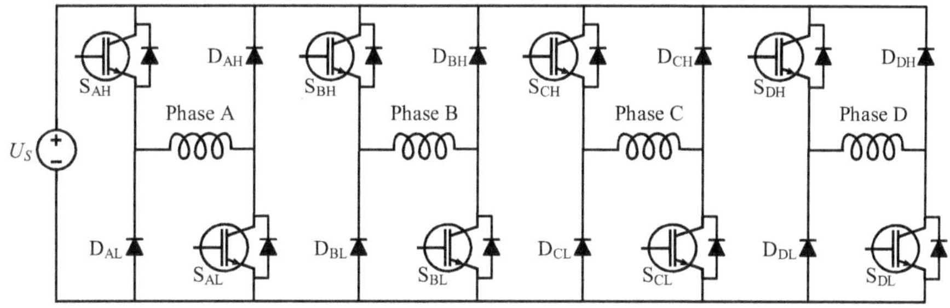

Figure 3.7 Asymmetrical H-bridge converter. Controller circuit with two active switches per phase

torque can be obtained whenever the rotor is between the unaligned and the aligned positions. Thus, each phase can produce a unidirectional torque during half of the rotor pole pitch, being the SRM controlled by supplying appropriately each phase between the ignition angle (θ_0) and the commutation angle (θ_C).

Having chosen the power electronics converter topology, it is now important to analyse the normal operation of the drive. Consider an 8/6 SRM powered by the circuit of Figure 3.7.

3.2.1 Single pulse operation

The analysis for phase A is particularly important since the processes for the other phases are similar, differing only in that they are time-shifted.

Single pulse operation mode is essentially used at high speeds when the inductance rate of change is high enough to promote a current decrease during the pole overlapping. In addition, single pulse operation mode presents a high torque ripple, which can be a major drawback for some applications.

In the single pulse operation mode, switches S_{AH} and S_{AL} are switched from OFF to ON when the rotor is at the turn-on position θ_0 and the supply voltage U_S is applied to the phase.

The same switches are switched from ON to OFF when the rotor reaches the commutation position θ_C. Due to the magnetic energy stored in the phase, the diodes D_{AH} and D_{AL} switch from OFF to ON and phase terminal voltage is reversed. The demagnetization of the motor phase starts by applying a terminal voltage equal to $-U_S$. The rotor position corresponding to the end of the demagnetization period is defined as extinction angle θ_q. Neglecting the ohmic losses, between the commutation position and the aligned position, part of the energy stored in the magnetic field is used to develop torque, and the remaining energy is returned to the source. As can be seen in Figure 3.8, if the demagnetization is not completed at the aligned position, after the aligned position and until the complete demagnetization, the drive starts to operate in generation mode were the energy returning to the source comes both from the shaft and the demagnetization phenomena. In fact, a proper set of both θ_0 and θ_C around the aligned position is used in a generator operation mode.

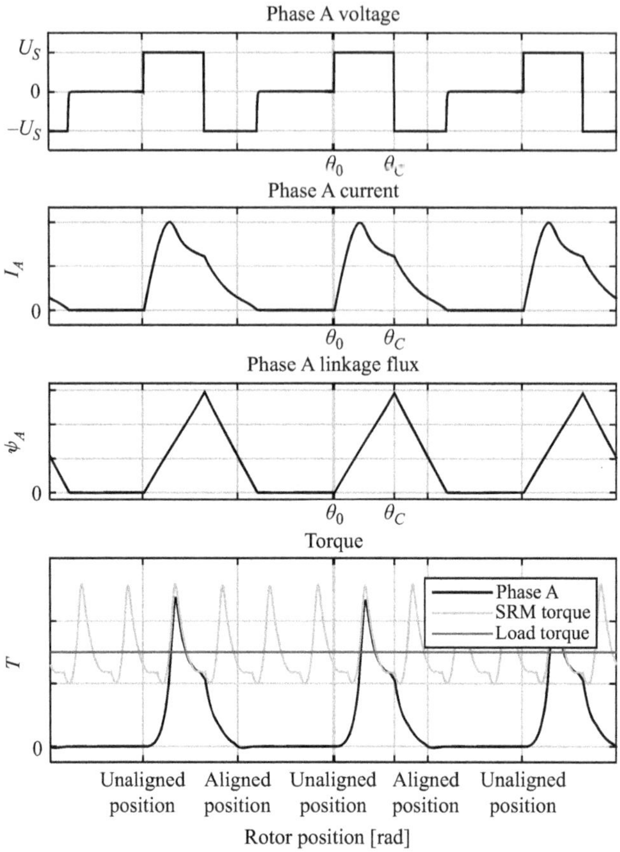

Figure 3.8 Single pulse waveforms

3.2.2 Voltage chopping

At normal operating conditions, and keeping the values of θ_0 and θ_C directly related to the speed, the phase currents are adjusted according to the load and/or other application requirements, such us torque ripple minimization.

To control the phase currents, and consequently to preform speed or torque control, the supply voltage between θ_0 and θ_C is chopped. Thus, the phase voltage is switched between a value that promotes the current increase and another one that promotes its decrease.

The commutation is usually grouped as *Soft Chopping* and *Hard Chopping*. However, for the SRM phase current control between θ_0 and θ_C, the hard chopping brings to higher switching frequencies presenting higher commutation losses [2].

3.2.2.1 Soft chopping

The soft chopping is characterized by applying a positive voltage U_S at the phase terminals to promote the current increase and, when the current reaches the desired

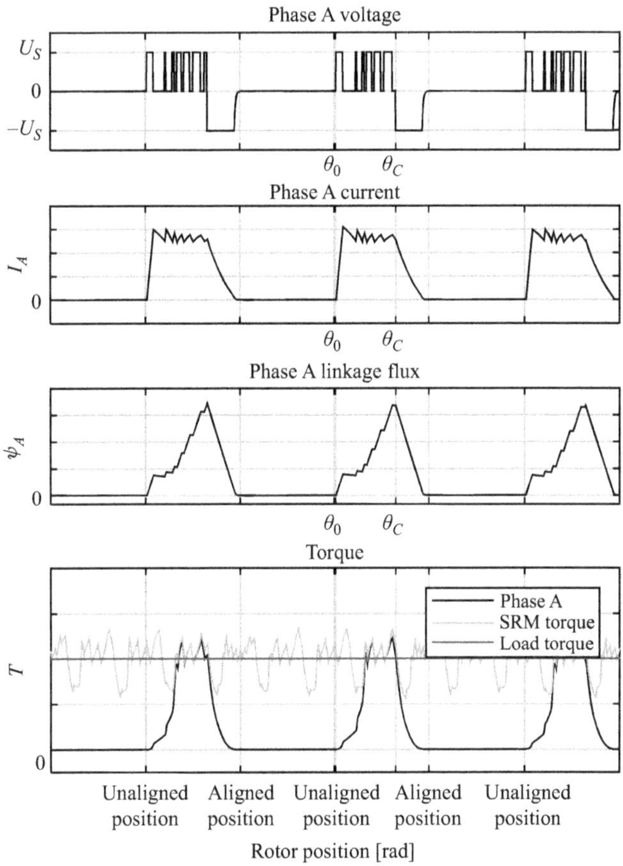

Figure 3.9 Soft chopping current control waveforms

amplitude, the phase terminals are short-circuited resulting in a null phase voltage, as can be seen in Figure 3.9.

To apply the positive voltage U_S, the switches S_{AH} and S_{AL} must be turned ON. To short-circuit the phase terminals, a freewheeling circuit is created by switching the designated chopping switch, S_{AH} or S_{AL}, from ON to OFF. Thus, if S_{AH} is the chopping switch, S_{AL} will remain ON, and the phase current is established by a freewheel circuit comprehending both S_{AL} and D_{AL}.

3.2.2.2 Hard chopping

Hard chopping can be used if a faster current fall is necessary. In this case, a negative voltage, $-U_S$, is applied at the phase terminals, instead of the freewheel short-circuit of the soft chopping.

To apply the negative voltage, $-U_S$, at the phase terminals, both S_{AH} and S_{AL} are switched from ON to OFF. In this case, the energy 'stored' in the winding will return to the source by means of the conduction of D_{AH} and D_{AL}.

3.3 Control of switched reluctance machine drives

The choice of the SRM control strategy can be considered as a multi-objective function which it is aimed at

- The production of a given torque, or a given speed movement.
- The maximization of the energy efficiency.
- The minimization of torque and/or speed ripple.

The second and third objectives are usually contradictory objectives and should, therefore, be balanced by taking into account the application requirements.

In such a system, the control variables are

- The control angles (the ignition angle θ_0 and the commutation angle θ_C).
- The duty-cycle PWM signal used to control the chopping switch, for voltage control.
- The current reference value and the width of the hysteresis band, for current control.

The maximum efficiency is usually obtained with a single pulse operation, since the switching losses are reduced and the SRM drive operates in its most favourable region for torque production. However, the control simplicity obtained by relating the control angles to the reference speed, and the need to control the waveform of the current to minimise the torque ripple and the starting current, may force another switching strategy, usually soft chopping [2].

The determination of the control angles θ_0 and θ_C is subjected to some constrains. Thus, θ_0 and θ_C must be such that the conduction period is less than the rotor pole pitch (τ_R), in order to establish the magnetic flux from zero, at each step:

$$\theta_q - \theta_0 < \tau_R = \frac{2\pi}{N_R} \tag{3.13}$$

It should also be avoided that θ_q lies beyond the aligned position, so that the phase current does not produce braking torque, decreasing the efficiency and increasing the level of the torque ripple.

In SRM operation, phase magnet circuit saturation is essential to achieve high efficient levels. For achieving magnetic saturation, the phase current must be sufficiently high. Thus, the phase current must grow enough before the overlapping between stator and rotor poles occurs. To achieve these current levels, it is sometimes necessary (mainly at high speeds) to start the phase conduction before the unaligned position. For the definition of the control angles, one must consider that

- For a constant average torque, an increase in speed is achieved by decreasing both θ_0 and θ_C, being the decrease of θ_0 greater than the decrease of θ_C.
- To increase torque, one can increase the conduction angle ($\theta_q - \theta_0$), keeping θ_q before the aligned position, preferably, based on the decrease of θ_0. However, the most usual approach is to index the control angles only to the speed and achieve the torque regulation by the control of the phase current reference value.

Whenever the torque ripple minimization is highly relevant for the application, torque control can be achieved with the definition of a proper phase current profile, based on a profound knowledge of each individual machine and requiring a current sensor per phase, since the current reference value is not constant along the magnetization period [8–12].

3.4 Fault analysis in switched reluctance machine drives

The application of a failure mode and effect analysis to a SRM drive system allows to conclude that potential faults can appear in the power converter, motor windings and bearings. Because of the intrinsic rotor robustness, it is not expected that any type of failure may occur in the rotor.

Bearing failures, on the other hand, are more likely to occur. Since the magnetic core flux distribution through the air gap is not uniform, the radial oriented forces between stator and rotor poles are high, highly concentrated and periodic. In addition, stator or power converter faults also increase these forces or their asymmetries.

Unlike other machines, SRM mechanical unbalances, due to manufacturing tolerances, eccentricity and poor shaft alignments, do not cause a significant increase of radial forces. Therefore, mechanical unbalances are not usually the cause of bearing faults.

However, it should be noted that bearing faults are a very peculiar type of fault, which has led to the scarce published work on bearing fault effects, diagnostics and corrective actions [13].

Power converter faults are almost exclusively semiconductor faults. As it is well known, semiconductors fault types can be classified as both open-circuit or short-circuit faults, being more usual in active switches than in diodes.

The SRM windings are composed of concentrated coils, one in each pole. As for other electric motors, the following SRM stator faults also tend to occur, mostly at the end windings region [13–16]:

- Disconnected phase, open-circuited phase.
- Disconnected phase, short-circuited phase.
- Disconnected phase's branch, in multiple branch phase windings.
- Short-circuited pole coil.
- Short-circuit to ground.
- Phase-to-phase short-circuit, usually between the mid-phase taps due to concentrated windings.
- Inter-turn short-circuits.

A large majority of the published work concerning both SRM fault detection and fault tolerance considers the series winding configuration, which is the most commonly found, and the one presented in Figure 3.10(a). In this case, the two coils wound on diametrically opposite poles of each phase winding are series connected. However, other configurations are possible, such as the parallel winding configuration, shown in Figure 3.10(b), which presents several advantages and drawbacks when compared with the series configuration [2,13,17].

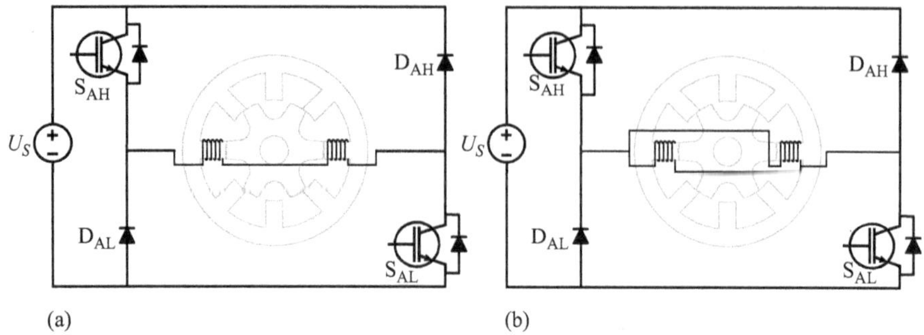

Figure 3.10 Stator winding configurations (a) series (b) parallel

3.4.1 Disconnected phase

A switched reluctance motor phase can become disconnected due to

- A wire break in the phase terminals or in the mid-phase tap of the series connected winding configuration,[1] causing an open-circuited phase.
- The action of the protection devices in the case of a phase terminal short-circuit.
- An open-circuit fault in a power converter switch.

When a phase becomes open-circuited, no current flows in the phase winding. If the motor is working at rated or maximum torque and rated or maximum speed (single pulse operation), a reduction in the average torque is obtained. In the case of an 8/6 SRM, the torque is approximately equal to 3/4 of its rated value, considering that the currents in the other phases remain unchanged [13–15,18–20].

If the drive is operated below the rated torque and with closed-loop speed control, to keep the average speed constant, the controller will increase the reference value of the current in the remaining phases, thus reacting to the fault with an increase of the torque developed by the healthy phases. As a consequence, both speed and torque ripple increase, as can be seen in Figure 3.11.

The system usually reacts to short-circuit faults by isolating the faulty motor phase, either through fuses or the control system. During the transient period between the occurrence of the fault and the complete demagnetization of the phase, the current in the faulty phase increases uncontrollably, which may cause a decrease in both torque and speed. The remaining phase currents tend to increase, to compensate for a possible brake torque created by the short-circuit current.

The short-circuit prevents the normal demagnetization of the phase. After the fault isolation, the short circuit current oscillates due to the generating and motoring modes of SRMs. However, the short-circuited phase current slowly decays according to the balance between the phase resistance and inductance. Due to current oscillation, a large torque ripple is expected [21].

[1]Concentrated winding disables end turns overlap, reducing wire vibration, and thus wire break is mostly expected at the coils terminals.

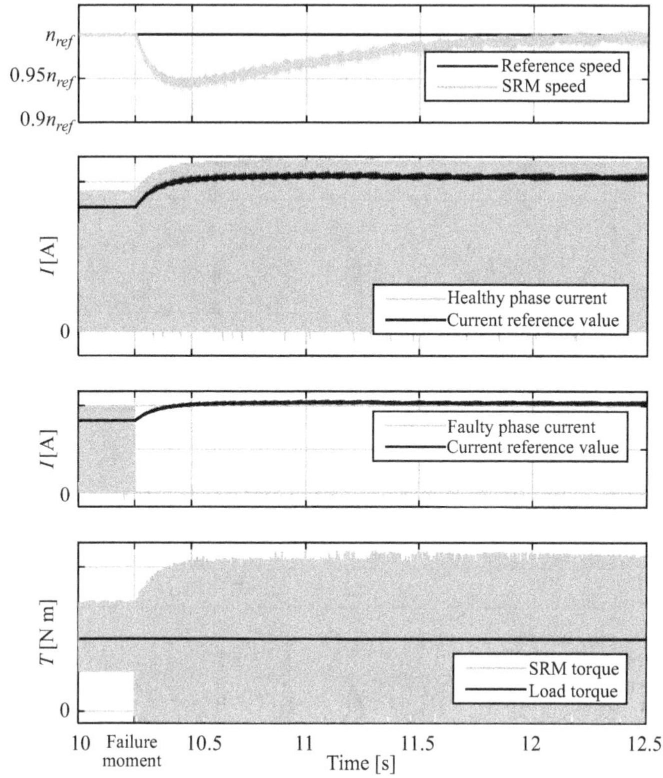

Figure 3.11 Transient of the SRM drive after a disconnected phase

As soon as the faulty phase is completely demagnetized, the system behaves as previously analysed, due to the low magnetic coupling between phases.

3.4.2 Disconnected phase branch

If the stator winding configuration presents two branches, each one is a pole coil of the phase, as it is the case of the parallel windings configuration presented in Figure 3.10(b). The open-circuit in the coil terminals may affect only one branch, and consequently a pole. In the case of an open-circuit in a branch, the closed-loop current control tends to double the current in the remaining branch, unless the SRM is in single pulse operation [14].

If the drive is in single pulse operation mode, the open-circuited branch fault will lead to a reduction in the average torque. For an 8/6 SRM with two branches per phase, the torque is approximately equal to 7/8 of its rated value and the phase current is reduced by half. The currents in the other phases remain unchanged.

The phase current will not present its normal waveform, as illustrated in Figure 3.12, and the radial forces created by the faulty phase current are highly unbalanced, causing serious noise and vibration

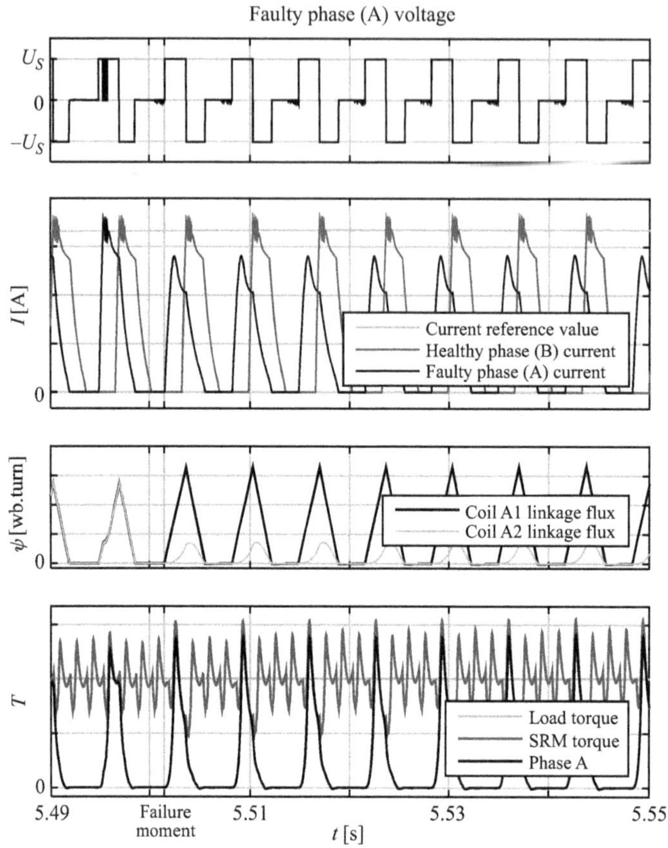

Figure 3.12 Waveforms for an open-circuited pole

3.4.3 Short-circuited pole

In the case of a short-circuited pole of a SRM with a series winding configuration, the phase inductance drops to about a quarter of the design value. The current control prevents an overcurrent situation in the healthy pole, but the rate of change of the current doubles, causing a similar increase in the switching frequency of the applied voltage.

The current in the short-circuited pole is induced by the current in the healthy pole. Due to the low magnetic coupling between the two poles, during the overlap, the short-circuited pole current presents a very low amplitude.

Due to the unbalance between magneto-motive forces produced in each phase's poles, the external vibration of the motor also increases considerably [13,22–27].

If the SRM presents a parallel winding configuration, the short-circuit between the pole coil terminals also shorts the phase terminals. In this case, the system evolves to a disconnected phase situation.

3.4.4 Short-circuit to ground

In the case of a short-circuit to ground, the most critical situation occurs when the following conditions are simultaneously verified [14]:

- The motor structure is connected to one of the supply buses (usually the negative bus).
- The short-circuit diverts the current from the current sensor used within the feedback control loop of the controller.

In this situation, the current can reach dangerous values leading to the destruction of the power converter switch or the action of the protection system, evolving to a disconnected-phase situation [14,16].

While the current continues to flow, it reaches a high value, and the motor experiences a considerable increase in mechanical vibration.

If the fault does not affect the feedback control of the phase current, the short-circuit effect depends on both the location and winding configuration, i.e. for a short-circuit between the ground and the mid-tap of a series winding, the fault will be very similar to a short-circuited pole.

3.4.5 Phase-to-phase short-circuit

In the case of consecutive phases, this fault starts to produce effects when both affected phases are being powered. A current flows through both phases in an unusual and undesirable way. The current in these two coils reaches relatively high values and an increase of the motor vibration is observed, but the torque and the speed of the machine do not suffer much variation [14,16].

3.4.6 Inter-turn short-circuit

Strongly related with aging effects, the inter-turn short-circuits are an evolving fault and are one of the main root-causes for electric machines stator winding failures.

Although the pole short-circuit is an inter-turn short-circuit, for the present analysis, inter-turn short-circuits refer to a situation where a portion of a pole coil is shorted. In this case, the phase current flows in the non-shorted turns of the pole and induces a current in the short-circuited turns, similarly to what happens in an autotransformer with the secondary in short-circuit [26].

Typically, the effects of an inter-turn short-circuit are strongly related with the portion of shorted turns as well as with the value of the short-circuit resistance. When a small portion of the pole is short-circuited, as presented in Figure 3.13, the motor performance is very close to normal. The decrease in both phase inductance and resistance has a small effect in the phase circuit time constant, and the current controller forces the phase current to maintain its value with a small change in the phase voltage.

Although the input phase current remains unchanged, the short-circuit current is high, and thus it might create a local hot spot.

If a large portion of a pole is in short-circuit, the motor behaviour resembles that of a short-circuited pole situation. The short-circuit current is low, and the most hazardous situation is due to the magnetic unbalance and vibration [13,14,16,21,26,27].

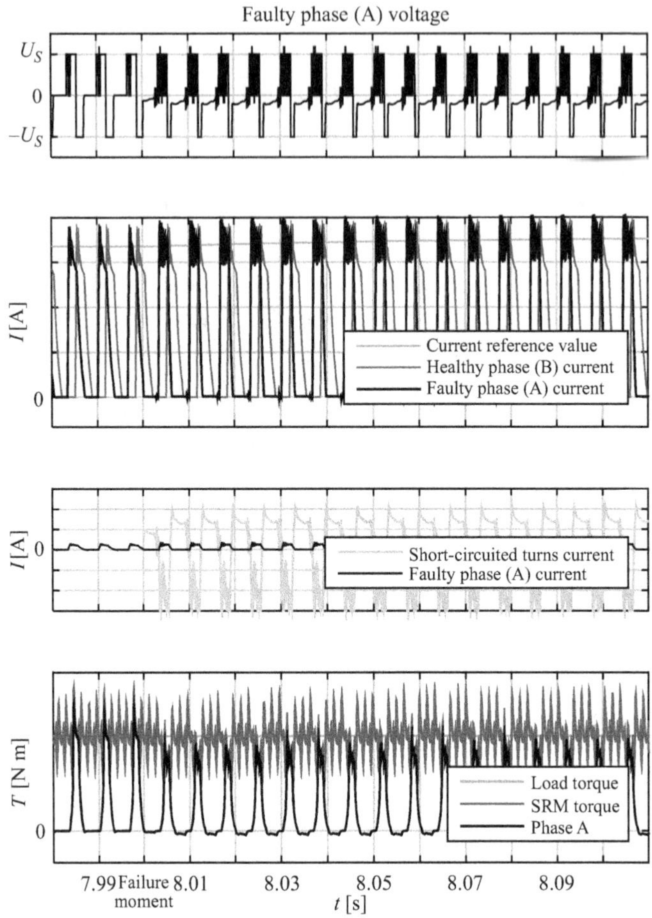

Figure 3.13 Waveforms for an inter-turn short-circuit situation

3.4.7 Power converter faults

When one of the power converter switches becomes defective, the faulty device becomes a permanent open-circuit or a permanent short-circuit.

In the case of converter diodes, the open-circuit fault disrupts the circuit path for phase demagnetization, enabling the appearance of high dV/dt values at the phase active switches terminals, causing their failure. The short-circuit fault of converter diodes causes a short circuit between the positive and negative DC supply bus and, hopefully, the protection system will rapidly isolate the faulty phase.

Active switches such as IGBTs or MOSFETs are less reliable and more prone to failure. A permanent open-circuit fault of these switches will have a similar effect as of an open-circuited phase.

A permanent short-circuit fault has several consequences, depending on the switch function, as can be seen in Figure 3.14, where the typical waveforms in the

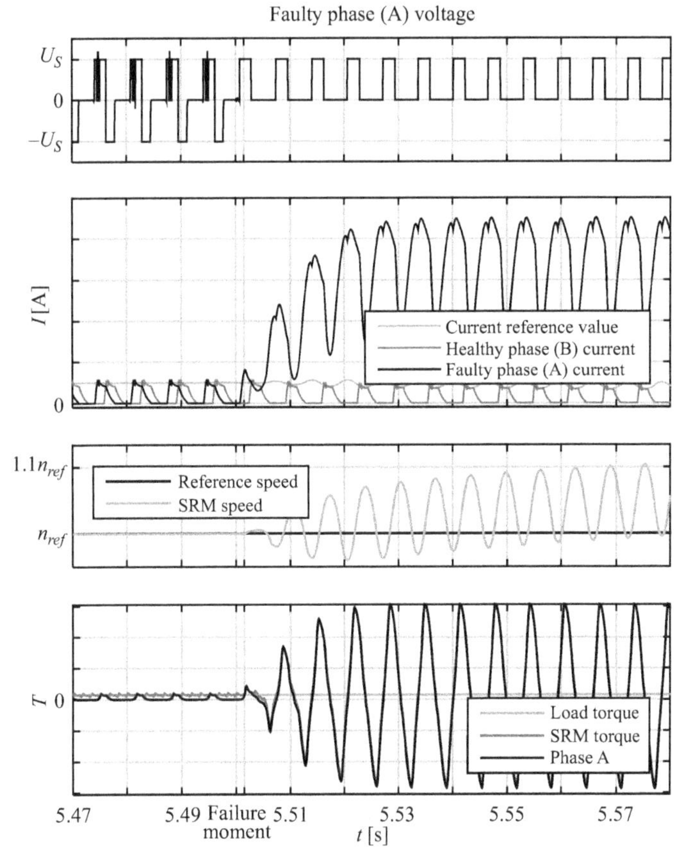

Figure 3.14 Waveforms for a short-circuit of the chopping switch

case of a short-circuit fault in the chopping switch are presented. In the case of a short-circuit fault in the other active switch, the current will be lower, thus less dangerous. In any case, it is impossible to apply negative voltages to the phase, making it impossible to perform the following actions:

- Fast demagnetization. The phase demagnetization is performed with a null phase voltage instead of a negative phase voltage.
- Current control, if the faulted switch is the chopping switch.

For both situations, the extinction angle will increase due to the non-application of a negative voltage between θ_C and θ_q. Consequently, the developed torque decreases and the torque ripple increases significantly.

When both phase switches are in a short-circuit fault condition, the phase will be driven by current all the time. Only the winding resistance will limit the phase current, and consequently an overcurrent situation will occur, which will, at least, lead to the protection fuse to actuate. Until the fuse acts, there will be a considerable decreasing of the drive speed and an increase of the current in the remaining phases [15,18,27].

3.4.8 Rotor-related faults

SRMs have an enviable rotor robustness, and it is not expected that any type of fault will occur in the rotor itself. However, eccentricity and bearing faults are expected.

Eccentricity in a rotating machine is characterized by a non-uniform air gap between the stator and the rotor. In the case of static eccentricity, the rotation centre, although coincident with the rotor geometric centre, is not coincident with the stator geometric centre.

The air gap of small power SRMs is of the order of 0.2–0.4 mm, thus smaller than that of both permanent magnet and induction motors [17]. Therefore, SRMs are even more sensitive to rotor eccentricity. Eccentricity levels up to 10% due to fabrication tolerances are common in 'healthy' motors, being the eccentricity level, e, defined as [28]

$$e = \frac{r}{g} \times 100\% \tag{3.14}$$

where g is the radial air-gap length at the aligned position of the motor with no eccentricity, and r is the displacement of the rotor in the radial direction.

Eccentricity, which is usually caused by misalignments with load or bearings degradation, can also appear if the rotor is sitting idle for some time. However, unlike other machines, the SRM mechanical unbalances, due to eccentricity and poor mechanical shaft alignment, do not cause a significant increase of the radial forces applied on the bearings. Thus, eccentricity is an unlikely cause of bearings degradation [17,28,29].

Since the SRM phase current ability to develop torque is maximum during the stator and rotor poles overlap, and in those positions the phase sees a reluctance corresponding to an air gap several times higher than g, an eccentricity level that does not cause friction between stator and rotor has a negligible effect on current, torque, speed and vibration. Thus, despite the fact of eccentricity being one of the most common faults, the large majority of the published works that focused on eccentricity fault diagnostics in SRM drives is related to off-line techniques.

Concerning bearing failures, SRMs are usually forced to operate with highly saturated cores, accompanied by high radial forces, causing vibrations, acoustic noise, abrasion and additional stress, and premature failures in bearings. In addition, the occurrence of both stator or power electronic converter faults increases those forces and their asymmetries.

Bearing faults usually lead to eccentricity. However, the very particular characteristics of bearing faults led to the scarce published work on

- Bearings fault effects and analysis.
- Bearings fault diagnostics.

3.5 Fault diagnostic techniques applied to switched reluctance machine drives

SRM drives present particular characteristics regarding the control strategies, power converter topologies and operating behaviour. Phase currents are usually unidirectional, and each motor phase is not continuously excited because its conductive state is dependent on the rotor position. These particular features of SRM drives lead to specific fault impacts on the behaviour of operating parameters such as, phase currents, electromagnetic torque, speed, etc. Therefore, most of the fault diagnostic techniques developed for traditional AC drives are not appropriate for SRM drives.

Fault diagnosis is extremely important to promote the isolation of the fault and thus ensuring the integrity of the remaining healthy motor windings and power electronic devices, allowing the continuous operation of the machine. An open-circuit fault in a phase winding or in the power converter typically prevents the operation of a single motor phase and does not affect the other motor phases. A short-circuit event commonly promotes the sudden increase of the electric phase-current magnitude. Thus, usually, the power switches of the faulty phase must be turned off, and the affected phase must stay out of service. However, a short-circuit occurrence in a power switch may not conduct immediately to an inappropriate increase of a phase-current magnitude. At low mechanical load and/or speed conditions, the faulty phase current may stay within normal amplitudes that do not put the affected phase winding in a dangerous situation. Keeping the faulty phase operating, by adjusting appropriately the respective control, can be very important at some operating conditions such as, during start-up, and under high load level, and/or speed requirements.

The identification of the faulty element permits the fast adjustment of the employed control strategy and/or power converter topology, in order to minimize the effects of the fault occurrence. Although, an open-circuit fault in a power switch leads to the inactivation of the affected phase, the identification of the faulty element is important to establish a rapid hardware reconfiguration, when a fault-tolerant converter is used, and quickly recover normal operating conditions. In a short-circuit scenario, the identification of the faulty element is also important to evaluate if the affected phase may or may not be magnetized and to establish an appropriate fault-tolerant control strategy.

Several devices and techniques have been developed for fault diagnosis in SRM drives, mostly based on the analysis of the motor phase currents. In order to minimize the requirements needed for a proper fault diagnostics process and keep it completely non-invasive, fault diagnosis is generally dependent on the employed control strategy and is based on the parameters measured and used in the control loop itself. For example, in some applications, torque ripple is not a concern, so the measurement of all phase currents is not necessary. In such applications, the voltage control strategy is commonly implemented and voltage and/or current sensors are not required. For those applications, fault diagnostic methods proposed in the literature are based on a single electric current which is generally the DC bus current. In applications where the torque ripple must be minimized, or where the fault tolerance is a main feature, all phase currents are measured and used for fault diagnosis.

3.5.1 Fault detection devices

Stephens [14] has proposed some fault detection devices for SRM drives. Those devices were intended to be used as additional elements, and to generate a signal whenever a fault is detected, in order to disable the power switches of the affected motor phase.

3.5.1.1 Overcurrent detector

This detector is easily implemented, but it is not fast acting, since the fault is detected when the electric current is already very high. It operates from the current sensor signal and sets a comparison with a threshold above the normal operating range of the phase currents.

The overcurrent detector is able to detect a short-circuit fault in a power switch, but some short-circuit faults, as for example, phase-to-phase short-circuits, phase-to-ground short-circuits or inter-turn short-circuits, cannot be detected if a phase current control is implemented.

Any overcurrent event due to a short-circuit fault easily causes a failure of a power switch, leading to an open-circuit fault occurrence.

3.5.1.2 Differential current detector

The differential current detector proposed in [14] uses an iron-core current transformer. The incoming and outgoing phase leads are passed through the core. Under normal operating conditions, the currents are equal, and no magnetic flux is set in the core. When there is a fault that bypasses the phase winding through an abnormal path, a difference between the incoming and outgoing phase current will be established.

This detector is fast acting and reacts to some short-circuit faults such as a phase-to-phase short-circuit and a phase-to-ground short-circuit.

3.5.1.3 Rate-of-rise detector

The rate-of-rise detector proposed in [14] consists of a linear magnetic coupler placed around the phase-winding lead. The change in the magnetic flux created by the phase current induces a voltage in the secondary of the magnetic coupler. The induced voltage is proportional to the rate-of-change of the phase current, in the range of operation where the magnetic material presents a linear feature. The fault detection is established when the induced voltage is greater than a threshold value. This threshold value is set to not respond to the highest rate-of-change of phase current that occurs at rotor position near to the unaligned position where phase inductance is at its lowest value.

Any short-circuit fault that promotes the by-pass of part of a phase winding, such as a short-circuit to ground or an inter-turn short-circuit, results in a decrease in the effective inductance, and the time rate-of-change of the phase currents exceeds normal levels.

This detector is fast acting, but its sensitivity is dependent on the restraining threshold accuracy.

3.5.1.4 Differential flux detector

The differential flux detector also proposed in [14] requires search coils wrapped around stator poles. The search coils of each motor phase are connected in opposing series in order to make a simple comparison between the voltages induced in each search coil. At normal operating conditions, the magnetic flux has equal behaviour at all stator poles of a particular motor phase, and the voltage at the terminals of the search coil series is zero. When a fault promotes a phase stator poles magnetic flux unbalance, a non-zero voltage will be induced at the terminals of the search coils series. This detector is able to detect phase-to-phase short-circuits, phase-to-ground short-circuits and inter-turn short-circuits. Despite the low magnetic coupling between phases, the detector is oversensitive and reacts to faults in other phases, able to produce unbalance magnetic fluxes in the phase poles. Thus, the fault pattern of each differential flux detector must be considered, only, when all other phases are demagnetized. Moreover, the machine must be appropriately assembled with additional coils which mean a more complex manufacturing process, as compared to traditional SRM drives. Additional connections to the motor are also a disadvantage of this detector.

3.5.2 Methods based on a single electric current

Some fault diagnostic techniques based on the observation of a single electric current can be found in the literature. Those techniques are intended to be used in applications where the phase currents are not measured, such as in applications with open-loop control of phase currents. Phase currents are usually indirectly controlled by the closed-loop speed control that establishes a duty cycle for the PWM voltage control of the motor phases. The techniques proposed are especially focused on the most common faults, which are power switches open-circuit faults, power switches short-circuit faults and inter-turn short-circuit faults. In an open-circuit fault scenario, none of the techniques proposed is able to identify the faulty element. As aforementioned, the open-circuit fault may either occur in one of the power switches or in the phase winding. All of these open-circuit faults make the magnetization of the affected motor phase impossible and, consequently, all these faults produce a similar SRM behaviour.

Gan *et al.* [30] presented a fault diagnosis scheme based in wavelet packet decomposition. Open and short-circuit faults in the power switches are addressed. The wavelet packet decomposition algorithm is applied to the estimated phase currents. The discrete degree of the wavelet packet node energy is the indicator for fault diagnosis and identification of the faulty power switch. A discrete degree is established for each phase. Each of the faults addressed presents a particular fault signature, which permits the fault diagnosis. The identification of the faulty power switch is only achieved in the case of short-circuit faults, because the respective fault signatures are quite different. The method is not able to distinguish between an open-circuit fault in one of the power switches and an open-circuit fault in the phase winding.

The current sensor used measures a specific electric current which is the sum of the electric currents that pass through the lower power switches. The power converter used is a traditional half-bridge converter, presented in Figure 3.7. The upper power switch is the chopping power switch used to control the phase current amplitude, and the lower power switch, where the current is measured, is permanently ON during the magnetization period, i.e. the time interval between the ignition angle and the commutation angle. This means that the phase current is not detected when the respective phase is being demagnetized, and both power switches are turned off.

The estimation of the phase current considers two different regions in each phase-magnetization period. When only one motor phase is being magnetized, the current sensor measures the respective phase current. When two motor phases are being simultaneously magnetized, two phase shifted PWM signals are used to turn off the respective lower switch, which permits to detect the phase current of the phase that keeps its lower power switch conducting. These PWM signals put the respective phase in freewheeling mode or demagnetization mode during a very short time interval and do not interfere significantly in the normal phase current behaviour.

The method proposed by Gan *et al.* [30] needs complex mathematical computation for wavelet packet decomposition which is a disadvantage. The fault signatures are based on normal and abnormal values of the discrete degrees. The authors do not explain if the values presented are suitable for other SRM drives with different rated parameters. Moreover, the authors do not present the evolution of the fault coefficients before and after the fault occurrence, and the time needed for fault diagnosis is not also established. In the particular case of a power switch short-circuit fault occurrence, a large time interval for fault diagnosis can be inappropriate to promote the isolation of the fault and thus ensuring the integrity of the remaining healthy motor windings and power electronic devices.

A different fault diagnosis scheme is proposed by Gan *et al.* in [31] based on the spectrum components of a single electric current. A fast Fourier algorithm with Blackman window interpolation is applied. The authors use a four-phase machine and an asymmetric half-bridge converter. Open-circuit faults in one or two motor phases are analysed. The fault diagnosis is established by the analysis of spectral components at particular frequencies, which are the fundamental frequency of a phase current and the respective second harmonic frequency. The following three different current sensor positions are considered, individually:

- The power supply current, which is the electric current that flows from the power supply to the several motor phases.
- The chopping bus current, which is the sum of the electric currents that pass through the upper and chopping power switches of each motor phase.

- The demagnetization bus current, which is the sum of phase currents at demagnetization mode.

The spectral analysis of one of these currents permits the fault diagnosis because their spectra present specific components during fault conditions, depending if there is an open-circuit fault in a single phase, in two consecutive motor phases, or in two non-consecutive motor phases.

In order to achieve comparable values at different rotor speeds, authors used normalized spectral components.

At normal and steady-state conditions, any of the analysed electric currents repeats its behaviour each time a motor phase starts operating. This means that its fundamental frequency (f_{bus}) is four times that of a phase current (f_1). The amplitude of the respective component at the frequency f_1 is negligible.

A single open-circuit fault causes an increase in the amplitude of the spectral component at the frequency f_1. The increase of that spectral component is higher if there is a fault in two consecutive phases and is not relevant when the phases are non-consecutive. The spectral component at the frequency $2f_1$ is used to diagnose an open-circuit fault in two non-consecutive phases. The identification of the faulty phase or phases is achieved through the analysis of the logic amplitude of the power supply current at particular rotor position ranges. The fault identification is based in the demagnetization process of each motor phase. When a motor phase starts its demagnetization, the energy stored in its magnetic field is recovered to the power supply and causes a power supply current with negative amplitude. The diagnosis time is within one period of a phase current.

As well as in the previous fault diagnosis scheme proposed by Gan *et al*, complex mathematical computation is needed to apply the fast Fourier transform. The authors have used a PWM voltage control strategy, and the impact of other control strategies and/or motor drives in the spectral components or in the fault identification is not addressed.

Hoseini *et al.* proposed in [32] an online method for inter-turn fault detection that uses an extended Kalman filter for winding resistance estimation. The achieved value is compared to the normal value and when the winding resistance is lower than a pre-set threshold a fault indicator is established. The method is model-based. The SRM is accurately modelled, using finite element software, to extract its inductance profile. The inputs of the extended Kalman filter are a specific electric current amplitude and rotor position. The unique current sensor used measures the sum of the several phase currents. The power converter employed is a buck-fronted Miller converter which has a main power switch that is used for all phases control and a dedicate power switch for each motor phase. The information of the measured electric current is used for closed-loop current control to define the duty cycle of the main power switch. The main disadvantage of the proposed method is the mandatory requirement for SRM model definition.

A single current sensor is used for fault diagnosis of open- and short-circuit faults in power switches, as proposed by Chen and Lu in [33]. The power converter used is the conventional converter applied in SRM drives. The control adopted is a

PWM voltage control, and when two phases are magnetized at the same time, the command signals of the respective power switches are the same. This means that both phases are being magnetized or in freewheeling mode at the same time. The upper power switch is used as the chopping power switch. The current sensor is especially assembled and measures the difference between the sum of the electric currents that flow from the upper diodes and the sum of the electric currents that flow to the lower diodes. The fault diagnosis method is based on the measured signal behaviour at particular regions and command signals states. The analysis is divided in two regions for each motor phase. The first region is bounded by the respective ignition angle and the commutation angle of the previous motor phase. The second region is bounded by the commutation angle of the previous phase and the ignition angle of the following phase. The positive, negative or zero amplitude of the measured current denotes a status of 1, -1 or 0, respectively. The fault occurrence conducts to a particular fault signature in the measured current status at specific regions and states of the upper power switch command signals. The different fault signatures achieved permits the fault diagnosis of open-circuit faults without switch identification and short-circuit faults with fault element identification.

The proposed method presents a high diagnosis time which is particularly inconvenient when in presence of a short-circuit fault, as known. To decrease the diagnosis time it is proposed, in the same paper, a second fault diagnosis method based on two current sensors. One measures the sum of the electric currents that flow to the upper power switches, and the other measures the sum of the electric currents that flow from the upper diodes. The theory applied is similar. A fault event results in a particular fault signature that is defined by the status of each measured electric current at each region and at each conductive state command of the upper power switches.

The second fault diagnosis method proposed by Chen and Lu in [33] presents a diagnosis time of two periods of the PWM control signal, after the fault event has produced an impact in the phase current behaviour. However, these fault diagnosis methods may not be applied in SRM drives with a different control strategy.

3.5.3 Methods based in all electric phase currents

The measurement of all electric phase currents is commonly used for accurate phase current control when a torque control strategy is performed to minimize the torque ripple. The adoption of electric current sensors in all motor phases gives important information for fault diagnosis.

The diagnosis of an open-circuit fault in a phase winding or in the respective power switch is easily established because the affected phase cannot be magnetized after the fault occurrence, and the respective phase current will permanently present zero amplitude. However, the identification of the faulty element is not so easy to achieve. For that, the fault diagnosis must be very fast, and an appropriate test must be performed.

Fault diagnosis of open- and short-circuit faults in the power switches can be established correlating the phase current behaviour with the command signals of the respective power switches, as it is assumed by Han *et al.* in [34].

An increase of the phase current amplitude when one or both associated power switches are turned off indicates a short-circuit event.

An open-circuit fault causes the decrease of the phase current amplitude when both associated power switches are turned on. However, as it can be seen in Figure 3.8, the decrease of the phase current may occur at healthy conditions, in particular, at high speeds, when the stator and rotor poles are partially overlapped, and the back electromotive force of the phase is higher than the supply voltage.

Marques *et al.* proposed in [35] a diagnostic technique for open- and short-circuit faults diagnosis in a power switch, based on average values of phase currents. In order to achieve comparable values, at different speed and/or mechanical load conditions, the authors used the average value of a normalized phase current taking into account the reference current signal provided by the main control system. The time interval for the average calculation corresponds to the period of each phase current which is dependent on the mechanical speed and the number of rotor poles.

At normal conditions, the average value of a normalized phase current can assume values within the range between zero and one, depending on both speed and mechanical load conditions. To overcome this problem, the authors proposed six diagnostic variables for a four-phase machine. Each diagnostic variable corresponds to the difference between two average normalized phase current values ($I_{A\mathrm{av}}-I_{B\mathrm{av}}$; $I_{A\mathrm{av}}-I_{C\mathrm{av}}$; $I_{A\mathrm{av}}-I_{D\mathrm{av}}$; $I_{B\mathrm{av}}-I_{C\mathrm{av}}$; $I_{B\mathrm{av}}-I_{D\mathrm{av}}$; $I_{C\mathrm{av}}-I_{D\mathrm{av}}$). All possible combinations are used for the diagnosis. The diagnostic variables are compared with threshold values, which generate a unique fault signature, according to the fault type and the faulty phase. Two threshold values are used; one for open-circuit fault diagnosis and another one for short-circuit fault diagnosis.

After the open-circuit fault diagnosis, a test is performed to identify the faulty power switch. The test forces one of the power switches of the faulty phase to turn on (or kept turned on), while the other one is turned off. Then, the demagnetization time is measured and related with the period of the phase current. If the relative demagnetization time is small, the faulty element is the power switch that has been forced to turn off. Otherwise, the faulty element is the other power switch. Nevertheless, the identification of the open-circuited element cannot be established if the fault occurs outside the magnetization period.

When a short-circuit fault is diagnosed, both power switches of the faulty phase are turned off in order to quickly decrease the phase current amplitude. The test is performed when the phase current has a small amplitude. The test is similar to the test adopted for open-circuit faults. If the phase current rises, the faulty element is the power switch that has been turned off. Otherwise, the faulty element is the other power switch. After the identification of the faulty power switch, both arm power switches are turned off.

The authors indicate that the time interval required for the fault diagnosis can be equivalent to 8% of the phase current period, which is a small time interval when compared to other methods. The fault diagnosis algorithm is not computationally demanding because it requires a few and basic mathematical operations. However, the method accuracy and diagnosis time is dependent on the two threshold values adopted.

3.5.4 Other methods

Some other fault diagnostic methods can be found in the literature, based on voltage or current sensors which are not directly necessary for the closed-loop control of the SRM drive.

Han *et al.* presented in [34] a phase current reconstruction strategy also suitable for fault diagnosis of open and short-circuit faults in the power switches. The strategy uses two current sensors specially assembled. One of the current sensors measures the sum of all phase currents. The second current sensor is also of the current transformer type, but each phase has a particular number of turns wound in the current sensor and/or is wound in opposite directions. The two current sensors permit to establish two linearly independent equations and allow phase current reconstruction because the authors consider that only two motor phases are being excited at a time.

The analysis considers different regions bounded by the ignition angle of the several phases. The fault diagnosis is based on the relationship between the phase current slopes and the command signals of the respective power switches.

An open-circuit fault is diagnosed when there is a negative slope, and both power switches are turned on. However, as aforementioned, the decrease of the phase current may occur at healthy conditions.

A short-circuit fault in the chopping power switch leads to a positive slope, while the command signal should disable the chopping power switch. The fault is then diagnosed. If the short-circuit affects the other power switch, the phase presents a normal behaviour until the commutation angle is reached. After that, the phase current decreases slowly and the demagnetization time is measured and compared with the demagnetization time of the previous motor phase. The increase of the demagnetization time permits the fault diagnosis.

The method presented allows the identification of the short-circuit power switch but is not able to do the same when there is an open-circuit fault. Moreover, the reconstruction strategy presented is not suitable when there is more than two phases conducting at the same time. This can happen at particular SRM drives or when there is a short-circuit fault in a power switch and the respective phase is always conducting.

The fault diagnostics method proposed by Gameiro and Cardoso in [36] uses an extra current sensor in addition to the phase current sensors. The extra current sensor measures the electric current provided by the power source to the machine. This measurement is used to check if the phase current path is consistent with the command signals of the power switches, or if the electric phase current path has suffered a change due to a fault occurrence in a power switch. The power converter used is the conventional asymmetric half-bridge converter. The fault diagnostic algorithm compares the DC bus current measured with an estimated value. The estimated DC bus current is calculated as a sum of the electric currents that flow from the power supply to each phase windings according to the command signals of the power switches and the measured phase current. If both power switches are turned on, the electric current that flows from the source to the phase winding is equal to the phase current. If both power switches are turned off, the electric current flows in the opposite direction, and the current that the power source provides to the phase winding is negative and equal to minus the phase current. If one of the power switches is turned on while the other one is turned off, there is no electric current flowing between the power source and the phase winding.

If any one of the power switches is faulty, the phase current of the affected phase has an unexpected path, and the fault is detected due to a difference between the measured and the estimated DC bus currents. A short-circuit fault conducts to a positive difference, while an open-circuit fault results in a negative difference. The fault is detected as soon as it forces an abnormal phase current behaviour. To avoid erroneous fault diagnoses, the fault is only considered if there is a difference between the DC bus currents at two successive sampling times. The diagnosis time is then of two sampling periods. The difference between the DC bus currents is compared to the phase currents which permits the identification of the faulted phase. The absolute value of this difference is equal to the faulted phase current amplitude. The identification of the faulted power switch is immediately established when the conductive command state of both power switches are different.

After an open-circuit diagnosis and if the faulted power switch has not been already identified, a test is conducted to identify the faulted power switch. During the test, one of the power switches is turned on, while the other one is turned off. If the estimated current equals the measured DC bus current, the current path is consistent with the command signals, and it can be concluded that the faulted power switch is the power switch turned off. Otherwise, the faulted power switch is the other one.

After a short-circuit diagnosis, the power switches of the faulted phase are both turned off to avoid an excessive and uncontrollable phase current. If the faulted power switch has not been already identified, the same test is conducted, as soon as the phase current presents a small amplitude. If the phase current increases, it means that the faulted power switch corresponds to the power switch that should be turned off. Otherwise, the faulted power switch is the one that has been turned on.

The main disadvantage of the diagnostic method proposed by Gameiro and Cardoso in [36] is the need for an additional current sensor. The diagnosis time is the smallest diagnosis time achieved when a fault diagnosis method based in current sensors is applied.

However, if the open-circuit fault in the power switch occurs outside the magnetization period, the proposed method cannot identify the faulty element. The fault is diagnosed, and the faulted phase is identified, due to the insignificant mean value of the affected phase current. The identification of such faults in such conditions is impossible if the diagnosis method uses only current sensors.

3.6 Fault-tolerant strategies

In a fault scenario, non-affected phases can keep their normal operating status due to electromagnetic independence between motor phases.

Most of the time, after a fault occurrence, the faulty phases may not or cannot be magnetized, and the machine's performance is clearly deteriorated.

Fault-tolerant strategies, based on software or/and hardware reconfiguration, can promote improvements in the motor dynamic behaviour. Fault-tolerant control strategies are prepared to adapt the control algorithm after fault detection. Those strategies can be easily implemented, but normal operating conditions cannot be restored if a single phase or multiple phases keep out of duty.

Fault-tolerant converters have been developed to completely or partially restore the operation of a faulted phase. For that purpose, additional components, that start operation after the fault occurrence, are introduced in the converter. Fault-tolerant converters are obviously more expensive than conventional converters and required more space available.

3.6.1 Fault-tolerant control

The most common result of electrical failures is the disability of one motor phase. Open-circuit events, in the winding or in a power switch, inhibit the use of the affected phase. When a short-circuit event occurs, the disconnection of the affected motor phase may be the best choice to take. That action can avoid uncontrollable and excessive phase currents and preserves the integrity of the non-affected drive elements, which permits the machine to keep working. When a motor phase is missing, motoring torque production may not be possible at all rotor positions or may be insignificant at some rotor positions. Therefore, after the fault occurrence, the torque ripple increases considerably. If no fault-tolerant control is taken, the mechanical rotor interval with no motoring torque is typically large. Additionally, depending on the adopted control strategy, the drive may register one or both of the following behaviours:

- At some rotor positions, the SRM drive produces a negative torque. During normal operating conditions, the negative electromagnetic torque contribution of one motor phase is compensated by the electromagnetic torque contribution of the neighbouring motor phase.
- Electric phase currents exhibit high magnitudes at rotor positions where the faulted phase should be working. The control loop may react to the absence of one motor phase and increase the reference control value of phase current, phase torque or phase voltage, during the time interval where the faulted phase should act, especially if a direct torque control strategy is adopted. The phase that precedes the missing phase may display a high electric current magnitude at the end of its magnetization period. This phase current can be excessive and conduct to a significant negative electromagnetic torque contribution at the end of the respective conduction period. The phase that follows the missing phase may present a high electric current magnitude at the beginning stage of its magnetization period. This can be useless because at those rotor positions, the phase torque production capability is typically low. However, an excessive increase of the phase current at those rotor positions may lead to a torque overshoot. When the rotor pole starts overlapping the stator pole, the phase torque production capability is higher and a lower phase current is needed.

A fault-tolerant control can reduce or eliminate the aforementioned problems and enhance the SRM drive performance. Changes in the control strategy are focussed on the adjacent phases of the faulted phase during the time interval where the missing phase should operate. Outside this time interval, control strategy can be equal to the control adopted at normal operating conditions. Due to the non-linear electromagnetic characteristic of SRM drives, it is not possible to establish a unique fault-tolerant strategy that is appropriate for all kinds of machines and operating conditions. The appropriate control actions are dependent on the electromagnetic characteristics of SRM phase windings. Thus, the magnitudes of phase currents are dependent on the mechanical load and specific for each machine. Despite these difficulties, it is possible to put in place some control actions to enhance the SRM drive performance.

3.6.1.1 Change the commutation angle of the phase that precedes the missing phase

Lagging the commutation angle of the phase that precedes the faulted phase promotes additional torque production at some rotor positions. This healthy phase can operate during a longer time interval trying to fulfil partially the lack of the electromagnetic torque contribution of the faulted phase. However, this action may not be adequate because lagging the commutation angle may lead to undesirable braking torque.

At normal operating conditions, and at some load and/or speed levels, motor phases contribute with a significant negative electromagnetic torque at its final conducting period. This braking torque is compensated by the positive and greater electromagnetic torque contribution of the following motor phase. This is usually adopted to optimize the drive efficiency, keeping each motor phase operating as long as possible when its electromagnetic torque capability is strong.

The optimization of the commutation angle of the phase that precedes the missing phase imposes a deep engagement between the electromagnetic torque production capability, nearby the commutation angle, and the consequent negative electromagnetic torque contribution.

At those operating circumstances, lagging the commutation angle may be inappropriate, and advancing the commutation angle may be a better choice.

3.6.1.2 Advance the ignition angle of the phase that proceeds the missing phase

The start over in advance of the phase that proceeds the faulted phase promotes also additional electromagnetic torque at some rotor positions. The time interval in which the mechanical torque of the SRM drive is negligible can be reduced. However, the improvements achieved may not be significant due to the weak electromagnetic torque production capability of one motor phase nearby its unaligned position. If the adopted ignition angle, using the usual control strategy, is not far away from the unaligned position, the additional effort of the motor phase may not be compensated by the small increase of its electromagnetic torque contribution.

3.6.1.3 Appropriately adjust the reference control parameter during the time interval where the missing phase should be in operation

It is common, at normal operating conditions, that two adjacent phases are being simultaneously magnetized. When one motor phase is missing, the adjacent phase may present a higher reference control parameter, during the time interval where that phase should be in operation, to compensate for the inexistent electromagnetic torque contribution of the faulted phase. This action may promote a constant mechanical torque during a longer time interval. However, the phase current magnitude needed for that may be inappropriate or useless.

Increasing the phase current magnitude of the phase that precedes the faulted phase may result in significant braking torque production.

Increasing the phase current magnitude of the phase that precedes the missing phase may be useless due to the weak electromagnetic torque capability of that phase at some rotor positions. Moreover, as previously explained, it may cause torque overshoot.

An optimized fault-tolerant control strategy adapts the control angles and adjusts the phase currents magnitudes, taking into account the open-phase fault. Dúbravka *et al.* introduced in [37] an approach for current profile computation that minimizes the torque ripple both under healthy and faulted conditions. The motor used is a three phase SRM. In one open-phase condition, two lookup tables $I=f(T,\theta)$ are used to define the electric current reference of the two healthy phases. Mir *et al.* proposed in [38] an adaptive fuzzy logic controller that adapts its properties to regulate the SRM torque as desired by the drive system even under fault conditions.

A short-circuit fault in a power switch leads to an overcurrent event. The behaviour of the faulted phase current is especially dictated by load and/or speed levels and is also dependent on the electromagnetic characteristics of the machine. If the phase current presents an extreme overcurrent magnitude, an open-circuit fault in the power converter or in the motor will occur as consequence. The short-circuit fault inhibits the power flow from the faulted phase to the source and/or to other motor phases. The decrease of the faulted phase current is always smooth. The control of the faulted phase current is hard to establish. After the aligned position, when the motor phase starts its generating mode, no actions can be implemented to decrease the phase current magnitude. Most of the time, keeping the faulted phase out of operation promotes a smaller torque ripple and a better drive efficiency [39].

The use of the faulted phase can be very important at start-up or at high load and/or speed levels. For that purpose, a different control strategy must be implemented into the faulted phase. Advancing the commutation angle of the faulted phase permits a higher demagnetization level at the aligned position and promotes smaller braking torque, as can be seen in [39]. Additionally, a smaller reference control parameter can be applied to the faulted phase, when compared to healthy phases. Reversing the role between the power switches of the faulted phase is imperative if the short-circuited switch is the chopping device [39,40].

3.6.2 Fault-tolerant converters

SRM intrinsic fault-tolerance capability and performance deterioration when a motor phase is out of service are evident. The impact of a fault can be reduced if a special converter and/or a special machine are used. The increase of fault tolerance of SRM drives, proposed in the literature, is based on one, or more, of the following characteristics:

- High number of motor phases.
- Redundant power electronic elements or power converter modules.
- Dissociation, or possible dissociation, of the coils that form a motor phase.
- Extra power electronic elements for hardware reconfiguration.

Ruba *et al.* proposed in [41] an unusual SRM structure with 12 stator poles and 14 rotor poles. The main goal of the proposed structure is the increase of the mean torque under normal and faulty operating conditions, when compared with a usual 12/8 SRM. The machine proposed has six phases which permits a high fault-tolerant capability. Each phase has two coils wound around two opposite stator poles. The proposed power converter has a single phase H-bridge inverter applied to every coil. This power converter topology allows the independent control of each coil.

In a recent study, Ruba *et al.* [42] proposed a separate asymmetrical half-bridge connection for each coil, reducing the number of power switches.

The application of separate half-bridge converters is also explored by Hennen *et al.* in [43]. The authors proposed a 20/16 SRM with an integrated and distributed power converter. The machine has five phases, and each motor phase is located in four stator poles. An asymmetrical half-bridge converter is used to supply every coil of a phase individually. The power converter of the SRM drive is composed of several power electronic modules placed nearby each stator pole. Each module is composed of an asymmetrical half-bridge converter and supplies a single coil.

The fault-tolerant SRM drives proposed by Ruba *et al.* and Hennen have a higher number of phases than traditional SRM drives and use a half or full bridge converter for each coil. The impact of the inactivation of a single coil, due to a fault occurrence in a power switch or in the coil, is then smaller than when a traditional power converter and SRM with three or four phases are used. However, the increase of motor phases and power electronic modules increases the cost of the machine drive as well as the complexity of the control. Each branch needs separate control devices and protection circuits.

The non-operation of a single coil leads to an unbalanced magnetic pull which is characterized by radial forces on the rotor, stressing the bearings [43]. Unbalance forces can be reduced if the opposite stator pole coil, of the faulted coil, is turned off. However, the use of a healthy coil, when the opposite pole coil is not working, may be an important help, especially at start-up or at high speed and/or mechanical load conditions.

To avoid unbalanced radial forces in the rotor due to the non-operation of a single coil, at faulty operating conditions, Ding *et al.* proposed in [44] a fault-tolerant converter decoupled in two channels. The machine used is a 12/8 SRM

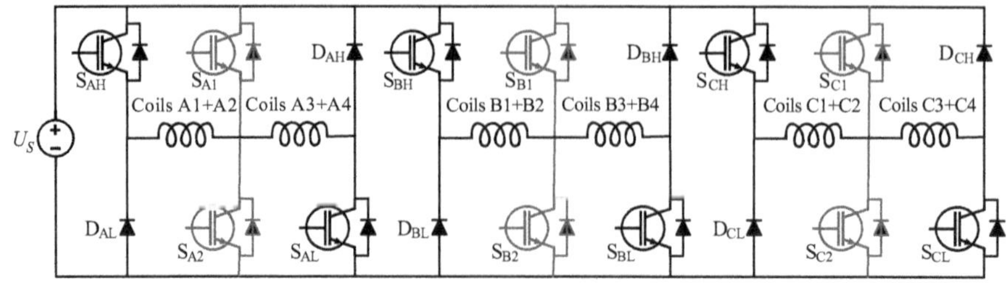

Figure 3.15 Fault-tolerant converter proposed by Hu et al. [45]

with three phases, four stator poles per phase, each pole having a single coil wound around. Each phase has two channels. A channel is composed by two coils, which are located in diametrically opposite stator poles, connected in series. Each phase channel is supplied by an asymmetric half-bridge converter. The SRM drive presents the same fault impact, due to the absence of a channel, as a six-phase SRM, when one of its phases is out of service. However, the control complexity is smaller in the three-phase machine because the control signals may be the same for the two channels of each motor phase.

Hu *et al.* proposed in [45] a fault-tolerant converter composed of a traditional asymmetrical half-bridge converter and a fault-tolerant module which is a common three-phase inverter (Figure 3.15). The machine used is also a 12/8 SRM with three phases. Each phase has four coils connected in series and a midpoint node of each phase winding which is electrically accessible. The windings are divided in two parts, each of them composed of two coils. The outputs of the three-phase inverter are connected to the midpoint node of each phase winding.

Under normal operating conditions, the three-phase inverter is not working, and each motor phase-magnetization state is dictated by the control of the respective upper and lower power switches (S_{AH} and S_{AL} for phase A), as it occurs using the traditional asymmetric half bridge.

If a fault takes place, either an open- or a short-circuit event, in the upper power switch of a phase, an upper power switch of the three-phase inverter starts operating. The fault is bypassed, the first part of the winding is electrically isolated, and the second part of the winding continues its operation. For example, when S_{AH} has a fault or there is a fault in coils A1 or A2, the control signal of S_{AH} is transferred to S_{A1}. Then, the first part of phase A winding, coil A1 and coil A2, stays out of service, and the second part, coil A3 and coil A4, can be working normally, using the power switches S_{A1} and S_{AL} and the diodes D_{AH} and D_{A2}. An identical procedure is used when the fault affects a lower power switch or the second part of a phase winding. In the case of phase A, the power switch S_{A2} starts conducting when there is a fault in the second part of phase A winding or in the respective lower power switch S_{AL}. The control signal of S_{AL} is transferred to S_{A2}.

The SRM drive fault-tolerant converter proposed by Hu *et al.* in [45] allows the operation of half of a faulty phase. However, a central node of the winding must be electrically accessible to be connected to the fault-tolerant module. The number of electrical connections, between the power converter and the machine, increases when compared to traditional connections of the two end-windings of a motor phase. When the fault-tolerant module is working, the DC bus voltage is applied to half of a phase winding. Then, the coils must be designed with a higher rated voltage.

Due to the inability of some coils, when a fault-tolerant procedure is implemented, the current phase measurement process must be different to the traditional one. The authors suggest two solutions. The easiest solution is the use of two current sensors per phase. One of the sensors measures the phase current at the beginning (or at the end) of the winding and the other sensor measures the electric current that flows to or from the fault-tolerant module. The second solution consists of a special connection of a common electric current sensor. The first part and the second part of the phase winding pass equally through the electric current sensor. Under normal operating conditions, the measured electric current magnitude is twice the phase current magnitude, and under fault-tolerant operating conditions, those magnitudes are the same.

Hu *et al.* [46] proposed a fault-tolerant converter made up of a traditional asymmetrical half-bridge converter and a fault-tolerant module which is a single-phase full bridge inverter (Figure 3.16). The theory applied is based on the same principles as the previous work, which is the promotion of a new path for the electric current.

In his work, a 12/8 SRM is also used, and the windings structure is the same. Each phase winding is electrically divided in three parts, and two inner electric nodes are established to promote their electric connection to the fault-tolerant module, when necessary. The first part of the winding is the first coil of the respective phase. The second part is composed of the two subsequent coils, and the third part is the last coil.

The fault-tolerant module is not permanently connected to the inner nodes. A relay is activated whenever a fault occurs. The fault-tolerant module has two bridge

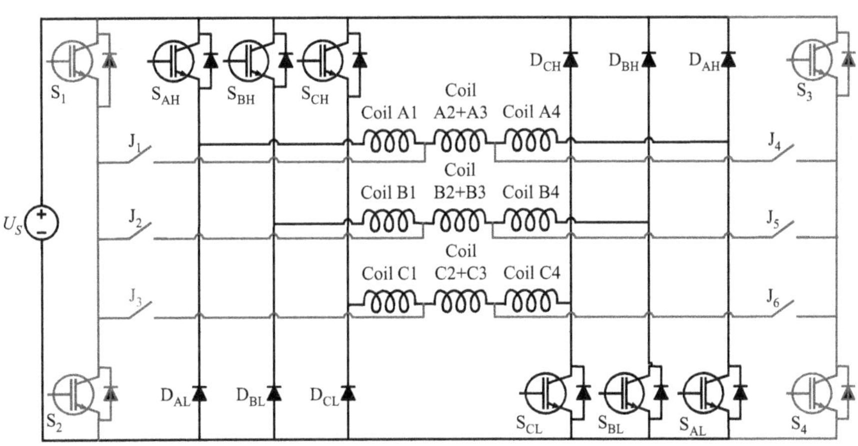

Figure 3.16 Fault tolerant converter proposed by Hu et al. [46]

arms, each one containing two switches and two diodes. In case of a fault, a bridge arm starts operating and does the work of one of the original arms, and one part of the affected winding stays out of work. Due to the number of new connections that can be made, the proposed fault-tolerant converter assures the operation of all motor phases in a lot of scenarios. For example, if there is a fault in the upper power switch of phase A, the relay J_1 is closed and the upper power switch of the left arm of the fault-tolerant module operates. Parts 2 and 3 of phase A keep operating, and part 1 is inactivated. The same is done if the fault occurs in the first phase coil.

The right arm of the fault-tolerant module is activated to fulfil the role of a lower power switch of the traditional asymmetric half-bridge converter. For example, the relay J_4 is closed when there is a fault in the lower power switch of phase A (S_{AL}) or in part 3 of the respective winding.

The lower power switch of the left arm of the fault-tolerant module can be used when there is an open-circuit fault in part 2 of a phase winding, as well as the upper power switch of the right arm of the fault-tolerant module. In these circumstances, there are two independent electric circuits for the same motor phase. One of them controls the magnetization of part 1, and the other one controls the magnetization of part 3. For example, after an open-circuit fault in part 2 of phase A, relays J_1 and J_4 are closed. The magnetization of part 1 is done turning on the power switches S_{AH} and S_2. The diodes D_{AL} and D_1 are conducting in freewheeling and/or demagnetization modes. The magnetization of part 3 is assured by controlling the conductive state of S_3 and S_{AL}. Identical analysis can be made for the others motor phases.

Beyond these fault scenarios, a power switch of the fault-tolerant module can be shared by two motor phases when there is a fault affecting two motor phases simultaneously. In these circumstances, there is some control dependence between those two phases.

The fault-tolerant converter proposed by Hu *et al* in [46] assures that at least half of a phase winding continues its operation after a fault occurrence. The disadvantages of this fault-tolerant power converter are similar to the disadvantages of the fault-tolerant converter proposed in [45]. The connections between the power converter and the machine are increased because two inner winding nodes must be electrically accessible. Moreover, if the measurement of electric phase currents are necessary for the control loop, special current sensors or additional current sensors must be used, due to the change of the electric phase current flow whenever a bridge arm of the fault-tolerant module is activated.

Special care must be taken when three of the four coils of a particular phase are being magnetized: this operating condition leads to unbalance radial forces in the rotor.

The fault-tolerant power converter proposed by Gameiro and Cardoso [47] has some additional power switches that must be activated for hardware reconfiguration after an open-circuit fault occurrence in a power switch (Figure 3.17). This work considers a four phases 8/6 SRM. At normal operating conditions, a traditional asymmetric power converter is used.

After the fault occurrence in one of the power switches of the traditional asymmetric power converter, an electrical connection is permanently established in

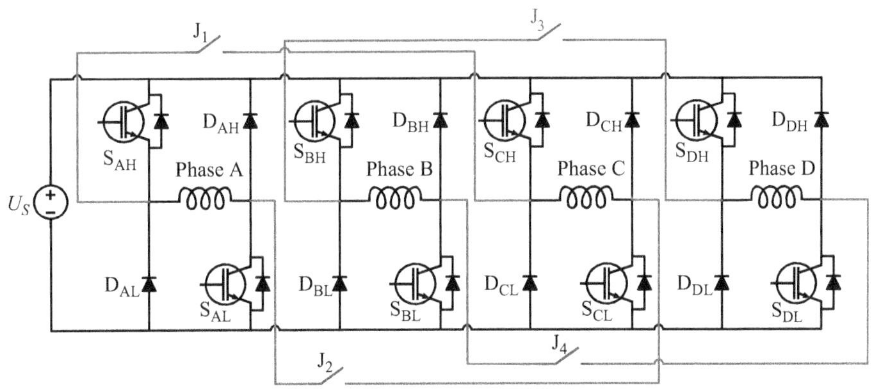

Figure 3.17 Fault-tolerant converter proposed by Gameiro and Cardoso [47]

order to keep the affected phase in operation. That electrical connection may be established by a simple electrical device, such as a relay.

Under fault conditions, a power switch is shared by two motor phases. For example, if the power switch S_{AH} fails, J_1 is activated and the power switch S_{CH} is used to magnetize both phases A and C.

There is some control dependence between the motor phases that share a power switch. However, the impact is not visible at low speed because the phases are not adjacent, and they do not work simultaneously. At higher speed or higher mechanical load levels, there are some time periods when both phases are conducting. It is convenient to define the shared power switch as the chopping power switch for current regulation proposes. If a voltage pulse control is used, a fault control strategy must be taken to avoid generative phenomena. The authors suggest, for those situations, to start the magnetization process of one of the phases that share a power switch only after the other one is completely demagnetized.

Despite the control dependence between those phases, the achieved performance is clearly better than when a phase is not working.

Since all power switches of the asymmetrical power converter may be shared by two motor phases, under faulty operating conditions, their rated electric currents must be twice the rated current of a power switch used by a single motor phase.

Oliveira *et al.* proposed, in [48], to apply a fault-tolerant converter based in a common three-phase bridge inverter in three-phase SRM drives (Figure 3.18). The phases are star connected and the neutral node is connected at the midpoint of two series connected capacitors, which permits the independent control of all motor phases. The electrical connections are the same either at normal or faulty operating conditions. The magnetization of each motor phase can be made by turning on any of the power switches of the respective bridge arm. The choice of the conducting switch, in normal operating conditions, depends on the voltage levels of each capacitor. This means that, during normal operating conditions, the phase current is bidirectional. The voltage applied to the phase winding is half of the DC bus

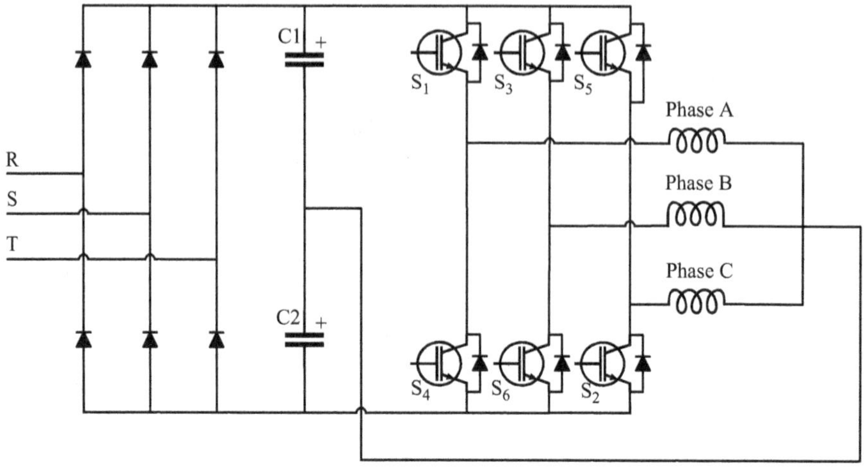

Figure 3.18 Fault-tolerant converter proposed by Oliveira et al. [48]

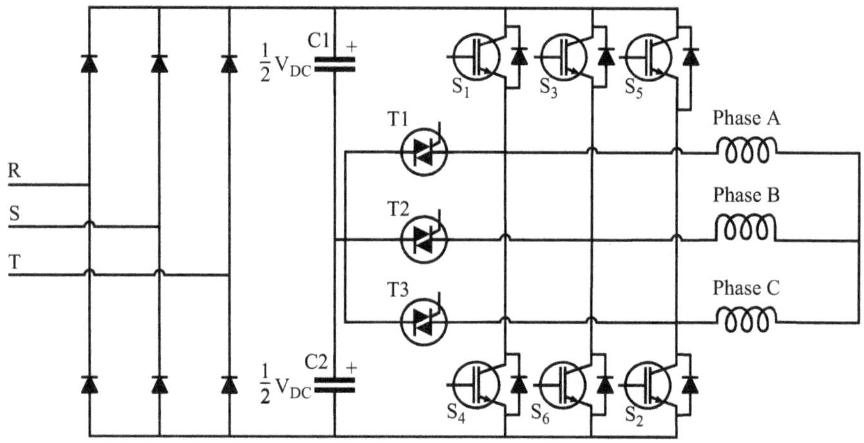

Figure 3.19 Fault-tolerant converter proposed by Lee et al. [49]

voltage, but all power switches must be prepared to support the DC bus voltage. The freewheeling mode is not possible.

After an open-circuit fault in one of the power switches, the healthy power switch of the affected motor phase will be the only one responsible for its magnetization, and the respective phase current is then unidirectional. The other phases must be appropriately controlled to guarantee the voltage balance at the terminals of each capacitor.

Lee *et al.* [49] proposed a fault-tolerant power converter also based on a common three-phase bridge inverter, similar to that adopted in AC drives (Figure 3.19) with additional power electronic devices and electrical connections.

When an open-circuit fault occurs in a power switch, the affected motor phase can keep its operation. For that, a TRIAC is activated and establishes the connection of the affected phase to a midpoint of the DC bus. Due to the star configuration, there are, always, two phases conducting, and the respective electric currents are symmetrical. This control dependency implies that each phase is in conduction over an extended time interval. When an asymmetric half-bridge converter is used, it is necessary half of that time interval to obtain equal torque output. This converter topology is, for this reason, criticized by several authors, taking as an example Clothier and Mecrow [50].

The adopted control strategy essentially divides each control cycle in three operating modes. In each mode there is a conducting phase which is responsible for the production of significant electromagnetic torque. The respective phase current is properly regulated. The other phase that is simultaneously in conduction produces a relatively small electromagnetic torque, sometimes positive and sometimes negative.

In the case of an open-circuit fault situation, in one of the power switches, the control signals of the power switches associated to the affected phase are inhibited. The control signals of the remaining power switches are identical to those of normal operation. This topology reduces the voltage applied after the activation of one of the TRIACs.

Acknowledgement

This work was supported by the European Regional Development Fund (ERDF) through the Operational Programme for Competitiveness and Internationalization (COMPETE 2020), under Project POCI-01-0145-FEDER-029494, and by National Funds through the FCT – Portuguese Foundation for Science and Technology, under Projects PTDC/EEI-EEE/29494/2017 and UID/EEA/04131/2013.

References

[1] T. J. E. Miller, *Brushless Permanent-Magnet and Reluctance Motor Drives*, Clarendon Press, Oxford, 1989.

[2] T. J. E. Miller, *Switched Reluctance Motors and their Control*, Magna Physics Publishing and Clarendon Press, Oxford, 1993.

[3] P. J. Lawrenson, J. M. Stephenson, P. T. Blenkinsop, J. Čorda and N. N. Fulton, "Variable-speed switched reluctance motors", *IEE Proceedings B*, vol. 127, no. 4, pp. 253–265, 1980.

[4] J. Čorda, S. Maši and J. M. Stephenson, "Computation and experimental determination of running torque waveforms in switched-reluctance motors", *IEE Proceedings*, vol. 126, no. 5, pp. 393–396, 1979.

[5] J. M. Stephenson and J. Čorda, "Computation of torque and current in doubly salient reluctance motors from nonlinear magnetisation data", *IEE Proceedings B*, vol. 140, no. 6, pp. 387–392, 1993.

[6] M. Barnes and C. Pollock, "Power electronic converters for switched reluctance drives", *IEEE Transactions on Power Electronics*, vol. 13, no. 6, pp. 1100–1111, 1998.

[7] S. Vukosavić and V. R. Stefanović, "SRM inverter topologies: a comparative evaluation", *IEEE Transactions on Industry Applications*, vol. 27, no. 6, pp. 1034–1047, 1991.

[8] J. Reinert, R. Inderka, M. Menne and R. de Doncker, "Optimizing performance in switched-reluctance drives", *IEEE Industry Applications Magazine*, vol. 6, no. 4, pp. 63–70, 2000.

[9] Y. A. Kwon, "Computation of optimal excitation of a switched reluctance motor using variable voltage", *IEEE Transactions on Industrial Electronics*, vol. 45, no. 1, pp. 177–180, 1998.

[10] N. C. Sahoo, S. K. Panda and P. K. Dash, "A current modulation scheme for direct torque control of switched reluctance motor using fuzzy logic", *Mechatronics*, vol. 10, no. 3, pp. 353–370, 2000.

[11] C. Moron, A. Garcia, E. Tremps, and J. A. Somolinos, "Torque control of switched reluctance motors", *IEEE Transactions on Magnetics*, vol. 48, no. 4, pp. 1161–1164, 2012.

[12] J. Bayless, N. Kurihara, H. Sugimoto and A. Chiba, "Acoustic noise reduction of switched reluctance motor with reduced RMS current and enhanced efficiency", *IEEE Transactions on Energy Conversion*, vol. 31, no. 2, pp. 627–636, 2016.

[13] T. J. E. Miller, "Faults and unbalance forces in the switched reluctance machine", *IEEE Transactions on Industry Applications*, vol. 31, no. 2, pp. 319–328, 1995.

[14] C. M. Stephens, "Fault detection and management system for fault-tolerant switched reluctance motor drive", *IEEE Transactions on Industry Applications*, vol. 27, no. 6, pp. 1098–1102, 1991.

[15] I. Husain and M. N. Anwa, "Fault analysis of switched reluctance motor drives", *IEEE International Conference Electric Machines and Drives 1999*, Seattle, WA, USA, pp. 41–43, 1999.

[16] B. Schinnerl and D. Gerling, "Analysis of winding failures of switched reluctance motors", *Record of the International Electric Machines and Drives Conference*, Miami, FL, USA, pp. 738–743, 2009.

[17] J. Li, D. Choi and Y. Cho, "Analysis of rotor eccentricity in switched reluctance motor with parallel winding using FEM", *IEEE Transactions on Magnetics*, vol. 45, no. 6, pp. 2851–2854, 2009.

[18] N. Saha and D. S. Choudhury, "Analysis of different types of faults exhibited in switched reluctance motor drives", *10th International Conference on Intelligent Systems and Control (ISCO)*, Coimbatore, India, pp. 1–5, 2016.

[19] A. A. Arkadan and B. W. Kielgas, "Switched reluctance motor drive systems dynamic performance prediction under internal and external fault conditions", *IEEE Transactions on Energy Conversion*, vol. 9, pp. 45–52, 1994.

[20] C. P. Weiss, M. Huebner, M. D. Hennen and R. W. De Doncker, "Switched reluctance machine model considering asymmetries and enabling dynamic fault simulation", *Record of the International Electric Machines and Drives Conference*, Chicago, IL, USA, pp. 979–985, 2013.

[21] W. Wand and B. Fahimi, "Short-circuit analysis of switched reluctance machines", *Proceedings of the 40th Annual Conference of the IEEE Industrial Electronics Society*, Dallas, TX, USA, pp. 863–868, 2014.

[22] A. A. Arkadan and B. W. Kielgas, "The coupled problem in switched reluctance motor drive systems during fault conditions", *IEEE Transactions on Magnetics*, vol. 30, no. 5, pp. 3256–3259, 1994.

[23] L. A. Belfore and A. A. Arkadan, "Modeling faulted switched reluctance motors using evolutionary neural networks", *IEEE Transactions on Industrial Electronics*, vol. 44, no. 2, pp. 226–233, 1997.

[24] A. A. Arkadan, M. Sidani and P. Du, "Characterization of SRM drives under normal and fault operating conditions", *Record of the International Electric Machines and Drives Conference*, Seattle, WA, USA, pp. 249–251, 1999.

[25] A. A. Arkadan, P. Du, M. Sidani and M. Bouji, "Performance prediction of SRM drive systems under normal and fault operating conditions using GA-based ANN method", *IEEE Transactions on Magnetics*, vol. 36, no. 4, pp. 1945–1949, 2000.

[26] B. Lequesne, S. Gopalakrishnan and A. M. Omekanda, "Winding short-circuits in the switched reluctance drive", *IEEE Transactions on Industry Applications*, vol. 41, no 5, pp. 1178–1184, 2005.

[27] P. Bogusz, M. Korkosz and J. Prokop, "Current harmonics analysis as a method of electrical faults diagnostic in switched reluctance motors", *Proceedings of the 6th IEEE International Symposium on Diagnostics for Electrical Machines, Power Electronics & Drives*, Cracow, Poland, pp. 426–431, 2007.

[28] J. Faiz and S. Pakdelian, "Finite-element analysis of a switched reluctance motors under static eccentricity fault", *IEEE Transactions on Magnetics*, vol. 42, no. 8, pp. 2004–2008, 2006.

[29] D. G. Dorrell, I. Chindurza and C. Cossar, "Effects of rotor eccentricity on torque in switched reluctance machines", *IEEE Transactions on Magnetics*, vol. 41, no. 10, pp. 3961–3963, 2005.

[30] C. Gan, J. Wu, S. Yang, Y. Hu and W. Cao, "Wavelet packet decomposition-based fault diagnosis scheme for SRM drives with a single current sensor", *IEEE Transactions on Energy Conversion*, vol. 31, no. 1, pp. 303–313, 2016.

[31] C. Gan, J. Wu, S. Yang, Y. Hu, W. Cao and J. Si, "Fault diagnosis scheme for open-circuit faults in switched reluctance motor drives using fast Fourier transform algorithm with bus current detection", *IET Power Electronics*, vol. 9, no. 1, pp. 20–30, 2016.

[32] S. R. K. Hoseini, E. Farjah, T. Ghanbari and H. Givi, "Extended Kalman filter-based method for inter-turn fault detection of the switched reluctance motors", *IET Electric Power Applications*, vol. 10, no. 8, pp. 714–722, 2016.

[33] H. Chen and S. Lu, "Fault diagnosis digital method for power transistors in power converters of switched reluctance motors", *IEEE Transactions on Industrial Electronics*, vol. 60, no. 2, pp. 749–763, 2013.

[34] G. Han, H. Chen, X. Shi and Y. Wang, "Phase current reconstruction strategy for switched reluctance machines with fault-tolerant capability", *IET Electric Power Applications*, vol. 11, no. 3, pp. 399–411, 2017.

[35] J. F. Marques, J. O. Estima, N. S. Gameiro and A. J. M. Cardoso, "A new diagnostic technique for real-time diagnosis of power converter faults in switched reluctance motor drives", *IEEE Transactions on Industry Applications*, vol. 50, no. 3, pp. 1854–1860, 2014.

[36] N. S. Gameiro and A. J. M. Cardoso, "A new method for power converter fault diagnosis in SRM drives", *IEEE Transactions on Industry Applications*, vol. 48, no. 2, pp. 653–662, 2012.

[37] P. Dúbravka, P. Rafajdus, P. Makyš and L. Szabó, "Control of switched reluctance motor by current profiling under normal and open phase operating condition", *IET Electric Power Applications*, vol. 11, no. 4, pp. 548–556, 2017.

[38] S. Mir, M. S. Islam, T. Sebastian and I. Husain, "Fault-tolerant switched reluctance motor drive using adaptive fuzzy logic controller", *IEEE Transactions on Power Electronics*, vol. 19, no. 2, pp. 289–295, 2004.

[39] S. Gopalakrishnan, A. M. Omekanda and B. Lequesne, "Classification and remediation of electrical faults in the switched reluctance drive", *IEEE Transactions on Industry Applications*, vol. 42, no. 2, pp. 479–486, 2006.

[40] H. S. Ro, D. H. Kim, H. G. Jeong and K. B. Lee, "Tolerant control for power transistor faults in switched reluctance motor drives", *IEEE Transactions on Industry Applications*, vol. 51, no. 4, pp. 3187–3197, 2015.

[41] M. Ruba, L. Szabo, L. Strete and I. A. Viorel, "Study on fault tolerant switched reluctance machines", *2008 18th International Conference on Electrical Machines*, Vilamoura, Portugal, pp. 1–6, 2008.

[42] M. Ruba, L. Szabo and F. Jurca, "Fault tolerant switched reluctance machine for wind turbine blade pitch control", *2009 International Conference on Clean Electrical Power*, Capri, Italy, pp. 721–726, 2009.

[43] M. D. Hennen, M. Niessen, C. Heyers, H. J. Brauer and R. W. De Doncker, "Development and control of an integrated and distributed inverter for a fault tolerant five-phase switched reluctance traction drive", *IEEE Transactions on Power Electronics*, vol. 27, no. 2, pp. 547–554, 2012.

[44] W. Ding, Y. Liu and Y. Hu, "Performance evaluation of a fault-tolerant decoupled dual-channel switched reluctance motor drive under open-circuits", *IET Electric Power Applications*, vol. 8, no. 4, pp. 117–130, 2014.

[45] Y. Hu, C. Gan, W. Cao, W. Li and S. J. Finney, "Central-tapped node linked modular fault-tolerance topology for SRM applications", *IEEE Transactions on Power Electronics*, vol. 31, no. 2, pp. 1541–1554, 2016.

[46] Y. Hu, C. Gan, W. Cao, J. Zhang, W. Li and S. J. Finney, "Flexible fault-tolerant topology for switched reluctance motor drives", *IEEE Transactions on Power Electronics*, vol. 31, no. 6, pp. 4654–4668, 2016.

[47] N. S. Gameiro and A. J. M. Cardoso, "Fault tolerant power converter for switched reluctance drives", *2008 18th International Conference on Electrical Machines*, Vilamoura, Portugal, pp. 1–6, 2008.

[48] A. C. Oliveira, C. B. Jacobina, A. M. N. Lima and F. Salvadori, "Startup and fault tolerance of the SRM drive with three-phase bridge inverter", *2005 IEEE 36th Power Electronics Specialists Conference*, Recife, Brazil, pp. 714–719, 2005.

[49] K. J. Lee, N. J. Park, K. H. Kim and D. S. Hyun, "Simple fault detection and tolerant scheme in VSI-fed switched reluctance motor", *2006 37th IEEE Power Electronics Specialists Conference*, Jeju, South Korea, pp. 1–6, 2006.

[50] A. C. Clothier and B. C. Mecrow, "The use of three phase bridge inverters with switched reluctance drives", *1997 Eighth International Conference on Electrical Machines and Drives (Conf. Publ. No. 444)*, Cambridge, pp. 351–355, 1997.

Chapter 4
High-power synchronous machine drives

Alberto Tessarolo[1] *and Adérito N. Alcaso*[2,3]

In this chapter, high power synchronous motor drives will be addressed. First of all, an overview will be provided of the main technologies and design features which characterize large synchronous machines (Section 4.1) and the relevant supplying converters (Section 4.2), also taking into due account their field of application. Subsequently, the attention will be placed on the major strategies intended to improve high-power synchronous machine drives fault tolerance (Section 4.3), acting on the system-level drive architecture as well as on the design and operation of the individual components (electric motor, converter, control system). Finally, the main diagnostics and condition monitoring techniques for high-power synchronous motor drives will be covered, describing the main methods to detect possible malfunctioning, anomalies and faults in drive operation before they result in serious damages or hazards.

4.1 High-power synchronous motors

The synchronous machines used in high-power drives can be mainly classified, based on their rotor technology, into the two categories of wound-field synchronous machines and permanent-magnet synchronous machines.

4.1.1 Permanent magnet motors

Permanent magnet machines are known to offer various advantages over their wound-rotor counterparts, such as higher efficiency due to almost absent rotor losses, no rotor electrical excitation and field supply equipment, high torque density, better capability of withstanding centrifugal stresses and, therefore, high rotor speeds. Their drawbacks mainly relate to the high and fluctuating cost of permanent magnets, possible rotor demagnetization risks and no possibility to control the rotor flux.

[1]Engineering and Architecture Department, University of Trieste, Italy
[2]Polytechnic of Guarda, Portugal
[3]CISE – Electromechatronic Systems Research Centre, Universidade da Beira Interior, Portugal

Traditionally, permanent magnet motor technology has been originated in the field of small-power applications and is still today relatively rare in large synchronous motor drives. Nevertheless, there are some remarkable examples, mainly in the field of ship propulsion and in high-speed drives for the oil and gas industry.

Regarding ship propulsion, we can mention the use of low-speed high-pole-count permanent magnet motors in the pods of large ships, with power ratings in the order of 4 MW or more and speeds around 100 rpm [1]. The main reason for using the permanent magnet technology is, in this case, the need to reduce the motor size as much as possible given the very small space available for the electric motor inside the pod structure. Other examples can be found in the electric propulsion of military vessels (frigates) [2].

Another important field where we can find permanent magnet motors in relatively large electric drives is the oil and gas industry, where built and tested permanent magnet machines with power ratings in the order of 8 MW at 15,000 rpm are reported in the literature [3]. In these applications, the reason why the permanent magnet technology is preferred evidently relates to the very high speeds (often above 10,000 rpm) which would make it difficult or prohibitive to mechanically secure a rotor excitation winding against centrifugal forces.

4.1.1.1 Rotor design

Both in low-speed and in high-speed permanent magnet synchronous motor drives for high power applications the most frequent rotor design is the surface-mounted permanent magnets (SPM) arrangement, where magnets are fixed on the external surface of a ferromagnetic rim (Figure 4.1). Permanent magnets are practically always made of rare earths (neodymium-iron-bore or samarium-cobalt alloys) to obtain good resilience to high temperatures and demagnetizing currents and to guarantee good torque density values.

In high speed motors, permanent magnets are usually secured against centrifugal forces by a high tensile strength carbon-fibre retaining sleeve wrapped around the rotor (Figure 4.2).

Conversely, in lower speed applications, a fibreglass tape wrapping is usually sufficient (Figure 4.3).

Figure 4.1 Example of a SPM rotor for a high-speed motor. © 2017 IEEE. Reprinted, with permission, from [4]

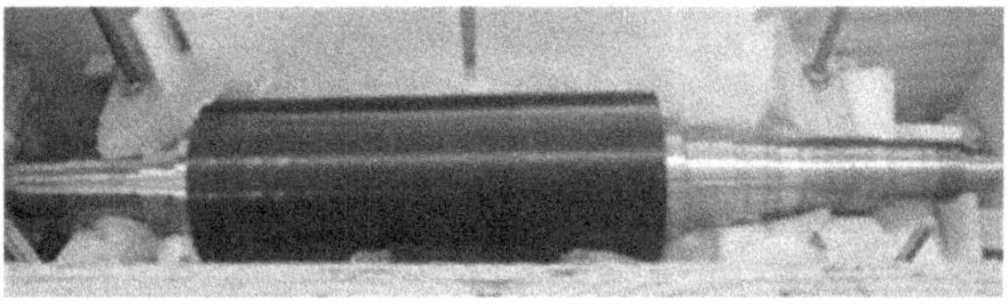

Figure 4.2 SPM rotor wrapped by a carbon-fibre retaining sleeve. © 2017 IEEE. Reprinted, with permission, from [4]

Figure 4.3 SPM rotor wrapped by a fibreglass

4.1.1.2 Stator design

As regards the stator technology, permanent magnet motors for high power applications rarely include random-wound stator windings made of wire, which are typical of low-voltage electric motors with power ratings of less than several hundred kW [5]. Wound-formed multi-turn coil windings are typically used, instead. According to this technology, peculiar to medium- and high-voltage machines, the stator winding is composed of several wound-formed coils (Figure 4.4), each including a certain number of turns in series.

The turn is made of various shunt-connected copper conductors of rectangular sections. The wound-formed coils are suitable to be fit, one by one, into the open slots of the stator laminated core to form a double-layer distributed winding (Figure 4.5). After all the wound-formed coils are mounted in the stator slots, series and parallel connections between coils are implemented (Figure 4.5).

In some high-speed high-power motor drives, like that reported in [6], the stator supply frequency can be in the order of some hundreds hertz. To reduce the eddy-current losses in the stator conductors at such high frequencies, Litz-wire technology is used to form each turn. This means that the turn is composed of a bundle of very thin shunt-connected wires twisted together so as to reduce circulating currents among them (Figures 4.6 and 4.7).

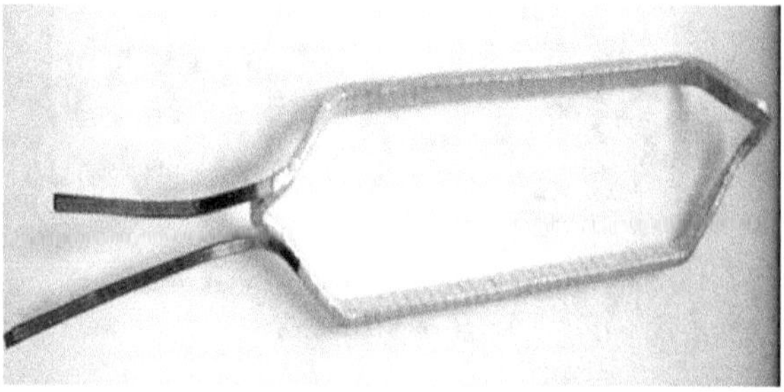

Figure 4.4 Wound-formed multi-turn coil

(a)

(b)

Figure 4.5 (a) Stator winding during the winding process and (b) finished stator winding with series and parallel connections between coils

Finally, there are low-speed applications, like large ship propulsion, where the pole count of the motor is so high that the number of poles is close to the number of slots. This may result in a fractional-slot concentrated winding (FSCW) design [8]. In this case, each coil embraces only one tooth and is, in fact, known as 'tooth coil'. An example of a portion (sector) of a modular concentrated winding stator is shown in Figure 4.8.

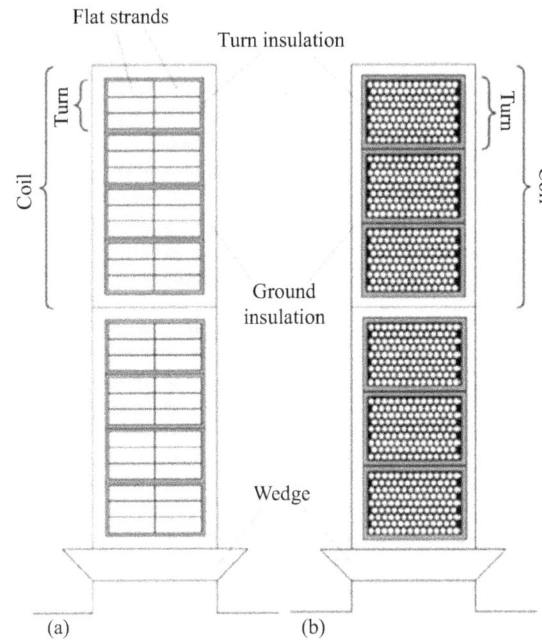

Figure 4.6 (a) Stator slot cross section in case of conventional multi-turn coil winding and (b) turn implementation with Litz wire technology.

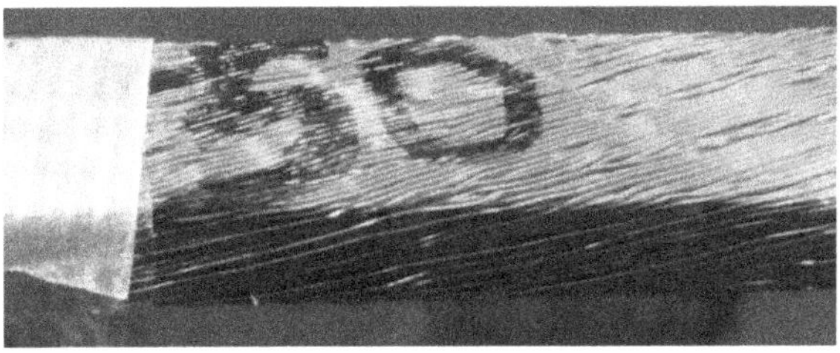

Figure 4.7 Part of a turn made of Litz wire

4.1.1.3 Bearings

Low- and medium-power electric motors mainly mount ball or rolling bearings. High-power machines, with power ratings above 750 kW, or even less in the case of high speeds, generally use journal bearings, which are characterized by no friction or contact between parts in relative rotation (Figure 4.9). A thin fluid film needs to exist between the revolving and the static parts of the bearing, which is assured by means of a pressurized lubrication system. Among the various advantages of journal bearings, we can mention beneficial damping effect for shaft lateral dynamics, especially during operation near critical speeds; reduced noise and vibration; and longer life under normal operating conditions.

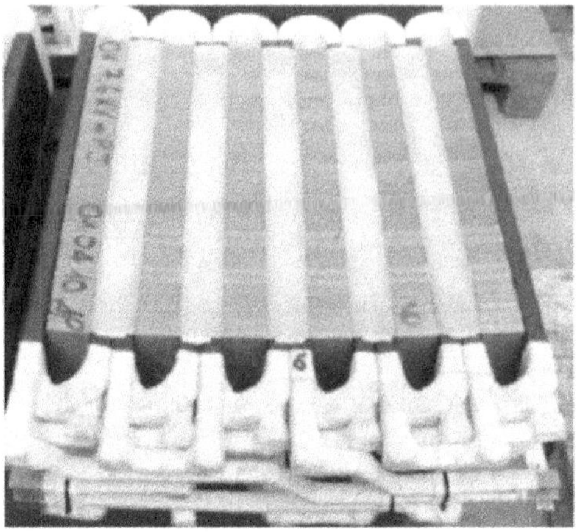

Figure 4.8 Example of a stator portion wound with tooth coils. © 2017 IEEE. Reprinted, with permission, from [9]

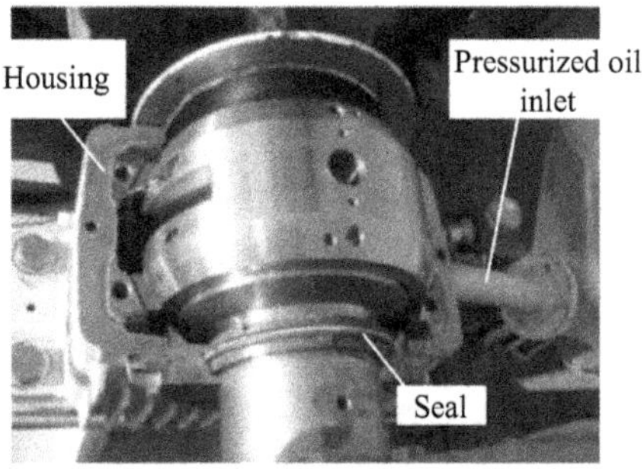

Figure 4.9 Journal bearing view after removal of the upper housing part

For some high-speed applications, with particular rotor-dynamics requirements, active magnetic bearings can be used as reported in [6].

4.1.2 Wound-field synchronous motors

Wound-field synchronous motors are the most traditional and widespread solution for high-power drives, both in low-speed and high-speed applications. Compared to permanent magnet machines, they make it possible to adjust the rotor excitation, which is paid in terms of need for additional equipment to supply energy to the rotating field winding and in terms of larger size and weight. This motor technology makes it possible to reach power ratings in the order of several tens of MWs [10] but is limited in terms of maximum speed (typically well below 10,000 rpm) due to the large radial forces acting on the field winding.

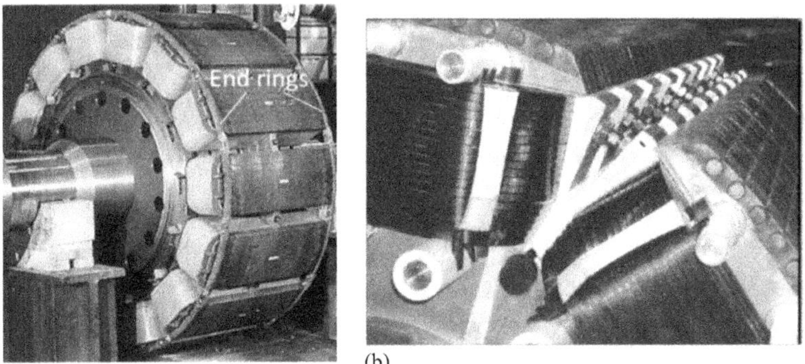

(a) (b)

Figure 4.10 (a) Rotors with salient laminated poles, damper bars short-circuited by end rings and (b) end plates (right)

4.1.2.1 Rotor design

For low-speed high-torque applications, the most suitable and convenient rotor design includes laminated salient poles (Figure 4.10). Example applications for this kind of rotor are high-power electric drives for rolling mills, grinding mills and large ship propulsion, with speeds of a few hundred rpm and power ratings of several MW's (sometimes above 20 MW). The number of poles is higher than six and, for example, is typically between 12 and 14 for large ship propulsion. In general, the pole count increases as the speed decreases. Rotors with high pole count usually feature larger diameters and smaller axial lengths.

Copper or aluminium damper bars are embedded in the salient pole laminations; the damper bars can be short-circuited by end rings or by end plates (Figure 4.10).

During steady-state operation, damper bars are useful to reduce the air-gap harmonics produced, for example, by stator current distortion or unbalance. During transients, dampers improve the synchronizing torque of the machine reducing pole-stepping risks.

When the motor speed is between 1,500 and 1,800 rpm, with power ratings of several MWs or a few tens of MWs, a four-pole rotor is adopted as a rule, either of cylindrical type or with salient poles. Cylindrical rotors for this range of power and speed are usually laminated (Figure 4.11). Damper bars are embedded in the laminations in this case, too, and are short-circuited by end plates.

The rotor excitation winding is composed of several (hundreds) of series-connected turns embedded in rotor slots and retained by conductive or non-conductive wedges (Figure 4.11).

The excitation winding overhangs are usually secured against centrifugal forces by wrapping them with a retaining sleeve (Figure 4.11).

As an alternative to laminated cylindrical rotor, for roughly the same speed and power levels, a four-pole solid salient pole construction can be used as depicted in Figure 4.12. The solid steel construction for the pole shoe relates to the high peripheral speed, for which a laminated structure would not be mechanically suitable.

(a) (b) (c)

Figure 4.11 (a) Laminated rotor core of a high-power four-pole synchronous motor; (b) excitation field winding (end coils) and slot wedges and (c) wound cylindrical rotor with retaining sleeve wrapped around excitation winding overhangs

Figure 4.12 Four-pole synchronous motor with solid salient poles

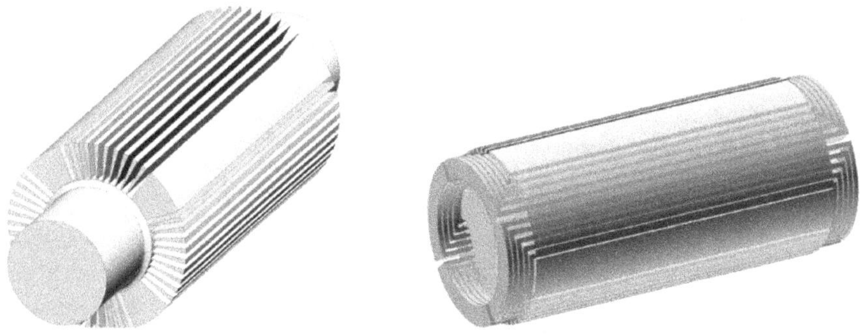

(a) (b)

Figure 4.13 3D rendering of a solid-steel cylindrical rotor: (a) without field circuit and (b) with field circuit. © 2013 IEEE. Reprinted, with permission, from [7]

Figure 4.14 End ring for a solid-steel rotor

Unlike laminated pole machines, this kind of motor does not need damper bars, the role of these being played by the eddy currents which arise in the solid pole shoe during transients or in presence of air-gap harmonic fluxes.

When the speed exceeds 3,000 rpm, a solid-steel rotor construction is adopted, with a technology which is very similar to that of turboalternators (Figure 4.13). The number of poles can be four or, more frequently, two. The design is roughly the same as that of cylindrical laminated rotors (Figure 4.13) except that, due to the higher peripheral speeds, forged steel need to be employed for the rotor body instead of the cheaper laminations. As a further difference, the field winding overhang need to be retained by a metal (non-magnetic steel) end ring (Figure 4.14) as tape wrapping would be mechanically inadequate.

A typical application field for solid-steel cylindrical rotor machines is constituted by high-power electric drives for the oil and gas industry, namely for gas compressors, natural gas liquefaction plants and pipelines. In this field, the highest power electric drives (with power ratings progressively approaching 100 MWs) are installed or under development.

4.1.2.2 Stator design

The most frequent stator winding technology for high-power wound-field synchronous machines is basically that based on wound-formed multi-turn coils, already described for permanent magnet synchronous motors (Figures 4.4 and 4.5).

On the other side, the use of Litz-wire to form stator turns (Figure 4.6) is generally not required in wound-field machine as the supply frequency (closely linked to the speed) is not as high as to produce dangerous eddy current losses in ordinary rectangular-shaped conductors.

Also the FSCW design (Figure 4.8) is infrequent if not absent in wound-field synchronous motor. In fact, the circumferential span of a wound pole (whether in a salient or cylindrical rotor construction) is such to typically cover multiple stator slots, so that the number of stator slot cannot be close to the number of poles.

In some high-power synchronous motors, like in the 35 MW gas compressor motor drive reported in [11], due to the very high stator current, it becomes convenient or necessary to adopt Roebel technology. The Roebel winding is a dual-layer distributed winding composed of bars connected in series at their ends as schematically depicted in Figure 4.15.

Each Roebel bar consists of a stack of shunt-connected strands which change their position inside the bar cross section along the bar length as illustrated in Figure 4.16. This particular construction guarantees that all the strands link approximately the same slot leakage flux, so that the occurrence of inter-strand circulating currents in prevented.

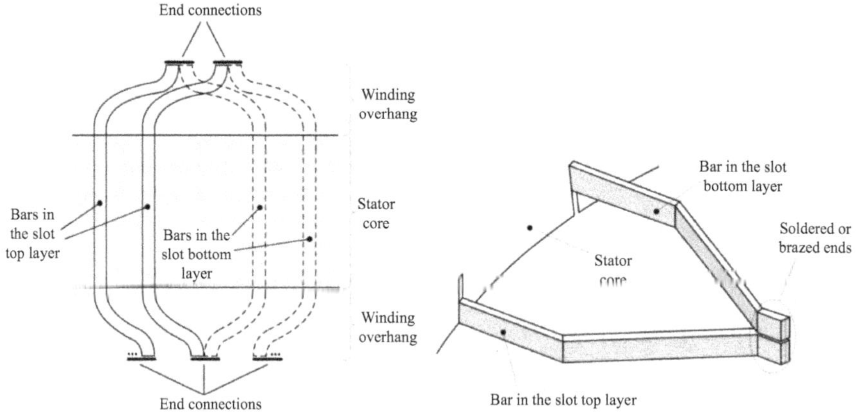

Figure 4.15 Schematic of a Roebel winding, composed of bars connected in series at their ends

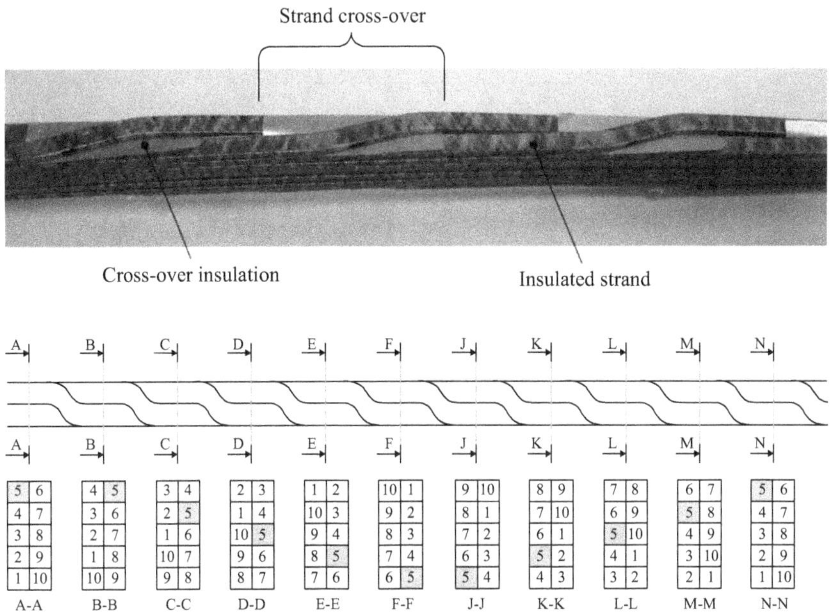

Figure 4.16 Top figure: portion of a Roebel bar. Bottom figure: various cross sections of a Roebel bar. The strand indicated with number 5 (like all the other strands) occupies all the possible positions inside the bar cross section. This applies to all the other strands

Large synchronous motors are typically medium-voltage or high-voltage machines, with a rated voltage between 3 kV and 13 kW. A very delicate aspect in this kind of machines is the stator insulation system. A schematic of the main insulation components in a stator coil is given in Figure 4.17. The ground-wall insulation may have thicknesses in the order of several millimetres and is composed of an appropriate number of turns of special insulating tapes. Modern ground-wall tapes are based on mica (a mineral with high-dielectric-strength characteristics) and glass fabric combined with thermally curable epoxy-based resins acting as an organic binder. Turn insulation tapes generally includes glass fabric with or without mica paper. For rated voltages above 4 kV, the coil is also covered with a semi-conductive coating in the slot region (Figure 4.18) so that the outer coil surface has nearly the same electric potential as the stator core, which reduces the risk of discharges inside the slot. Where the conductive coating ends, in the end-coil region, some electric field intensification may occur [Figure 4.18(b)] and, to avoid surface (corona) discharges, an additional 'field-grading' semi-conductive tape needs to be applied [Figure 4.18(c)].

Special attention, in the stator coil winding design and construction, is paid to minimize the presence of air-filled voids inside the coil insulation. In fact, if a void or gap exists and an electric field greater than 3 kV/mm arises in it (which is likely to occur in medium- and high voltage machines), partial discharges occur through

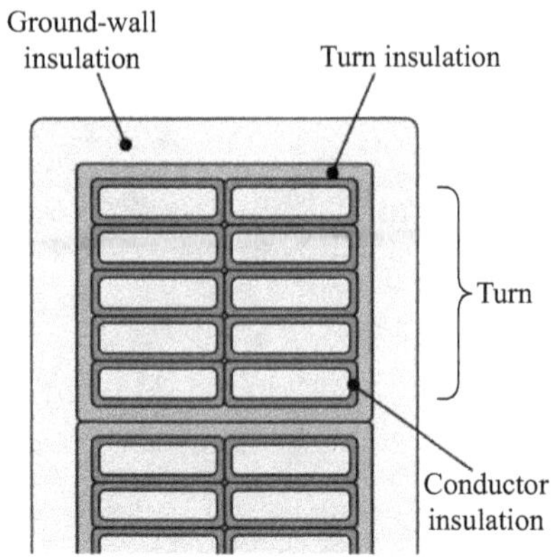

Figure 4.17 Main insulation components in medium-voltage coil

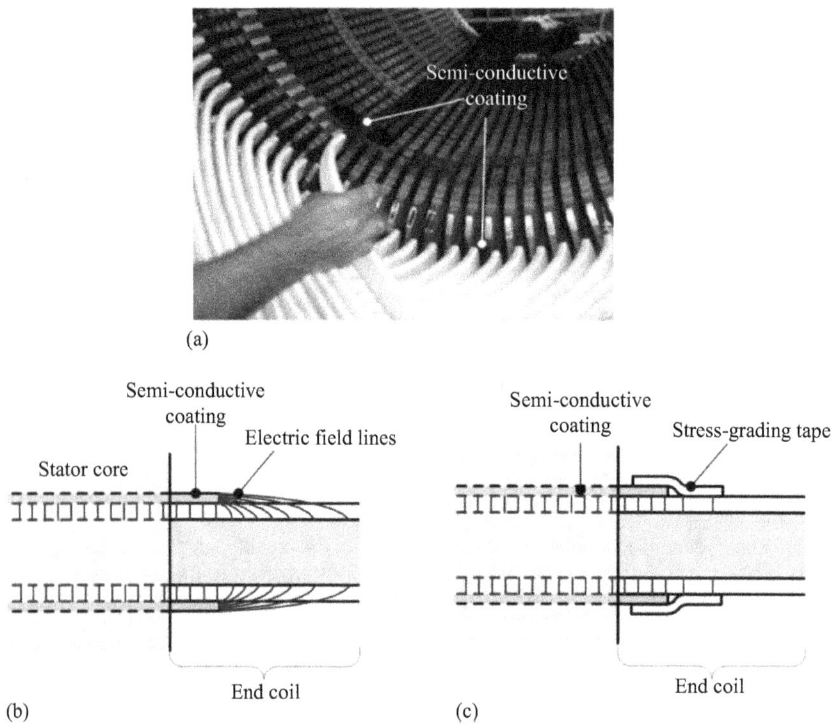

Figure 4.18 (a), (b) Semi-conductive coating and (c) field grading tape

the air of the void leading to progressive deterioration of the surrounding insulation [5]. For this reason, after assembly the winding is impregnated with epoxy resin. The usual process to do this is called 'vacuum pressure impregnation' [5] and is intended to fill all the possible voids and gaps in the insulation system with cured

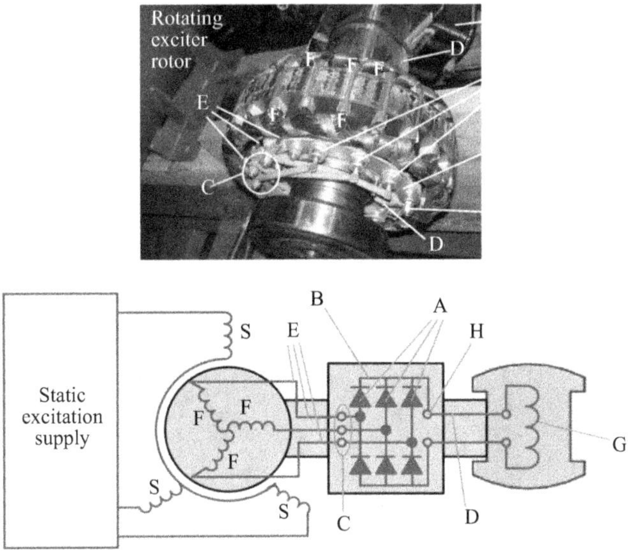

Figure 4.19 Rotating exciter structure

resin (having a higher dielectric strength than the air) so as to prevent the occurrence of partial discharges.

4.1.2.3 Rotor excitation system

There are two main methods to energize the excitation circuit of wound-field synchronous motors: using sliding contacts and brushes or using a brushless system based on a rotating exciter.

The former method is old-fashioned and is rarely employed in modern drives for its apparent drawbacks like: need for heavy maintenance, mechanical wear of the brushes and sliding contacts, possible sparkling issues, release of conductive particles (from brushes) which can sediment on winding insulation giving rise to local electric field intensification and, therefore, to additional dielectric stresses.

The basic structure of a rotating excitation system is illustrated in Figure 4.19; it consists of a rotating exciter with a three-phase stator and rotor, the latter mounted on the main motor shaft. The stator is supplied by an external AC source producing a rotating field in the exciter air gap. Such rotating field induces an electromotive force (EMF) in the exciter rotor phases. In order to maximize the amplitude of the induced EMFs, the stator exciter phase sequence is set so that the rotating field revolves in the opposite direction with respect to the rotor. Due to the induced EMFs, a three-phase system of currents is induced in the exciter rotor phases. Such currents are rectified by the rotating diodes and transformed into the DC current which is fed to the main rotor field.

4.1.2.4 Bearings

As regards bearings, the same considerations made for high-power permanent magnet motors apply to wound-field synchronous ones, except for magnetic

(a)

(b)

Figure 4.20 (a) Bearings integrated in the motor frame and (b) bearings installed on dedicated supports

bearings which are required only for very high speeds and are then peculiar to permanent magnet machines only.

The rotor of wound-field synchronous machines may have a very significant weight (several tens of tons); in this case, it is a common practice not to integrate the bearings in the motor frame but to place them on dedicated supports as shown in Figure 4.20.

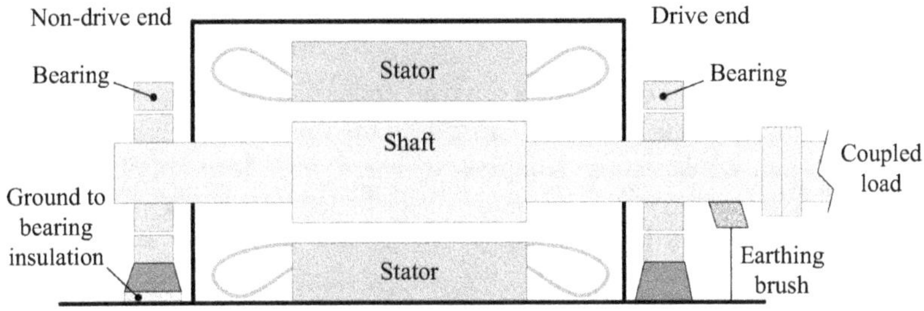

Figure 4.21 The non-drive-end bearing is usually insulated from the ground while the shaft is grounded

From an electrical point of view, the non-drive end bearing is usually insulated from the ground, while the drive-end bearing is grounded, as well as the shaft (Figure 4.21). All these provisions are meant to avoid the shaft from being electrically charged and to avoid parasitic currents, due to the so-called shaft voltages, to circulate through the bearings [12].

4.2 High-power converters

In high-power synchronous motor drives, the electric machine is supplied by medium-voltage AC/AC converters, featuring a typical voltage rating between 3 and 11 kV and covering a power range up to almost 100 MW.

Medium-voltage converters can be basically classified based on their inverter stage, which is the converter section directly connected to the motor, into the following main categories and subcategories:

- Voltage source inverters (VSIs), which include
  - Neutral point clamped (NPC) inverters
  - Flying capacitor (FC) inverters
  - Cascaded-cell or serial cell H-bridge (SC-HB) inverters
- Current source inverters (CSIs), which include
  - Self-commutated inverters
  - Load-commutated inverters (LCIs)
- Cycloconverters

The topologies and main features of the above-listed converters will be briefly discussed next.

4.2.1 Voltage source inverters

Nowadays, multi-level medium-voltage VSIs represent the most advanced high-end solution for large drives thanks to their superior performance in terms of output voltage quality, reduced grid-side harmonic pollution and power factor control in the case of active front end (AFE) employment, reduced motor losses and torque pulsations, possibility to operate the motor at unity power factor, good dynamic performance and high output frequency capability. They are progressively replacing the more traditional CSI's and cycloconverters in many applications, although their massive diffusion is still often limited by the relative high cost as well as by their design and construction complexity which may lead to possibly poorer robustness and reliability. Furthermore, VSIs include large DC link capacitors, which can cause more safety issues than the DC link inductors used in CSIs.

VSIs are suitable for supplying both wound-field and permanent magnet synchronous motors.

The power ratings of each individual three-phase converter unit reach few tens of MWs for NPC and FC VSIs and can exceed 100 MW for SC-HB VSIs. Voltage ratings are typically up to 7 kV for NPC and FC VSIs, while they can reach 13 kV for SC-HB VSIs. The output frequency is usually from 0 to around 250 Hz for all kinds of VSIs.

4.2.1.1 Front-end or AC/DC rectifier stage

Out of the three types of VSIs, the first two topologies (NPC and FC VSIs) are connected, by means of a single voltage DC link, to the AC/DC rectifier stage. The rectifier stage is, in turn, connected to the feeding transformer which interfaces it to the grid. The rectifier stage is therefore also known as the 'front end' of the converter. There are two types of front end, namely the AFE and the diode front end (DFE). The latter case is illustrated in Figure 4.22, in its six- or twelve-pulse configuration, where the twelve-pulse arrangement is obviously more complicated but allows for a smoother DC link voltage and less grid-side voltage harmonic distortion.

In the case of AFE, the arrangement is basically that shown in Figure 4.22(a) but with the inner structure of the rectifier including controlled-switching devices, such as integrated gate-commutated thyristors (IGCTs) or insulated-gate bipolar

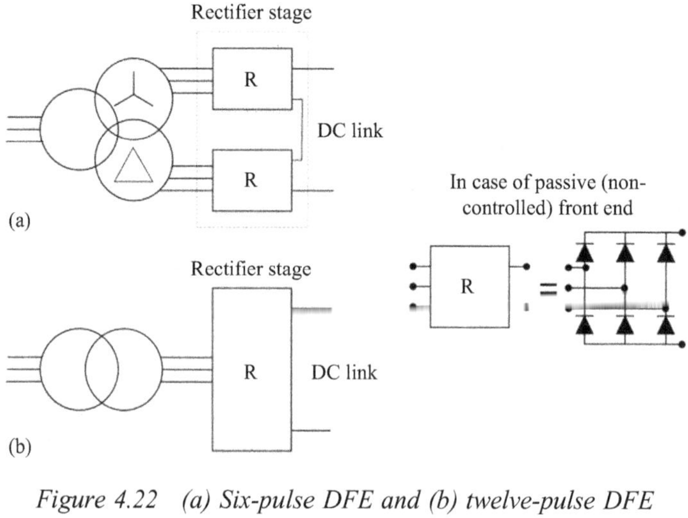

Figure 4.22 (a) Six-pulse DFE and (b) twelve-pulse DFE

transistors (IGBTs) instead of diodes. In this case, the rectifier has usually the same topology as the inverter although in a reversed arrangement (namely with an AC input and a DC output). The AFE is much more expensive and complex than the DFE, but it makes it possible to improve grid-side current and voltage waveforms (then reducing the harmonic pollution due to converter switching) and to control the grid-side power factor close to unity. The presence of an AFE makes it unnecessary to use traditional LC filters.

Regarding the SC-HB VSI topology, this is more complicated as it includes a distributed DC link and several rectifier stages as discussed later.

4.2.1.2 Multi-level voltage output

As a general feature, medium-voltage VSIs are multi-level inverters. This means that, compared to the simple two-level low-voltage inverters used in low-power drives (Figure 4.23), they can produce much better voltage waveforms, where the voltage transitions (voltage levels) associated to each commutation are a small fraction of the total DC link voltage. Figure 4.24 shows the multi-level VSI output

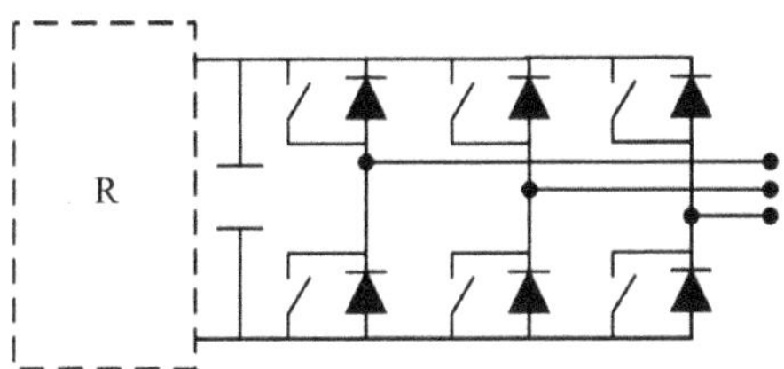

Figure 4.23 Two-level VSI generally used in low-voltage drives

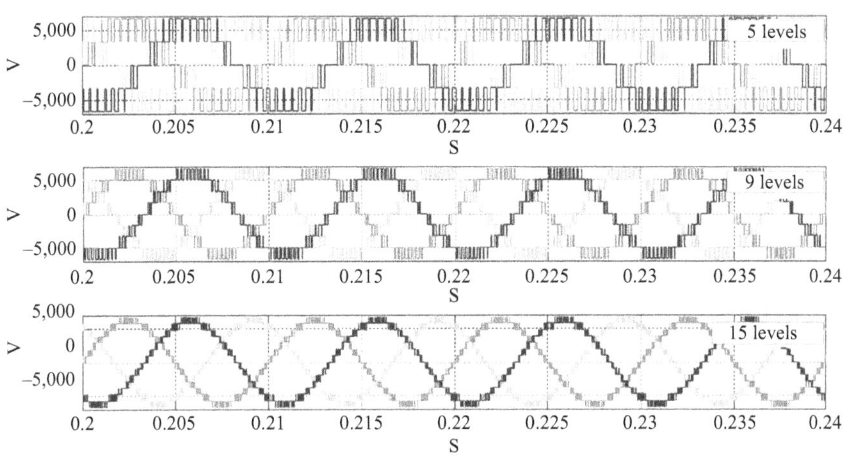

Figure 4.24 Multi-level VSI output voltage in the case of different number of levels

voltage waveforms in the case of different number of levels with the same voltage fundamental amplitude. Having a relatively high number of voltage levels means having relatively small voltage steps for each commutation, and this is essential to reduce the electric stress imposed on the electric motor insulation in terms of voltage spikes and of *dv/dt* stresses [5]. On the inverter side, multi-level topologies allow the use of semiconductor switching devices, such as IGBTs and IGCTs, with a smaller blocking voltage than the DC link voltage.

In general, a larger number of voltage levels can be achieved at the expenses of a more complicated VSI structure.

4.2.1.3 Inverter topologies

The standard NPC and FC VSI topologies are illustrated in Figures 4.25 and 4.26, where the rectifier stage may have one of the configurations shown in Figure 4.22. Common semiconductor devices used in these VSIs are IGBTs and IGCTs.

In the standard arrangement shown in Figures 4.25 and 4.26, the NPC VSI produces a three-level voltage output, while the FC VSI generates a four-level voltage output. The relatively low number of voltage levels, when combined with high-voltage ratings, may place significant electric stresses due to voltage spikes and *dv/dt* issues on the motor insulation. Therefore, an output RC filter between the VSI output and the motor terminals may be necessary. Of course, the number of levels can be increased by using more switches in each inverter leg, obviously leading to more complicated (although conceptually identical) structures. In practice, the risk of unbalance in the two DC-link capacitor voltages makes the NPC topology unsuitable to achieve more than three levels. Conversely, up to five voltage levels can be reached with the FC topology (further increasing the number of levels would excessively increase the number of required storage capacitors).

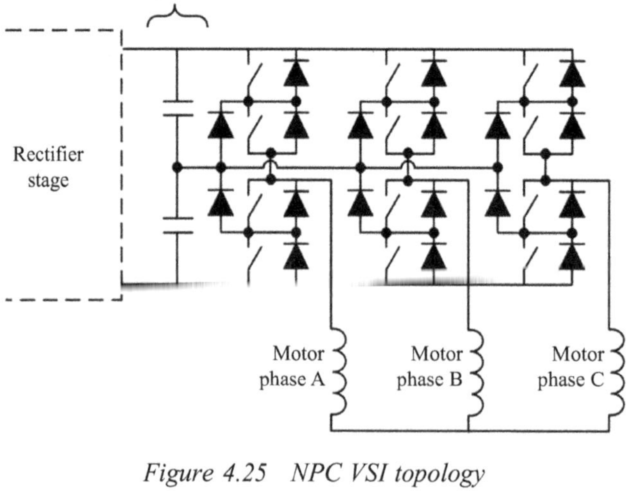

Figure 4.25 NPC VSI topology

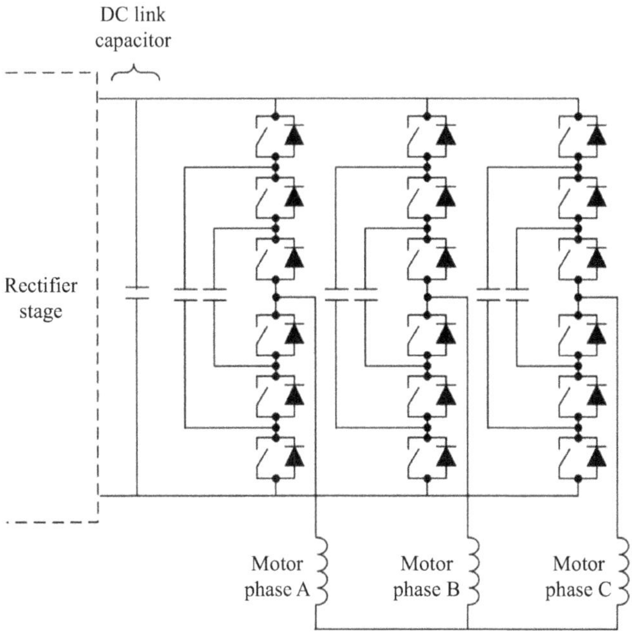

Figure 4.26 FC VSI topology

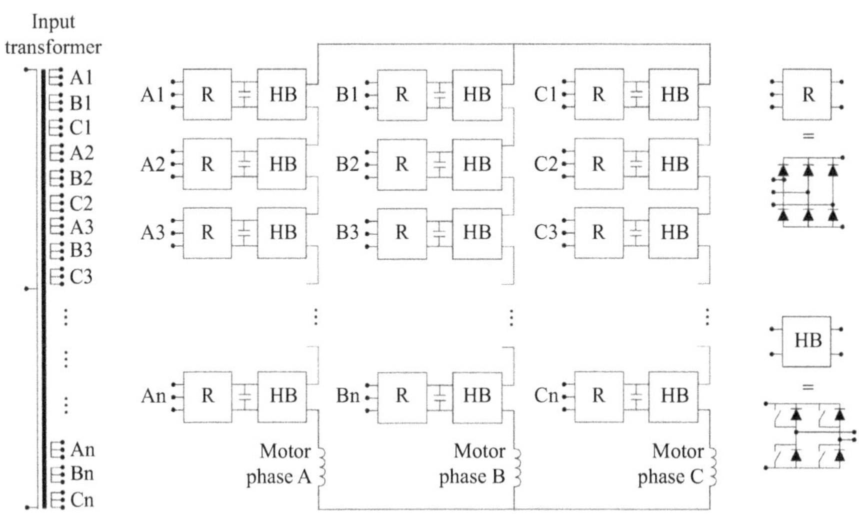

Figure 4.27 SC-HB VSI topology in the case of n cells per phase

The topology of a SC-HB VSI, in its simplest form, is shown in Figure 4.27. Each inverter phase is composed of several (n) low-voltage series-connected cells, each one including an individual rectifier stage, a DC-link capacitor and an H-bridge single-phase inverter. The use of low-voltage cells makes it possible to

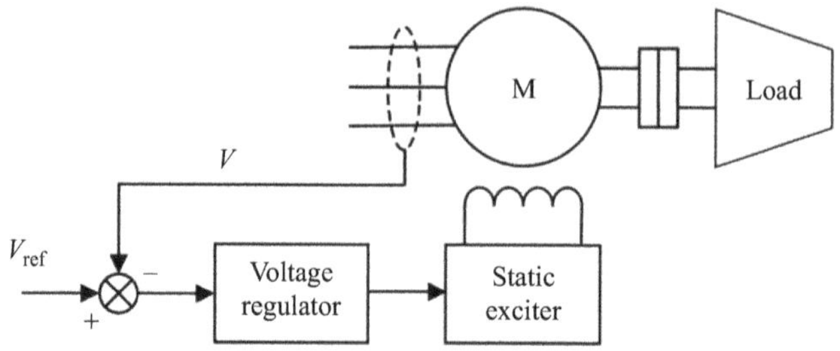

Figure 4.28 Voltage control loop acting on the rotor excitation

implement them with low blocking voltage IGBTs. Every cell is supplied by an individual three-phase secondary winding of a multi-secondary transformer. The output voltage contains a number of levels equal to $2n + 1$, being n the number of cascaded cells. It is known that, using a suitably high number of cells per phase, an outstanding output voltage waveform can be obtained (e.g. the 15-level voltage waveform in Figure 4.24 is achieved with seven cells per phase). This usually makes it unnecessary to use output filters between the inverter and the motor. Furthermore, it is usually guaranteed that an ordinary motor designed for direct-on-line supply is suitable for being fed by a SC-HB VSI with no need for insulation reinforcement. Of course, the high-end performance of this kind of converter is paid in terms of high cost, complexity and device components count.

There exist more advanced variants to the structure shown in Figure 4.27. For example, the cell rectifier stage can be implemented with controlled switches, constituting a cell AFE, which endows the converter a regenerative capability (the power flow can be from the mains to the motor but can be also reversed allowing for regenerative braking or electric machine generating mode operation). Another enhancement could be a hybrid structure where the H-bridge section of the cell is replaced by more complicated structures, like NPC or FC topologies.

4.2.1.4 Control features

Case of wound-field synchronous motors

When VSIs are used to supply wound-field synchronous motor, the control system includes the following main loops:

- Voltage control loop through the rotor excitation current (Figure 4.28)
- Stator flux control loop
- Speed control loop (possibly including an inner torque control loop)

The voltage control loop acting on the field excitation regulates the output current of the static exciter so that the motor terminal voltage equals the reference one. Due to the large field circuit time constant, this voltage regulation is relatively slow in the sense that it takes a long time (in the order of seconds) for the field current to increase or decrease following an input control step.

The other two control loops (stator flux and torque loops) act on the supply converter determining the instantaneous voltage that it applies to the motor

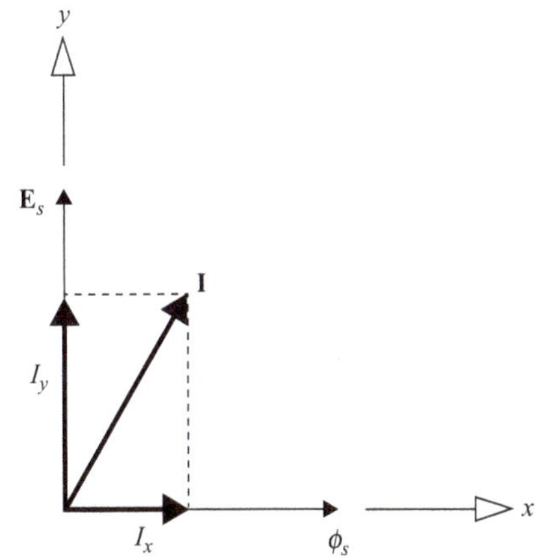

Figure 4.29 Decomposition of the stator current in the stator flux reference frame

terminals. To implement these loops, the stator flux vector ϕ_s (Figure 4.29) is generally estimated based on voltage and current measurements taken at motor terminals and an orthogonal xy reference frame, with the x axis aligned to the stator flux, is considered. Neglecting resistive drops (which are actually negligible in medium-voltage machines), the stator flux vector is orthogonal to the stator voltage vector $\mathbf{E}_s$. The stator current vector **I** is then decomposed into its component I_x and I_y, of which I_x regulates the stator flux magnitude and I_y regulates the torque. Therefore, the stator flux control loop acts on I_x and the torque control loop acts on I_y. The reference value for I_x is usually set to zero, which means that, at steady state, the machine is forced to operate at unity power factor. During transients, the flux control loops uses I_x for fast adjustments of the stator flux and the torque control loop uses I_y for fast adjustments of the torque.

The control approach described so far constitutes a field-oriented control (FOC) performed in the stator flux reference frame and is the most frequently implemented methodology. Nevertheless, some drive manufacturers also adopt a direct torque control (DTC) strategy, which establishes the voltage vector to be applied to the motor without the need for current (I_x, I_y) controllers, leading to a possibly faster torque dynamic response [13].

Case of permanent magnet synchronous motors

In permanent magnet synchronous motors, there is no excitation circuit. Hence, there are only two control loops, one controlling motor flux and the other controlling motor speed. The most frequent approach is the FOC performed in a dq rotor attached reference frame, where the d axis is aligned to the rotor pole axis (Figure 4.30).

The speed loop control (possibly including an inner torque loop) acts on the I_q current, while the flux control loop acts on the I_d current. In SPM motors (which are

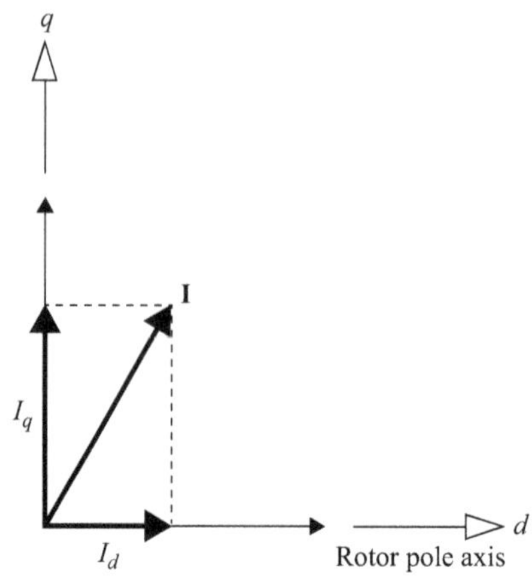

Figure 4.30 Decomposition of the stator current in a dq rotor-attached reference frame

the vast majority in high power applications), it is a common practice to set the reference value of I_d to zero, so as to minimize the stator current and hence Joule losses. Of course, if a flux weakening operation is needed above a certain speed, the I_d is set to an appropriate negative value in the flux weakening region so as to prevent the stator voltage from exceeding the supplying VSI capability.

Modulation strategies

The most popular method to synthetize the output voltage in industrial VSIs is the sinusoidal carrier-based PMW. Space vector modulation (SVM), originally conceived for low-voltage inverters, can be also used, especially in three-level NPC VSIs. Finally, when DTC is used to control motor torque and flux, this naturally yields a dedicated modulation strategy.

4.2.2 Current source inverters

CSIs are characterized by the presence of a current DC-link, equipped with an inductor meant to reduce the DC-link current ripple. Compared to VSIs, they generally feature a simpler and more robust structure, offer an intrinsic protection against short-circuit currents, naturally allow for a bi-directional power flow and produce a smoother voltage with no electrical stresses on motor insulation. On the other side, they suffer from a slower dynamic response and generally have a bulky overall size due to the need for large smoothing inductors.

There are two main variants of CSIs, namely the self-commutated CSI and the LCI, whose basic topologies are shown in Figures 4.31 and 4.32, respectively.

4.2.2.1 Self-commutated CSI

In the self-commutated CSI (Figure 4.31), gate turn-off devices with reverse voltage blocking capability, like symmetric gate commutated thyristors, are used. A pulse width modulation (PWM) is implemented in the rectifier stage in order to

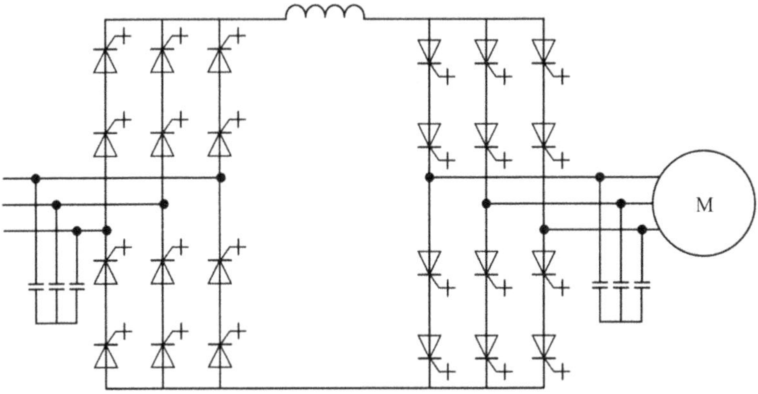

Figure 4.31 Self-commutated CSI, in a six-pulse configuration on both line- and motor-side

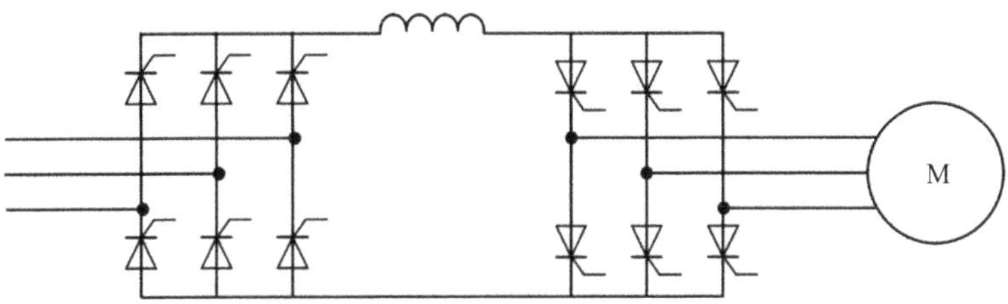

Figure 4.32 LCI in its simplest six-pulse configuration on both line- and motor-side

control the DC-link current and in the inverter stage to control the output current fed to the motor. Output capacitive filters, capable of carrying the DC-link current, need be used to enable the commutation of the turn-off switches. Capacitors also bring the benefit of improving the output voltage waveform, making it free from spikes and high *dv/dt* transitions. Nearly unity power factor performance can be achieved on both grid- and motor-side.

The self-commutated CSI is suitable for driving both wound-field and permanent magnet synchronous machines, with power ratings usually lower than 10 MW and voltage ratings below 7 kV, in those applications where very low dynamic responses can be accepted, like electric drives for blowers, pumps and extruders.

4.2.2.2 Load-commutated inverter

The LCI (Figure 4.32) substantially differs from the self-commutated CSI because its semiconductor devices are conventional thyristors, also known as silicon-controlled rectifiers (SCRs), which cannot be switched off before the current flowing through them becomes zero. Line-side SCRs are then commutated by the grid voltages, while motor side SCRs commutate thanks to motor phase back-EMFs. Because of the reactive power needed to commutate SCRs, LCIs can only be used to supply over-excited wound-field synchronous machines, while they are unsuitable for permanent magnet synchronous motors. SCRs are available at very high current and voltage ratings, and this leads to LCIs with an overall power exceeding 70 MW and voltages above 10 kV.

Well-known merits of LCIs are their simplicity, robustness and reliability, together with a quasi-sinusoidal input and output voltage waveform and an intrinsic self-protection from short-circuit over-currents; conversely, their major drawbacks are low power factor (usually below 0.92) on both motor- and grid-side; highly distorted input and output current waveforms causing important harmonic pollution on the grid side and large torque pulsations at the motor shaft; limited frequency, which cannot usually exceed 100 Hz; critical start-up, as below nearly 10% of the rated speed motor EMFs are insufficient to commutate SCRs and a special 'pulsed operation,' causing large torque pulsations, is required; frequent need for a special electric motor design with low sub-transient reactance to reduce commutation intervals and relatively poor dynamic performance.

From a control point of view, LCIs are much simpler than self-commutated CSIs. All SCRs are simply controlled adjusting their commutation delay angle, namely the time interval at which they are turned on with respect to the zero-crossing instant of the relevant commutating voltage. For the grid-side SCRs, this angle is determined by a speed control loop to obtain the desired DC-link current and, therefore, the desired motor torque; the commutation delay angle of the motor-side SCRs, instead, is either maintained constant or slightly adjusted based on the speed. As regards the motor flux, it is simply controlled by acting on the rotor excitation system (Figure 4.28).

LCIs are still today widely used in such application fields such as oil and gas industry, for large compressor drives; marine propulsion, for large ships; and starting drives, for large turboalternators and hydrogenerators. However, their diffusion is being limited by growing employment of high-power VSIs which may be preferred especially in those applications where low torque ripple, higher dynamic response and low current harmonic pollution are required.

4.2.3 Cycloconverters

The typical cycloconverter topology is illustrated in Figure 4.33. It differs from both VSI and CSI topologies because there is neither current nor voltage DC-link with any energy storage device. Like LCIs, it employs thyristors (SCRs) as semiconductor switching devices. SCRs are commutated by the line-side voltages, and this poses a serious limitation on the output frequency, which cannot be higher than half of the line frequency. One more disadvantage is the large voltage and current harmonics, both on the input and the output sides. Conversely, points of strength are a high overload capability thanks to the use of SCRs and a good dynamic response.

Cycloconverters are used in the cement industry, large ship propulsion, rolling mills and grinding mills, with typical power ratings up to 25 MW.

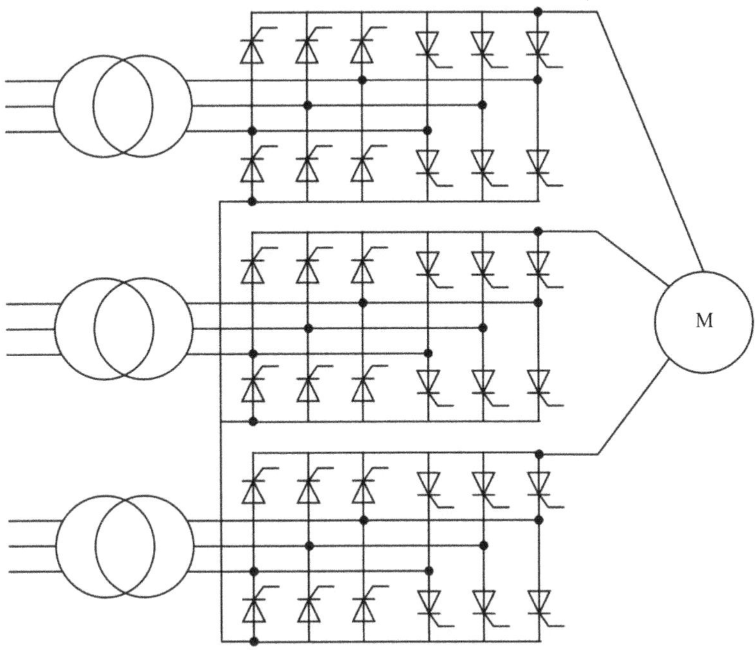

Figure 4.33 Cycloconverter topology

4.3 System-level fault-tolerant drive architectures

A basic strategy to improve the fault tolerance of a drive system is to act on its system-level design, including appropriate arrangements and interconnections of its main components, namely the electric motor and converter. Two main strategies can be identified in this sense, the former employing redundancy and the latter based on multi-phase drive arrangements. In the redundancy approach, a conventional three-phase design is preserved for each component (motor and converter) and the fault tolerance is pursued by using multiple motors or converters, or even converter sections in different combinations. In the multi-phase design approach, on the other side, fault tolerance is pursued by suitably increasing the number of motor and/or converter phases above three.

4.3.1 Redundant drive architectures

Redundancy, applied to the motor or to the converter section, or even to both of them, is probably the most intuitive way to increase the fault tolerance of a drive system. It consists of duplicating (or multiplicating) the whole drive or parts of it, making sure that all duplicated items can operate independently and that a fault in one of them does not harmfully impact on the others. This approach suffers from obvious drawbacks as it leads to increased system size, weight and cost. From a reliability point of view, it increments system complexity and parts count, which

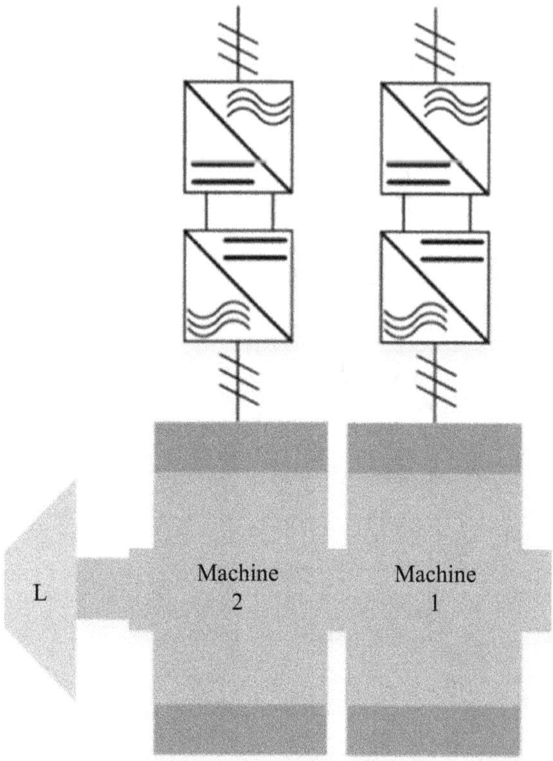

Figure 4.34 Drive redundancy with series mechanical connection of two motors and distinct converters. © 2015 IEEE. Reprinted, with permission, from [17]

can be paradoxically detrimental because the probability of a fault grows as the number of system components increases. Hence, redundancy should be very carefully employed always weighing its pros and cons [14,15].

Depending on which parts of the drives are duplicated, different kinds of redundancy can be identified, mainly distinguishing the case when multiple motors or a single motor with multiple converters is adopted.

4.3.1.1 Multi-motor redundant configurations

A relatively rough and cost-expensive approach to redundancy is to use multiple motors (typically two) mechanically coupled in either a series or parallel fashion. In the series mechanical arrangement, the two motor shafts are directly coupled (Figure 4.34), while in the parallel arrangement, a multi-pinion gear-box is needed (Figure 4.35). The series and parallel mechanically coupled dual-motor arrangements are sometimes used in fault-tolerant shipboard propulsion drives [16].

Regarding the power electronics supply section, the most redundant solution includes distinct complete converters to independently feed the electric motors

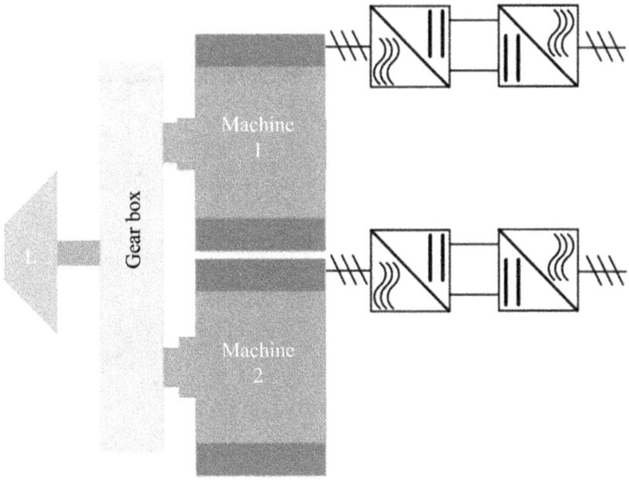

Figure 4.35 Drive redundancy with parallel mechanical connection of two motors and distinct converters. © 2015 IEEE. Reprinted, with permission, from [17]

(Figures 4.34 and 4.35). As an alternative, each of the motors can have its own inverter with a common DC-link connecting the distinct inverters to the same rectifier stage, as depicted in Figures 4.36 and 4.37.

It may be worth noting that the redundant configurations with common DC-links (Figures 4.36 and 4.37) are unsuitable in the case of CSIs (Section 4.2.2) because of current sharing and DC-link current control issues. Furthermore, it can be implemented only with those VSIs which are equipped with a concentrated DC-link, i.e. in the case of NPC and FC VSI topologies, and not with SC-HB VSIs (Section 4.2.1.3).

4.3.1.2 Single-motor design configurations

To reduce the cost, size or weight of the drive system it may be decided to use a single three-phase motor supplied by shunt-connected redundant converters (Figure 4.38). As shown in Figure 4.38, the redundancy can be either applied to the whole converter or be limited to the inverter stage, in the latter case making use of a common DC-link arrangement.

The shunt-connection of the inverter to the motor or among different DC-links would give rise to current control and current sharing issues if CSI's were used; this kind of redundancy is therefore applicable only if VSIs are used. In particular, the whole-drive redundancy [Figure 4.38(b)] can be achieved with any kind of VSI, while the common DC-link arrangement [Figure 4.38(a)] can be implemented only with VSI featuring a concentrated DC-link design, namely with NPC and FC VSIs, and not with SC-HB VSIs.

The parallel connection of different inverters to the same three-phase motor terminals can cause circulation current issues. To prevent them, it is usually necessary to adopt decoupling reactors at the output of each inverter.

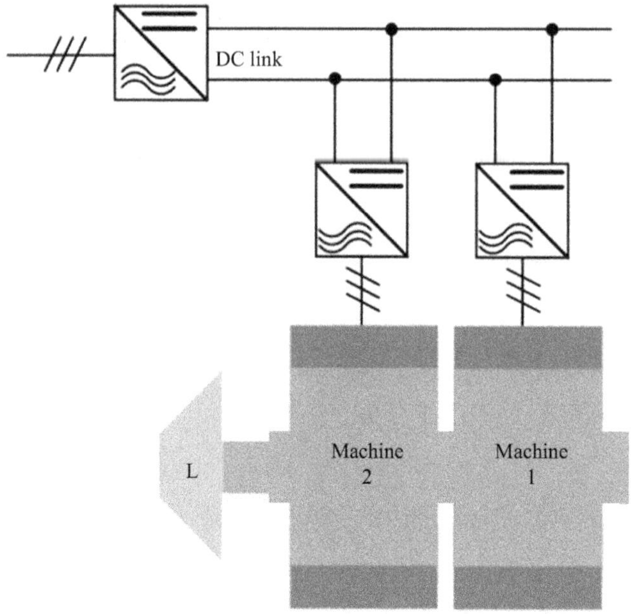

Figure 4.36 Drive redundancy with series mechanical connection of two motors and common DC-link converters

An example of industrial implementation of the single-motor redundant drive design is reported in [11]: a three-phase wound-field synchronous motor, rated 35 MW and used to drive a compressor for the oil and gas industry, is supplied by four shunt-connected NPC converters as illustrated in Figure 4.39.

A whole converter redundancy is adopted, with each converter including an AFE having the same NPC topology as its inverter stage, which endows the system with bidirectional power flow capability. The four converters are independently controlled, although the four switching patterns are designed according to a so-called interleaved technique which generates a voltage waveform with a high number of levels although each inverter outputs a three-level voltage typical of the NPC topology (Section 4.2.1). Choke inductors are used at each inverter output in order to prevent circulating current issues, although the inclusion of these inductors is paid in terms of drive size increase and efficiency decrease.

4.3.2 Multi-phase drive architectures

The use of multi-phase architectures (or, more specifically, both motors and converters comprising more than three phases) is a widely recognized way to increase the fault tolerance of both wound-field and permanent magnet synchronous motor drives. The basic idea is that the motor can continue operating, although at reduced power and with degraded performance, even if one of its phases (or group of phases) is out of service following a fault in either the motor or the supplying power

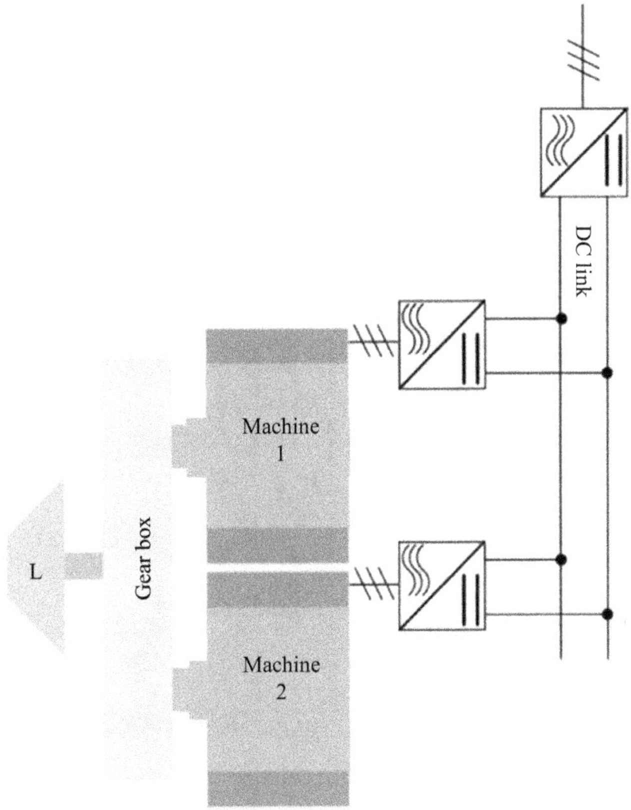

Figure 4.37 *Drive redundancy with parallel mechanical connection of two motors and common DC-link converters*

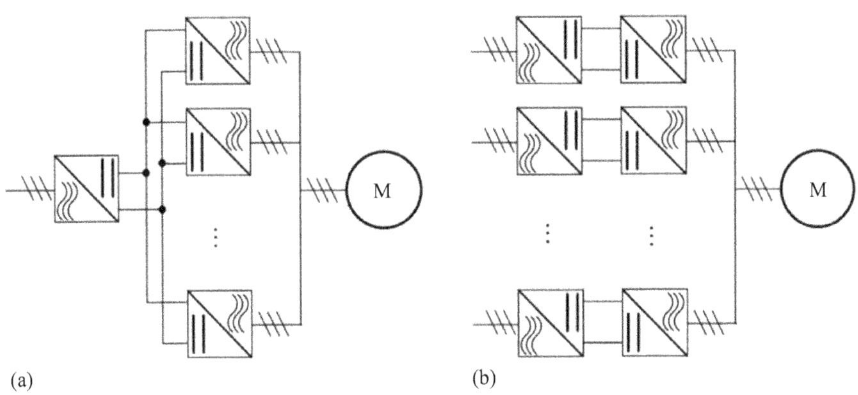

Figure 4.38 *Single-motor with redundant converters configuration: (a) partial and (b) total*

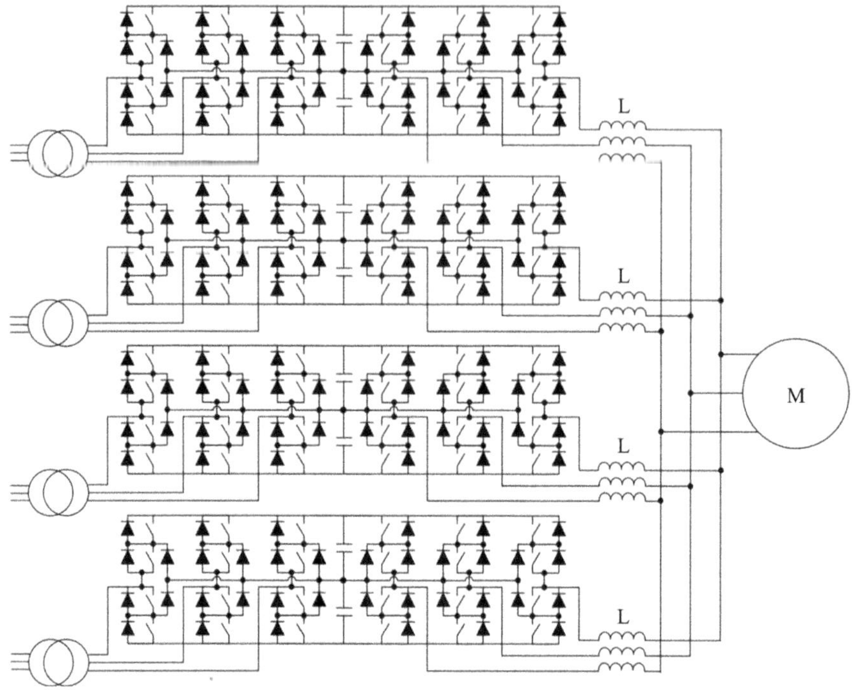

Figure 4.39 Single-motor redundant configuration including four NPC converters with AFEs

electronics equipment. This service continuity is essential in many safety-critical applications (like ship propulsion) and in those fields (like the oil and gas industry) where a temporary drive unavailability results in large economic losses.

From the electric motor design and construction point of view, the use of more than three stator phases does not introduce any particular cost and weight increase, but only some complications in terms of a larger number of connections between the stator winding and the terminal box (Figure 4.40), and a larger terminal box size due to the need for more phase leads (Figure 4.41).

Different possible multi-phase stator winding designs exist and can be basically grouped into the following categories:

- symmetrical multi-phase designs (Figure 4.42);
- asymmetrical multi-phase or split-phase or multi-three-phase designs with displaced winding sets (Figure 4.43);
- multi-phase designs with in-phase three-phase sections (Figure 4.44).

4.3.2.1 Symmetrical multi-phase architectures

In a symmetrical n-phase drive (Figure 4.42 provides an example for $n = 5$), phases are equally distributed in the electric motor stator winding with a spatial phase shift of $\alpha = 360/n$ electrical degrees. It is observed that the symmetrical n-phase motor

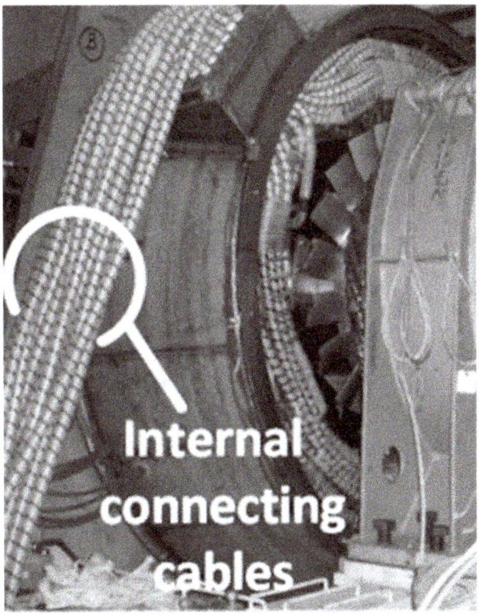

Figure 4.40 Internal connection cables (from winding to terminal box) in a 45-MW 12-phase machine. © 2015 IEEE. Reprinted, with permission, from [17]

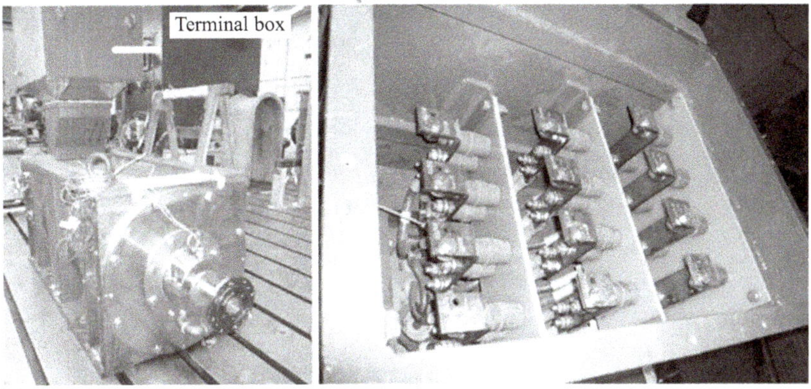

(a) (b)

Figure 4.41 A 12-phase machine (a) with detailed terminal box (b). © 2015 IEEE. Reprinted, with permission, from [17]

needs to be supplied by an n-phase inverter (either of VSI or CSI type), whose steady-state output is constituted by a symmetrical n-phase set of voltages (in case of VSI supply) or currents (in case of CSI supply), shifted by $360/n$ electrical degrees apart in time. The implementation of a symmetrical n-phase stator winding in the electric motor does not pose any challenge as the n-phase winding retains the

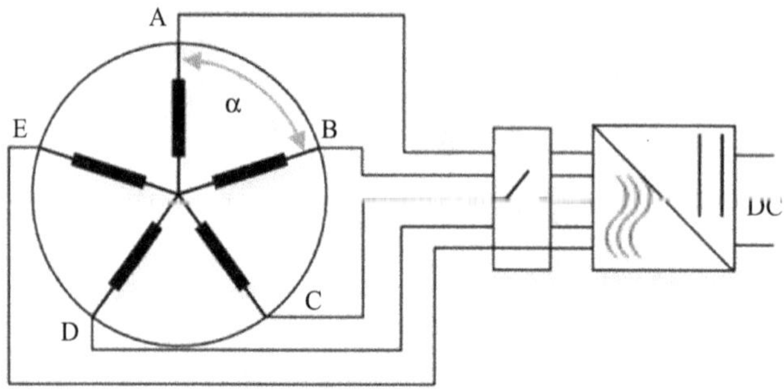

Figure 4.42 *Example of symmetrical multi-phase drive. © 2015 IEEE. Reprinted, with permission, from [17]*

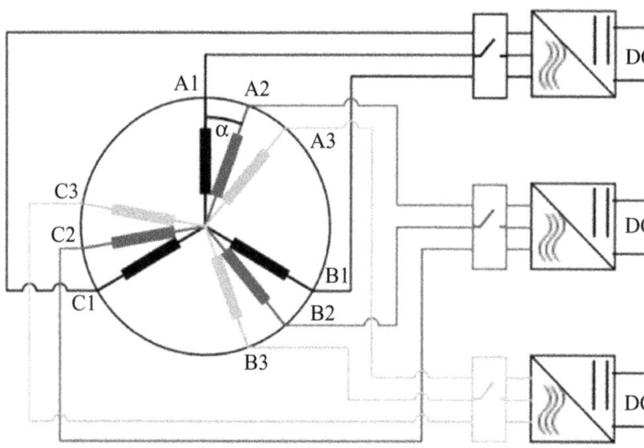

Figure 4.43 *Example of asymmetrical multi-phase drive scheme with displaced winding sets. © 2015 IEEE. Reprinted, with permission, from [17]*

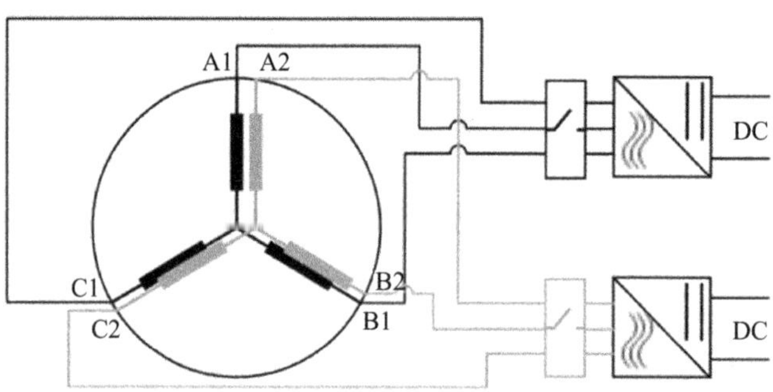

Figure 4.44 *Example multiple-three-phase drive scheme with in-phase winding sets. © 2015 IEEE. Reprinted, with permission, from [17]*

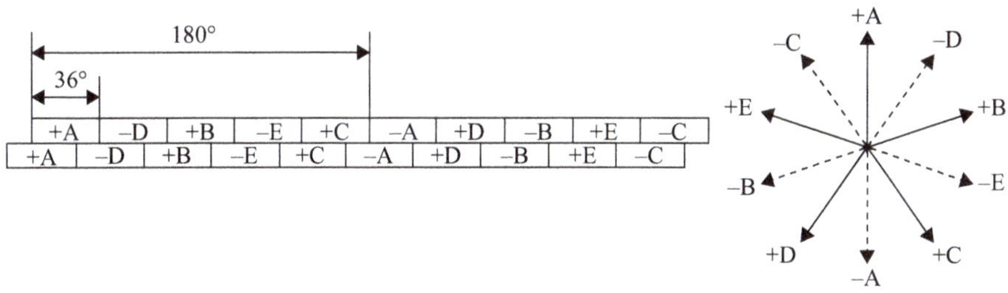

Figure 4.45 Left: phase belt arrangement for a five-phase symmetrical short-pitch stator winding. Right: corresponding phasor diagram

same structure as an ordinary three-phase one except that each phase belt spans over $180/n$ instead of $180/3 = 60$ electrical degrees. As an example, Figure 4.45 shows the phase belt arrangement for a five-phase winding ($n = 5$) where phases are denoted with letters A, B, …, E and signs '+' and '−' denote the conventional conductor direction.

The control and modulation strategy of medium-voltage inverters are consolidated for the three-phase case, while the extension to a generic number of phases n is still immature from an industrial point of view. Therefore, since reliability and risk mitigation are key factors in the development of large drive systems, it is often found a safe option to implement multi-phase drives configuration by suitably combining well-proven and tested three-phase converter modules, which leads to the asymmetrical or multi-three-phase designs being discussed in the next section.

We can therefore say that symmetrical n-phase drive designs have been receiving large attention in small synchronous electric machines [18–20], while their adoption in high power applications looks promising and attractive at a research and concept design level [21], but not yet industrially consolidated. The interest for the development of symmetrical multi-phase architectures in medium-voltage drives is not only related to fault-tolerance purposes but also to the possibility of exploiting higher order harmonics for torque production [22], so as to improve motor torque density, and of implementing multi-motor drives [19], where a single converter can supply multiple electric machines.

From a fault-tolerant perspective, it can be envisioned that symmetrical multi-phase designs offer better potential compared to the multi-three-phase architectures discussed in Section 4.3.2.2. In fact, if one of the n phases in a symmetrical n-phase drive is disabled due to a fault, the remaining $n - 1$ phases can remain in service and the drive can virtually continue working with a power output reduced by a factor $(n - 1)/n$ with respect to its healthy rated conditions. Also, remedial control strategies can be implemented, as investigated in the small-power drive field [23], to modify the n-phase inverter control to mitigate the performance degradation following the phase loss. Conversely, in a multi-three-phase drive architecture, if one phase needs to be disabled due to a fault, it is a common practice to disconnect the entire three-phase section including the faulty phase, resulting in service continuity with a more pronounced power reduction.

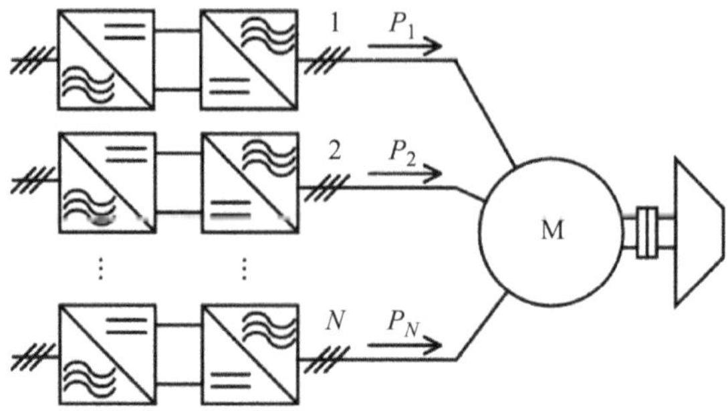

Figure 4.46 Schematic of an N-three-phase drive. © 2010 IEEE. Reprinted, with permission, from [24]

4.3.2.2 Asymmetrical multi-phase architectures

In asymmetrical multi-phase (or shifted multi-three-phase) designs, the electric motor winding is split into N sets (usually three-phase ones) displaced by $\alpha = 60/N$ electrical degrees apart. Each set is connected to an inverter (Figure 4.46).

This design is very frequently used in high-power applications because, unlike the symmetrical multi-phase configuration, it permits the use of consolidated three-phase inverters, with standard control features. Apart from the fault-tolerance benefits, the design also allows for power segmentation, resulting in smaller size converters and power electronics switches. The power rating of some drives is, in fact, so high (several tens of MWs) that the use of multiple inverters may be mandatory due to the unavailability of a single converter capable of delivering the whole power required [10]. Examples of large multi-three-phase with phase shift drives can be found in several high-power applications, as for oil-and-gas compressors [10] and in ship propulsion [16,25].

In motors with shifted multiple three-phase winding sections, each phase belt has a spatial span of $60/N$ electrical degrees. For instance, in the case of a triple three-phase drive shown in Figure 4.43, the motor winding, generally of distributed short pitch type, is arranged as illustrated in Figure 4.47.

A highly remarkable feature of multi-three-phase architectures with phase shift is their capability to yield a good motor air-gap flux waveform – and therefore a good output torque quality – even in presence of time harmonics in the phase currents. To explain this point, let us consider a high-power drive where the electric motor is supplied by a LCI (Section 4.2.2.2). It is well known that this kind of inverters produces output currents with a quasi-rectangular waveform as schematically shown in Figure 4.48. Each phase current is highly distorted and, in particular, includes significant fifth-order and a seventh-order time harmonics.

If the drive arrangement has a conventional single-phase architecture (Figure 4.32), the fifth- and seventh-order time harmonics in the phase currents give rise to large fundamental air-gap fields. In particular, the fifth-order time harmonic generates a fundamental air-gap field wave revolving in the opposite direction compared to the rotor at a speed equal to five times the rotor speed and

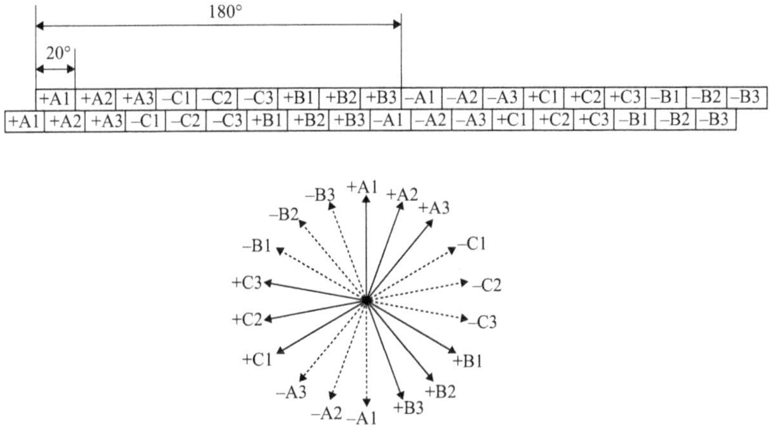

Figure 4.47 Left: phase belt arrangement for a triple-three-phase short-pitch stator winding. Right: corresponding phasor diagram

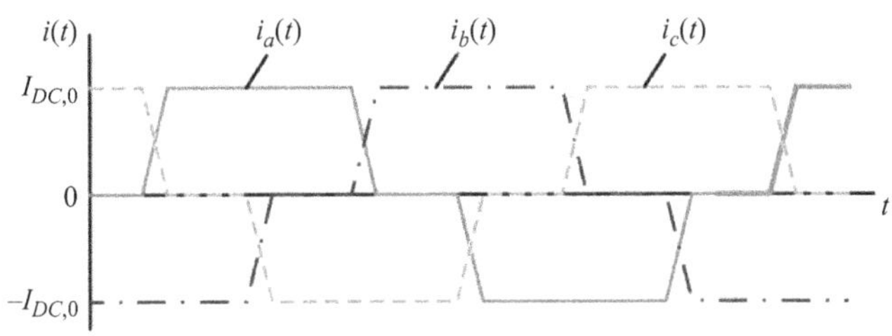

Figure 4.48 Current waveform for a LCI. © 2010 IEEE. Reprinted, with permission, from [24]

the seventh-order harmonic generates a fundamental air-gap field wave revolving in the same direction as the rotor at a speed equal to seven times the rotor speed. Both these fundamental air-gap fields interact with the rotor excitation giving rise to a sixth-order torque harmonic, i.e. to a torque ripple having a frequency equal to $6f$ being f the fundamental supply frequency. Such torque harmonics can be very dangerous as they may produce shaft vibrations and even possible mechanical resonance issues. If a dual-three-phase drive arrangement with phase shift is used (as shown in Figure 4.46 with $N=2$), the fifth- and seventh-order time harmonics still exist in the stator phase currents, but the air-gap fields they produce are shifted so that they mutually cancel out. As a consequence, no sixth-harmonic torque ripple appears. If a higher number N of shifted three-phase sections is used in the drive, then the described beneficial cancellation effect extends to higher order time harmonics. For example, if $N=3$, also the 11th- and 13th-order time harmonics in the phase currents give no air-gap field fundamental and, therefore, no torque ripple contribution. In general, considering a motor with N three-phase stator windings, each supplied by an inverter, we can state that the lowest order harmonics in the stator phases capable of contributing to the air-gap field and torque ripple production have order $6N \pm 1$ and the consequent lowest frequency harmonics which appear in the output torque have frequency $6N$ times the fundamental supply frequency. A formal explanation of the harmonic cancellation phenomenon can be found in the literature [24].

When it comes to evaluate drive fault-tolerance features, however, it is important to note that all the described harmonic cancellation effects in multi-three-phase drives with shifted sections take place only in the hypothesis of balanced operation, i.e. under the assumption that all the powers P_1, P_2, ..., P_N (Figure 4.46) supplied by the individual inverters to the motor are the same. In the case of unbalanced operation, instead, the air-gap field waveform worsens and as well as the torque quality. Finite element analyses or analytical formulas can be used to predict how the machine performance deteriorates under unbalanced supply conditions [24]. A particular case of unbalanced operation occurs, of course, when one or more of the powers P_1, P_2, ..., P_N is zero, i.e. when the corresponding inverters are disconnected from the motor (typically due to a fault). The deterioration of machine performance under such circumstances needs to be carefully evaluated in terms of additional torque ripples and additional rotor losses. As a consequence of the evaluation, it might be found safe to apply a suitable power derating factor when some of the motor three-phase sets are not supplied. Hence, if m of the N inverters are out of service, each of the active inverters should operate at its rated power multiplied by an appropriate derating factor $k < 1$; the consequent power capability of the overall drive is accordingly reduced to $P_n \times (N - m)/N \times k$. The derating factor should vary with m (growing as m increases) because the number and amplitude of the air-gap harmonics grows with the number of out-of-service inverters [24].

A further remarkable issue with multi-three-phase drive architectures with phase shift is the possible occurrence of current harmonics in the case where the N motor winding sets are independently supplied by VSIs, even in case of high-quality multilevel voltage waveforms. An example of the phenomenon is reported in [10] regarding a gas compressor wound-field synchronous motor, rated 45 MW at 3,000 rpm and 100 Hz, equipped with $N = 4$ stator winding sets displaced by 15 electrical degrees apart. Each winding set is supplied by a SC-HB VSI (Section 4.2.1.3, Figure 4.27) comprising $n = 7$ H-bridge cells per phase. The output voltages applied to the motor include 15 levels and therefore exhibit an almost sinusoidal waveform [Figure 4.49(a)]. In spite of the excellent voltage waveform, phase currents have a noticeable distortion [Figure 4.49(b)], mainly due to fifth and seventh harmonics.

The phenomenon has been studied in the literature and found to result from the back-EMF harmonics induced in the stator phases as a consequence of even slight distortions in the rotor excitation field [26,27]. Based on both experiments and theoretical analysis, it has been proved that the current harmonics in issue do not increase if any of the N supplying inverters is disconnected; furthermore, they have been shown not to cause significant torque ripples [26], their only negative effect being a slight increase in stator Joule losses. In small-sized VSI-fed synchronous

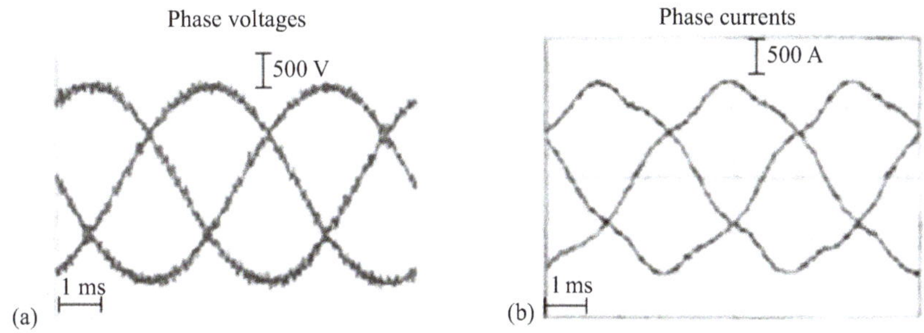

Figure 4.49 *(a) Phase voltages and (b) currents for a winding three-phase set of a quadruple-three-phase VSI-supplied motor. © 2011 IEEE. Reprinted, with permission, from [10]*

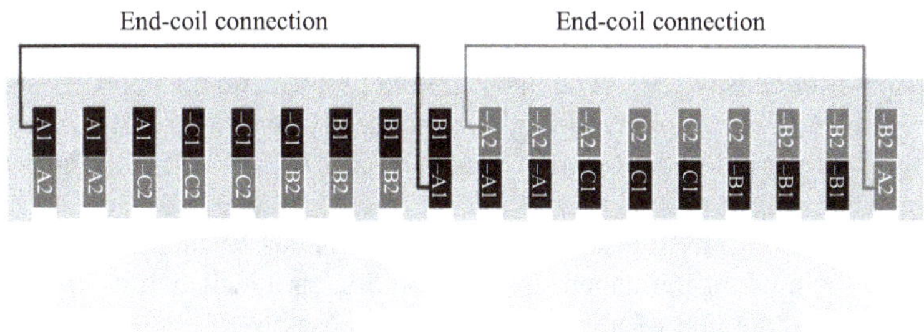

Figure 4.50 *Possible winding implementation for a dual-three-phase winding with in-phase winding sets A1, B1, C1 and A2, B2, C2*

motor drives with shifted multi-phase sections, the problem of phase current distortion has been effectively solved [28] using special control strategies (based on the vector space decomposition theory [29]) which involve the complete set of stator currents as an alternative to the wholly decoupled and independent control of the *N* supplying VSIs. The extension of these control strategies to high-power industrial or propulsion drives has not been carried out yet.

4.3.2.3 Multi-three-phase architectures with in-phase three-phase sections

In the multi-phase drive design with multiple in-phase sets (Figure 4.44), the electric motor winding is split into *N* three-phase sections in which identical EMFs are induced. The winding sections are connected to separate inverters (of either VSI or CSI type) which produce in-phase voltage or current systems. As an example and for the sake of clarity, Figure 4.50 shows (over a two-pole span) a possible implementation of the dual-three-phase motor winding for the drive system illustrated in Figure 4.51. The two winding sets are respectively composed of the phases A1, B1, C1 and A2, B2, C2.

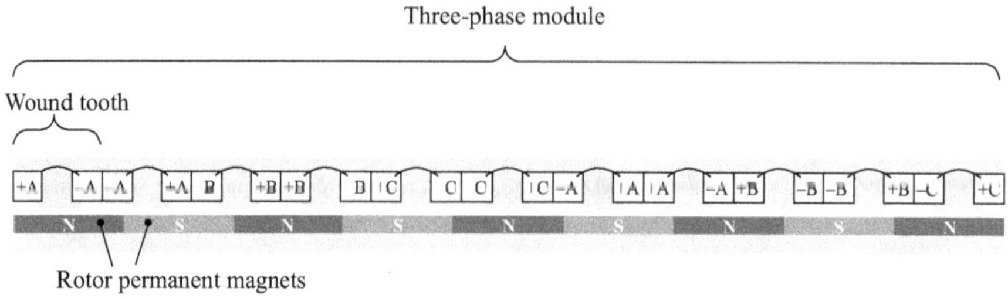

Figure 4.51 Example of a machine module comprising 12 wound teeth and 10 rotor poles

In-phase multi-three-phase configurations are not as widespread as those with phase shift (Section 4.3.2.2), although references can be mentioned where this kind of design is proposed for large wind generators [30]. In fact, during operation in healthy conditions, the systems under study behave as a usual three-phase one and the potentials of symmetrical and asymmetrical multi-phase designs in terms of control [22], increased power density [29], better air-gap field [29], etc. do not apply. On the other side, even in the case of supply through independently controlled VSI converters, the circulating current issues discussed in Section 4.3.2.2 are avoided or mitigated [27].

An interesting potential prospective is offered by FSCWs (Section 4.1.1.2), which are, however, rarely feasible in large synchronous machines, except for low-speed high-pole count permanent magnet motors, as it will be discussed in Section 4.4.1. The adoption of an FSCW makes it possible to implement a modular stator design where each stator module includes a three-phase stator winding section [30–32], such as in the example shown in Figure 4.51.

The three-phase modules have little or no magnetic coupling among them (i.e. a three-phase current inside a module produces little or no flux linkage in the other modules [32,33]) and are characterized by in-phase back-EMFs. Therefore, modules are suitable for being possibly connected in parallel, if necessary. Possible modular system architectures which can be built with this kind of machines are shown in Figure 4.52, where the case of four stator modules is represented.

In the configuration shown in Figure 4.52(a), the switches S1, S2, S3 and S4 are normally closed so that there are two couples of parallel-connected modules (1 and 3, 2 and 4, respectively). Each couple of modules is connected to an inverter. In the case of a fault in one of the four sections, the relevant couple of switches are opened. For example, if a fault occurs in module 1 (or in module 2), the switches S1 and S2 open at the same time; if a fault occurs in module 3 (or in module 4), the switches S3 and S4 open at the same time. In this way, any faulty condition includes supplied modules which are shifted by 180° so as to preserve symmetry and reduce the occurrence of large unbalanced magnetic pull (UMP). Such a configuration is implemented, for example, in the machine described in [32,33]: the machine is intended for operation as a wind generator but has a fully reversible operation capability as proved during its testing; thus, it could be used in a very low-speed synchronous motor drive as well.

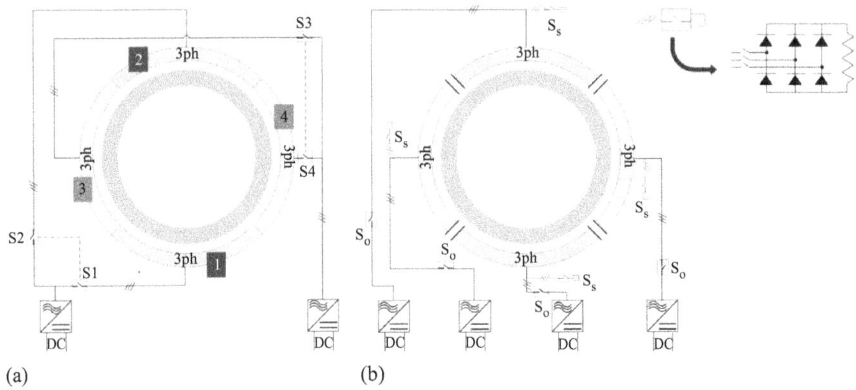

Figure 4.52 Examples of modular architectures with four in-phase stator modules: (a) partial magnetic decoupling and (b) full magnetic decoupling

The system's arrangement shown in Figure 4.52(b) exhibits much better fault tolerance potentialities. It features four magnetically decoupled in-phase modules which, in normal operation, are independently supplied by four inverters. In the case of an open-circuit fault in one of the modules, the corresponding breaker S_0 is opened. As for the arrangement in Figure 4.52(a), it may be necessary to disconnect the opposite module, too, to avoid excessive UMP issues. The design depicted in Figure 4.52(b) is suitable for coping not only with open-circuit fault, but also with short-circuit faults. If a short-circuit occurs in one of the modules, the relevant S_0 switch is opened and the relevant S_s switch is closed so that the entire three-phase module is closed on a small resistor load [34]. Of course, the ability of the machine to withstand the short-circuit fault strongly depends on the machine design and, in particular, requires the following conditions to be met [35,36]: the three-phase modules are magnetically decoupled so that the short-circuit current in one of them do not weakens the flux produced by the healthy modules [31]; the per-unit reactance of each winding module is less than or equal to one [35,36] so that the short-circuit current amplitude is comparable to the rated current. These conditions are not easily met in large synchronous machines as they typically require an FSCW, as discussed in Section 4.4.1.

4.4 Fault-tolerant electric motor design

So far, it has been discussed how the fault tolerance of a drive can be improved acting, at a system level, on the drive overall architecture. Of course, it is also important that all the drive components (i.e. mainly the electric motor and the converter) can be individually designed so as to enhance their capability of remaining in service in presence of a fault. In this section, the attention is being placed on some design provisions that can be used to improve the fault tolerance of large wound-field and permanent-magnet synchronous motors.

4.4.1 Fault-tolerant solutions in the stator design

4.4.1.1 Distributed vs concentrated stator windings

It is well known that the stator of a large synchronous machine can be equipped with two main types of windings, namely the distributed winding (Figure 4.5) and FSCW (Figure 4.8). In the former, each coil embraces several slot pitches and the number q of slots per pole per phase is either an integer or fractional number greater than one; in the latter, instead, each coil is wound around a single tooth, thus taking the form of a 'tooth coil' (Figure 4.8) [37] and the number q of slots per pole per phase is less than one.

Large synchronous motors typically feature a distributed dual-layer stator winding (Figure 4.5). The basic reason why FSCW designs are little suited for large synchronous machines (especially for wound-field ones) is that, to implement an FSCW, the slot (or tooth) pitch must be comparable to the pole pitch: such a condition is feasible (and often convenient) when the pole pitch can be in the order of a few centimetres as it may happen in small permanent-magnet synchronous machines [37] but is much more difficult to meet in large synchronous machines where the circumferential span of each pole is usually much larger than the maximum slot pitch which can be reasonably obtained (around 50 mm).

The mandatory adoption of a distributed-winding design in large synchronous motors has a strong impact on motor fault tolerance, with particular respect to phase short-circuit faults. In fact, the adoption of an FSCW is necessary to meet the key requirements for motor fault-tolerance to short-circuit faults (steady-state short-circuit current equal to or less than the rated current; magnetic decoupling between winding sections), making it possible to implement such drive system configurations as shown in Figure 4.52(b).

In the rare cases of large permanent-magnet synchronous motors where an FSCW can be implemented, the highest fault tolerance could be achieved using a multi-phase single-layer stator winding (Figure 4.53) where stator teeth are alternately wound so that each slot includes only one coil side.

Such a winding design is certainly critical for its very large space harmonic content [38] (often including large subharmonics) but can guarantee a magnetic, mechanical, electrical and thermal decoupling among phases. Figure 4.53, for example, shows the arrangement of a single-layer six-phase FSCW in its two

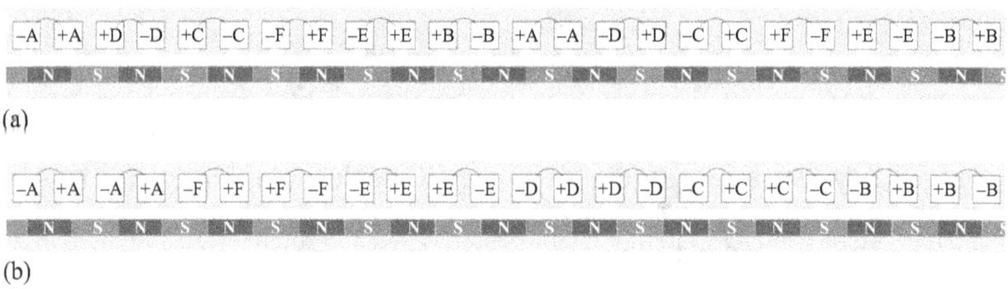

Figure 4.53 Example of a six-phase single-layer FSCW in its two possible implementations. © 2015 IEEE. Reprinted, with permission, from [17]

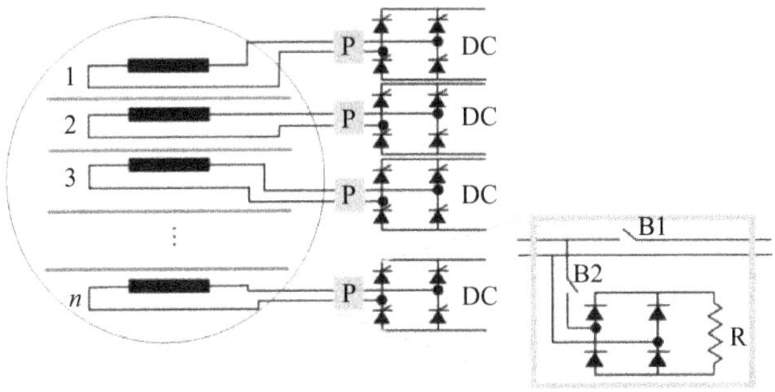

Figure 4.54 Ideal fault-tolerant drive based on an n-phase machine with decupled phases. © 2015 IEEE. Reprinted, with permission, from [17]

possible variants, of which the one shown in Figure 4.53(b) guarantees a better magnetic decoupling among phases compared to the more conventional configuration shown in Figure 4.53(a) [39]. Magnetic decoupling among phases makes it possible that the short-circuit current flowing through a faulty phase do not prevent the healthy phases to generate torque. This permits such drive structures as shown in Figure 4.54, which is recognized by many authors as the arrangement yielding the highest level of fault tolerance [35,36,40].

The solution includes a motor with n decoupled phases, each independently supplied by a single-phase converter (e.g. with a H-bridge structure). The protection device 'P' interposed between each single-phase converter and the relevant phase is provided with at least two breakers B1 and B2: the latter is normally closed and opens in case of any fault (of either open-circuit or short-circuit type); B2 is normally open and closes in the case of a short-circuit fault (including a turn-to-turn short fault, [37]) in order to close the faulty phase onto the low-value resistance R through a diode rectifier [41]. It may be worth noting that the arrangement shown in Figure 4.54 differs from that given in Figure 4.52(b) because the single phases are decoupled, not only the three-phase modules (which can be generally achieved only through a single-layer winding layout). So, in the case of a fault on one phase, the motor can continue working with all the other $n - 1$ phases.

It should be noted that the full fault-tolerant configuration shown in Figure 4.54, particularly suited for safety-critical applications, is much far from finding an industrial implementation in high-power drives, but it is worthwhile being mentioned as a target for future possible developments in those permanent-magnet motor drives where the use of an FSCWs (particularly single-layer ones) can be implemented.

4.4.1.2 Design for improving machine resilience to eccentricity faults

Equipotential connections (which are frequently employed especially in large synchronous motors) are additional low-impedance wires which short-circuit winding points that, in absence of eccentricity or construction asymmetries, would be exactly at the same potential. As a consequence of rotor eccentricity [42] and/or manufacturing imperfections [43], an UMP arises, together with circulating currents between the parallel paths of a phase. The UMP is harmful in terms of vibration and possible bearing damages. As discussed in [44], the addition of equipotential connections provides further circulating current paths which help reduce UMP effects.

4.4.2 Fault-tolerant solutions for the rotor design

4.4.2.1 Wound-field synchronous motors

The removal of brush-ring systems for well-known maintenance, wear and environmental issues makes brushless excitation (Section 4.1.2.3) the preferred choice for modern wound-field synchronous motors. While avoiding sliding contacts, this however introduces a certain level of complexity in the machine design, which requires a rotating exciter combined to a rotor-mounted rectifier (Figure 4.19).

Field protection against overvoltages

In normal steady-state operation, the field winding of large wound-field synchronous motors has a relatively low voltage (generally of a few hundreds of volts) applied to it through the diode rectifier. However, there are various transient conditions (some of which relate to faults) that can result in significant voltages being induced across the field circuit [45].

Different kinds of faults have strongly different impacts in terms of field overvoltages. For example, in three-phase, phase-to-phase and single-phase short-circuit faults the excitation system control loop responds to the sudden decay in machine voltage and flux by forcing the exciter supply to its maximum ('ceiling') value. During all stator short-circuit transients, a very high current is induced in the field but, in any case, the transient field current flows in the positive direction, that is in the same direction as it flows during normal steady-state operation. This causes the rotating rectifier diodes to act as free-wheel diodes so that the field current is never forced to reverse due to the blocking action of the diodes and no important overvoltage occurs. Damages to the rectifier diodes for excessive heating are not expected either. In fact, the extremely high fault currents appearing in both the main stator phases and in the exciter stator phases cause a rapid intervention of machine protections before any detrimental effect can be produced.

A completely different behaviour can be observed in the case of a pole-slipping or out-of-synchronism fault. Typically, this kind of fault may occur in a synchronous generator as a consequence of a sudden variation in the grid voltage (like voltage sags) or in the mechanical load. However, it can also happen in VSI-fed synchronous motors (Section 4.2.1) when the voltage applied by the inverter has a wrong phase relationship with the motor back-EMF, i.e. in the case of sudden excessive changes in the load angle. In these circumstances, the synchronizing torque naturally developed by the machine (also thanks to damper reaction currents) may not be sufficient to keep the rotor in synchronism with the externally applied voltage and a pole-slip event occurs. Possible faults of this kind are possible, in particular, during the starting process of sensorless VSI-fed motors due to the difficulty of precisely identifying the current rotor position. As an effect of a pole-slip event, the field current is caused to decrease abruptly and, especially if the

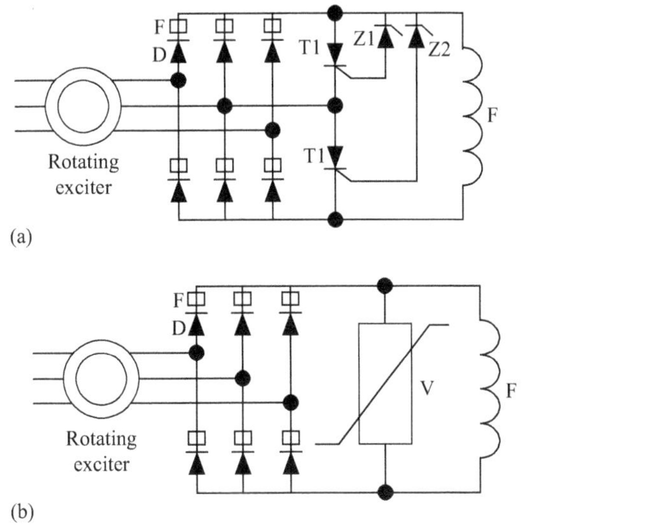

Figure 4.55 Rotor field protection through: fuses (F) and (a) thyristors fired by Zener diodes and (b) metal-oxide-varistors. © 2015 IEEE. Reprinted, with permission, from [17]

machine is operating at low loads or at no load (so with low field currents), it is likely to become zero. Each time the field current tries to reverse, it encounters the blocking effect of the rectifier diodes, and this causes a high voltage surge to occur across the field circuit. Simulations and various on-field experiences [10,45] prove that such field over-voltages can reach thousands of volts and, in absence of suitable protections, would certainly cause either the field or the rotating rectifier to discharge and breakdown.

A possible protection device is shown in Figure 4.55(a): two cascaded tyristors T1 and T2, connected in parallel to the main field, are fired, in case of an over-voltage, by Zener diode systems (Z1 and Z2) connected between their gates and the rectifier DC terminals. The connection of the mid points between T1 and T2 with an exciter rotor phase terminal is essential from a fault-tolerance viewpoint as it enables tyristors to be switched off in case they have been accidentally fired.

When the voltage across the field exceeds a limit depending on Zener diode sizing, thyristors become conductive, short-circuiting the field. The fault current is then allowed to reverse, flowing though the thyristors, without encountering the blocking action of the rectifier diodes.

Further protection devices, shown in Figure 4.55, are fuses F mounted in series to rectifier diodes D. Such fuses can be necessary because a rectifier's diode can fail into either an open-circuit or short-circuit [46]. While the former is compatible with a safe continued operation of the machine (although with a reduced rotor overload capability), the latter is likely to quickly cause severe effects due to exciter current overload [46]. The fuse is therefore useful especially to protect exciter phases against overheating by turning the diode short-circuit into an open-circuit fault, which can be subsequently detected as discussed in [47,48].

As an alternative to thyristors, metal-oxide-varistors (MOVs) can be used, in parallel with the field (Figure 4.55). MOVs are special resistors having non-linear voltage-vs-current characteristics such that, when subjected to normal field voltage, they behave like an open-circuit, while their resistance drops to low values when the voltage across the field increases due to a fault or abnormal transient.

The design of field protections against over-voltages is challenging from various points of view. From a mechanical standpoint, the design needs to guarantee perfect rotor balance, which often requires installation of the devices in pair (i.e. a couple of devices displaced by 180°) to prevent the occurrence of asymmetrical centrifugal forces and consequent vibrations; from an electrical point of view, the devices must be carefully selected so that

- they do not activate during normal motor operation, e.g. due to the normal voltage spikes following each rectifier diode commutation for reverse recovery effects [49];
- they clamp any overvoltage that could be dangerous for field winding insulation or rectifier diodes;
- they can withstand a discharge, dissipating the relevant energy, without an excessive temperature increase.

Examples of protection device design and selection based on dynamic simulations are given in [45].

Rotating excitation system fault-tolerant designs

One further approach to increase motor fault tolerance through rotor design is acting on the rotating excitation system.

For instance, the exciter rotor can be wound with a symmetrical n-phase winding (Section 4.3.2.1) and connected to a high-pulse-count rectifier composed of n legs ($2n$ diodes). An example of such a topology, with $n = 11$, is discussed in [46,50] showing how it can strongly help increase machine operation resilience to both open- and short-circuit excitation system faults. Otherwise, the exciter rotor can be designed with N three-phase winding sections (Section 4.3.2.2), each supplying a diode rectifier (REC), and the N rectifiers can be shunt-connected and protected with voltage-clamping devices (PROT), as shown in Figure 4.56.

Finally, redundant design arrangements have been recently proposed and implemented for on-board synchronous motor-generators [51] where the machine is equipped with two distinct exciters, each mounted on the same rotor and each feeding a different rectifier; the two rectifiers are shunt connected to supply the main field as shown in Figure 4.56(b). In addition to increasing fault tolerance, the design serves the purpose to let the machine operate with the DC-supplied exciter when operating as a generator and with the AC-supplied one when working as a propulsion motor, leaving the two control systems totally independent and decoupled. The obvious drawback of the solution is a noticeable increase of machine size, especially in the axial direction.

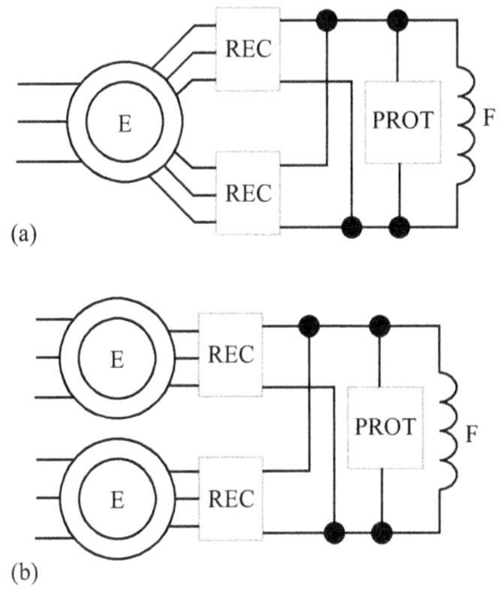

Figure 4.56 Fault-tolerant design solutions for the rotor excitation system: (a) with redundant rectifiers and (b) with redundant rotating exciters. © 2015 IEEE. Reprinted, with permission, from [17]

4.4.2.2 Permanent magnet motors

The margins to improve permanent-magnet synchronous motor fault tolerance through appropriate rotor designs seem relatively small and questionable. Some authors have actually proved that some interior permanent magnet (IPM) topologies (like the V-type) seem to exhibit a better resilience to demagnetization during short-circuit events if compared to others (like the spoke-type) [52]. As to the comparison between SPM and IPM topologies in terms of demagnetization withstand capabilities, works arriving at opposite conclusions can be found in the literature [53,54]. For sure, an accurate finite-element analysis of permanent magnet field distribution during short-circuit faults at the highest admissible temperature is essential to assure a safe design [41]. Other useful provisions include axial magnet segmentation to reduce eddy currents (and the consequent PM temperature) along with the selection of high-end materials like samarium–cobalt [41].

4.5 Fault-tolerant power converter design

The medium-voltage power converter, as an individual component of the drive included in either standard or redundant drive system layouts, can be endowed with fault-tolerance features acting on its design in terms of topology and/or control strategy. The provisions which can be taken to improve converter fault tolerance strongly depend on the converter type as discussed next.

4.5.1 Fault-tolerant VSIs

As discussed in Section 4.2.1, the VSI used for medium-voltage high-power applications are multi-level inverters. Compared to low-voltage two-level inverters, these topologies include a large number of switches which lead to better intrinsic fault-tolerance capabilities. In low-voltage two-level VSIs, the achievement of fault tolerance implies the addition of a fourth leg to the three legs typically employed in the ordinary three-phase design [55]. Conversely, in multi-level VSIs, interesting fault-tolerant features can be obtained even without the addition of a fourth leg [55,56].

The faults of interest are short-circuit and open-circuit faults affecting any of the switching devices included in the VSI. The main target is to maintain a balanced three-phase voltage system with acceptable multi-level output waveform. The degree of fault tolerance and the ways to achieve it vary from one VSI type to the other, as well as the industrial maturity of the fault-tolerant solution proposed in the scientific research literature, as discussed in various comprehensive surveys on the subject [55,56] and summarized next.

4.5.1.1 NPC VSIs

The ordinary NPC topology (Section 4.2.1.3) is not able to operate with a faulty switching device unless suitable provisions are taken to modify its switching algorithms and its hardware layout. Fault-tolerance features can be achieved basically in the following ways:

1. acting on the control strategy only, and leaving the hardware unchanged;
2. acting on the control strategy and including additional devices to the three VSI legs;
3. acting on the control strategy and including a fourth leg, plus possible additional devices.

The additional devices to be added can be thyristors, IGBTs and/or fast fuses in proper locations so that the VSI can work with acceptable performance also in case one or more switching devices undergoes either an open-circuit or short-circuit fault. Remedial switching control strategies after the fault are necessary both in the case of carrier-based PWM and of SVM: in the former case, appropriate phase switch changes need be applied to the modulating signals associated with the various device switches to guarantee balanced output voltages and to maximize the voltage output amplitude and quality; in the latter case, the basic idea is to use those switching patterns which may remain available to realize a certain instantaneous space vector after the fault of a switch. In fact, the NPC topology makes it possible, in general, to synthetize a given space vector with a variety of open-close switch combinations; of course, the presence of a faulty switch makes some of these combinations unfeasible but leaves the freedom to choose among the remaining ones. The selection of the switching patterns to be applied is, of course, a matter of optimization to be carried out assuming output voltage balance, amplitude and harmonic quality as the main objective functions.

While the target of retaining output voltage balance is considered mandatory, in all the proposed fault-tolerant solutions, the output power of the VSI in faulty conditions may undergo a severe derating (typically of 50%) due to a modulation index reduction, in some of the strategies based on a three-leg topology. However, a rated output power and voltage amplitude can be achieved even without adding a fourth leg, on condition that thyristors and fast fuses are suitably added to the standard VSI topology and considering that switching devices must be oversized to withstand the whole DC link voltage in faulty conditions. Conversely, the inclusion of a fourth VSI leg, besides additional thyristors and fuses, can guarantee full-power full-voltage post-fault operation capability even without the need for switching device oversizing. In any case, endowing the VSI with fault-tolerance capability is paid in terms of higher complexity and, therefore, higher predicted failure rates and higher cost due to the increase in the component count [55,56]. Most of the techniques proposed for NPC VSI fault-tolerance performance enhancement are relatively complicated both in terms of software and hardware implementations and have therefore a still limited diffusion in industrial drive applications.

4.5.1.2 FC VSIs

Compared to the NPC VSI, the FC topology (Section 4.2.1.3) exhibits superior fault-tolerance features, at least with respect to switch short-circuit faults. In fact, even with no addition of a fourth leg to the ordinary three-phase configuration and with no change in the conventional topology, the FC VSI can continue operating at its full voltage (with no modulation index decrease) even in the case of a faulty (short-circuited) switch. Post-fault remedial actions (including a suitable phase shift of carrier signals) can be implemented in order to maintain a certain output voltage quality. However, despite any post-fault control strategy, the occurrence of a short-circuit fault generally poses significant stresses in terms of over-voltages applied to the VSI capacitors and healthy switches, as well as in terms of additional switching losses. As a result, an appropriate reduction of the power output after the fault is highly recommended for a fail-safe continued operation [56].

If an inverter switch fails in an open-circuit, the effect on the FC VSI operation is much more severe [56] and usually leads to a trip. To improve the fault-tolerance capability of FC VSIs, even with respect to switch open-circuit faults, the inverter topology needs to be changed with addition of new hardware components. For example, the use of low-frequency switches in parallel to each IGBT or IGCT and in series to the FCs is mentioned in [55]. These additional switches make it possible to bypass the faulty device on one leg side as well as the corresponding device on the opposite leg side and, at the same time, to isolate the connected FC. After this topology reconfiguration, the leg including the faulty switch can continue operating as the leg of an ordinary low-voltage two-level inverter. Of course, the post-fault operation performance increases if a high number of switches per leg is adopted.

4.5.1.3 SC-HB VSI

The SC-HB VSIs are probably the medium-voltage VSIs which exhibit the highest and more industrially exploited fault-tolerance potential. This results from the inherently redundant architecture of the VSI, where each phase voltage is built up

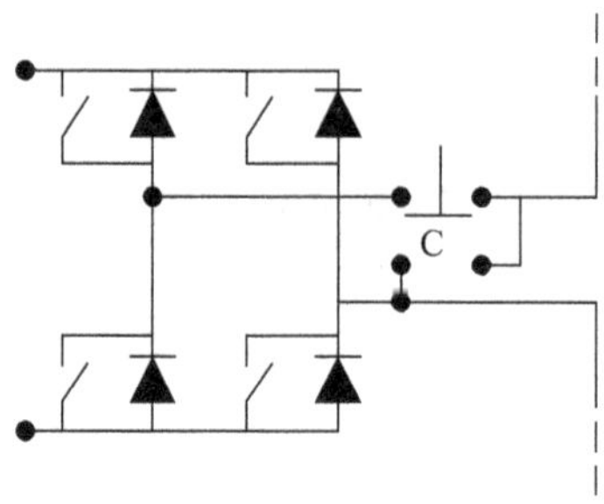

Figure 4.57 Mechanism for cell-bypass operation in SC-HB VSIs

from the contribution of several series-connected H-bridge low-voltage cells (Section 4.2.1.3). The basic idea to cope with a fault in a given cell is to bypass it through the conceptual scheme illustrated in Figure 4.57: as soon as a fault is detected in the H-bridge cell, the contactor C is moved from its upper position (which is its usual position during normal operation) to the lower position so as to short-circuit the faulty cell.

If no provision is taken, the cell-bypass produces an imbalance in the output voltages. However, there are post-fault control techniques which make it possible to preserve the balance and symmetry of line-to-line voltage systems [55]. A simple strategy to retain line-to-line symmetry and balance is to bypass the healthy cells in some phases so that each phase has the same number of active cells [56]. Of course, these countermeasures imply a reduction in the VSI output voltage amplitude. Such issues can be effectively avoided if the cells are connected to active rectifier stages (see Figure 4.27). In fact, this makes it possible to control the DC-link voltage reference of each cell in order to compensate for a bypassed cell loss [57]. For example, if a cell is bypassed in a certain phase including n series-connected cells, the DC-link voltage of the $n-1$ healthy cells of the same phase can be increased by a factor $n/(n-1)$ so that the output voltage amplitude can be maintained with no control algorithm modifications. The possibility to share the DC-link voltage increase among all the healthy cells of the phase minimizes the need to oversize cell devices. However, a certain oversizing is necessary depending on the maximum allowed number of cells that can be bypassed.

A further fault-tolerance feature which can be found in some industrial implementation of SC-HB VSIs [10,58] is its resilience to supply grid failure. In fact, a potentially dangerous event which may happen during a large drive operation is the temporary grid voltage reduction or loss. In this event, the SC-HB VSI can be designed so that the DC-link voltage during the grid fault reduces but to a limited extent thanks to the kinetic energy coming from the motor deceleration. In this way, the inverter can maintain the FOC of the motor during the fault (Section 4.2.1.4). When the grid voltage is restored, a DC-link current flow from the grid is produced in a controlled way so as to avoid severe inrush currents and progressively recharging the DC-link capacitors to their nominal voltage. Because

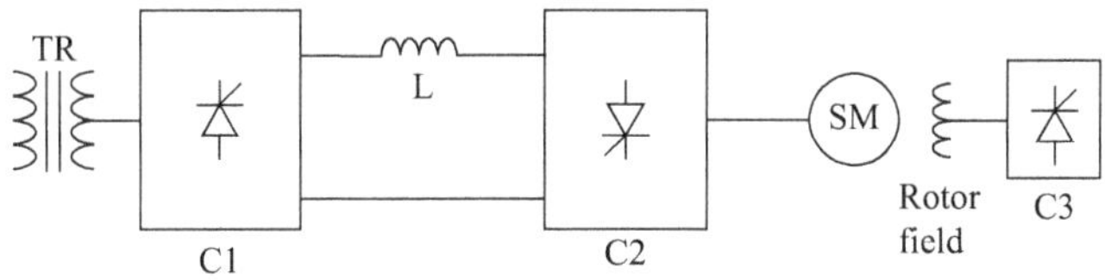

Figure 4.58 Basic LCI drive structure

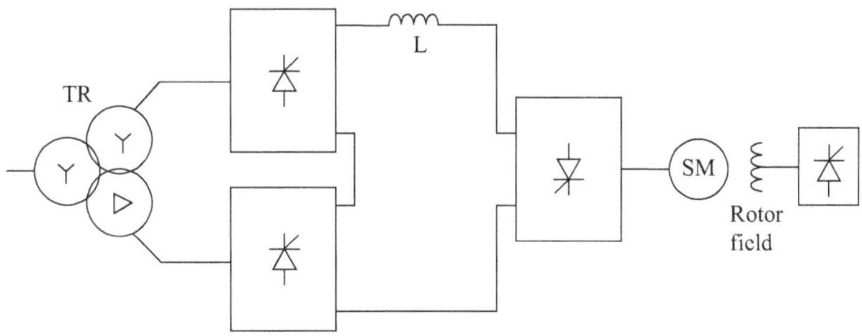

Figure 4.59 LCI drive arrangement with a 12-pulse structure on the grid-side and a 6-pulse one on the motor-side

the FOC of the motor is maintained throughout the event, the motor can be 'caught on the fly' reaccelerating it promptly when the voltage is restored [58].

4.5.2 Fault-tolerant CSIs

As mentioned in Section 4.2, the most commonly used CSI is the LCI for its simplicity, robustness and relatively low cost. The basic and simplest LCI drive arrangement is depicted in Figure 4.58.

In order to improve the drive performance on the grid- and/or motor-side, and also to endow the drive with fault-tolerance capabilities, it is a common practice to use two converters, instead of one, on the grid-side and/or on the motor-side. Various possible arrangements can result, the most significant being shown in Figures 4.59–4.62.

Each time a dual converter arrangement is used, we have a '12 pulse' instead of a '6 pulse' configuration, depending on the number of current 'peaks' per period produced in the DC-link current. The increase in the number of pulses on the line side (i.e. the use of two line-side converters) is highly beneficial as it reduces the amplitude of the harmonics injected into the grid; similarly, the increase in the number of pulses on the motor side (i.e. the use of two motor-side converters) is highly beneficial as it reduces the amplitude of motor air-gap MMF harmonics and, hence, the motor torque ripple.

For the mentioned benefits to occur, it is necessary that the two grid-side converters be supplied by three-phase sources shifted by 30 electrical degrees apart; similarly, the electric motor is to be equipped with two three-phase sets (according

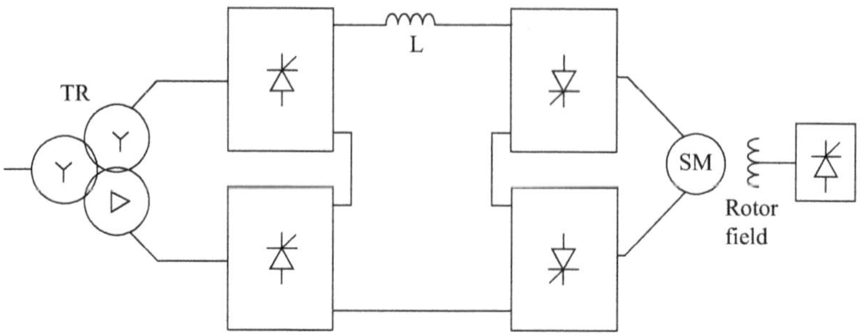

Figure 4.60 12/12 pulse configuration with LCI series connection

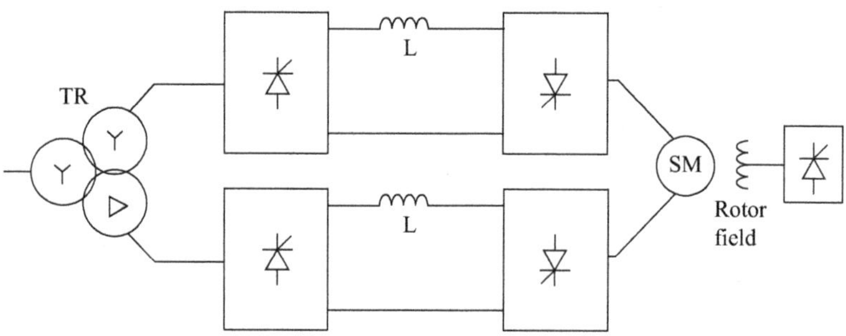

Figure 4.61 12/12 pulse configuration with parallel connection of LCIs

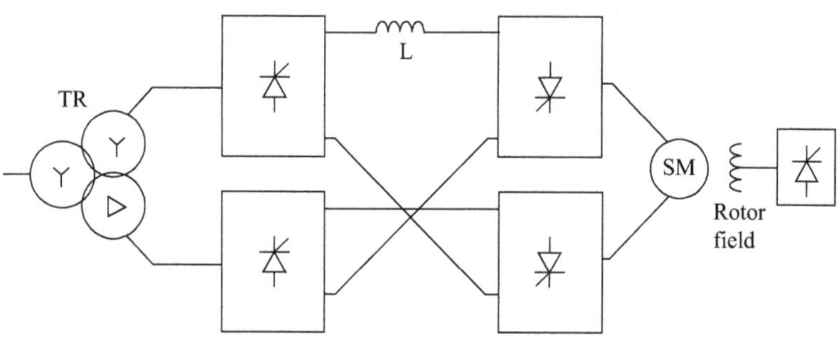

Figure 4.62 12/12 pulse configuration with cross series connection of LCIs

to a multiple-three-phase arrangement as discussed in Section 4.3.2.2) displaced by 30 electrical degrees apart.

The dual-three-phase configuration shown in Figure 4.62, in particular, is useful to reduce the so-called sub-harmonics which may arise from the interaction of the motor-side and grid-side converter operations. However, the solution should be carefully designed considering that it increases the voltage stress between phases belonging to different motor three-phase sets as investigated in [59].

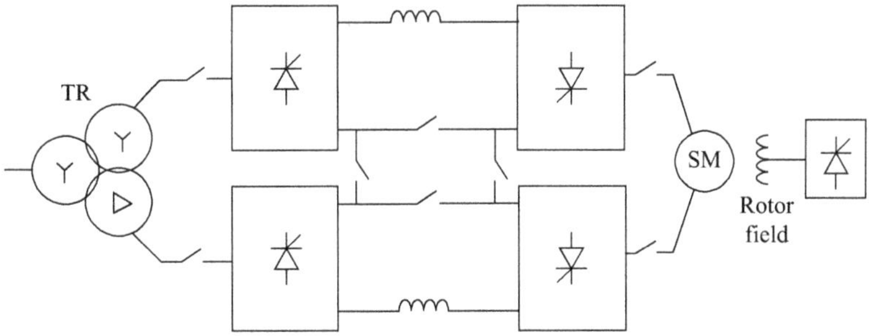

Figure 4.63 Schematic for LCI drive reconfiguration following a fault

In addition, to improving drive performance in healthy conditions, the multi-three-phase (12-pulse) arrangements are very beneficial in terms of fault tolerance. The most common faults involve single converter switches (SCRs), which can undergo either an open-circuit or a short-circuit fault. Additionally, the open-circuit fault may involve either a motor-side or line-side converter. All these fault scenarios can be managed differently and will be separately discussed below.

4.5.2.1 SCR short-circuit fault

A possible way to make the drive resilient to SCR short-circuit faults is to adopt a redundant configuration by employing multiple SCRs in series instead of a single one. This, however, increases the component count and, therefore, the system complexity and fault rate.

As an alternative, if an SCR fails in a short-circuit and a 12-pulse (dual bridge) configuration is adopted on the converter side where the fault occurs, it is possible to reconfigure the 12-pulse converter structure and control into a 6/6 pulse one (Figure 4.58) by disabling the faulty converter and the associated connections.

A possible schematic for the reconfiguration of a 12/12 pulse converter is shown in Figure 4.63, where the additional breakers added with respect to the ordinary structure (Figures 4.60 and 4.61) are to be closed or opened depending on the status (healthy/faulty) of each converter.

Of course, when turning a 12-pulse drive into a 6-pulse form, the power output decreases (by one half or more according to what is discussed in Section 4.3.2.2) and an increase in the motor vibrations is expected, possibly causing a drive trip, due to the increase in the torque pulsations.

4.5.2.2 SCR open-circuit fault in a line-side converter

An SCR open-circuit failure in the line-side converter for a 6/6 pulse structure (Figure 4.58) causes an instantaneous decrease in the DC-link voltage and the appearance of a DC component in the line current, resulting in a saturation of the input transformer. A possible way to prevent this is to control the faulty bridge as a single-phase one. However, this approach leads to very large DC-link current oscillations and therefore to large grid-side harmonic pollution and motor torque ripple.

As in the case of SCR short-circuit fault mode, the use of a 12-pulse (dual converter) topology allows for better fault tolerance. For example, if a parallel 12/12 pulse configuration is used (Figure 4.61), the entire converter section including the faulty bridge can be disconnected so that the drive reduces to a basic 6/6 pulse form; only one of the two motor windings is supplied and the power output can be at most one half of the rated power. In case of series connection (Figures 4.59 and 4.60), the faulty bridge can be bypassed so that the drive structure is converted to a 6/6 or a 6/12 one. A possible solution to bypass one of the faulty grid-side bridges with no need for additional switches is to control it as a single-phase bridge, firing SCRs with a delay angle close to 180° so that the bridge produces a quasi-null voltage. Also, this solution implies that the drive can continue operation with an output power not higher than half of the rated value. The output power during post-fault operation can be increased if the faulty bridge is controlled as a single-phase one, but firing the SCRs with a delay angle higher than 180° so as to increase the DC-link current and, hence, the motor torque. In any case (whether the faulty bridge is bypassed or partly used for current and torque production), the benefits of the 12-pulse configuration are lost, leading to significant growth in the grid-side harmonic pollution and motor torque ripple. Possible ways to mitigate the issue through suitable post-fault control algorithms are discussed in [60]: the two proposed strategies include either managing the faulty bridge as a diode bridge and controlling the healthy one or managing the healthy bridge as a diode bridge and controlling the faulty one. If controlling the healthy bridge, it is possible to trigger SCRs in the usual way, i.e. with equal firing angles for the three-phases, or to suitably change the firing angles of the phases so as to minimize the DC-link current ripple and, hence, the motor torque ripple. This asymmetrical firing pattern can effectively help mitigate the effects of the fault in terms of DC-link current and motor torque waveform [60].

4.5.2.3 SCR open-circuit fault in a motor-side converter

If the open-circuit fault affects an SCR located in the motor-side converter, the DC-link current after the fault behaves as exemplified in Figure 4.64.

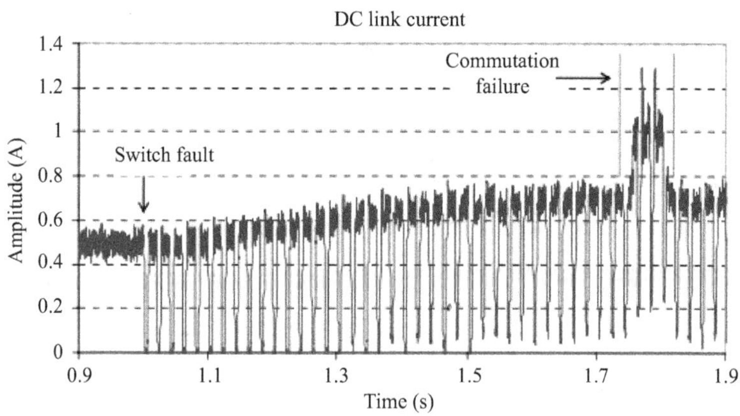

Figure 4.64 Experimental evolution of the DC-link current with an open-circuit fault affecting a motor-side SCR

It can be seen that the fault produces a reduction in the mean value of the DC-link current and the control tries to compensate for this by increasing the DC-link current reference. The increase in the DC-link current peak value finally results in a commutation loss event.

A post-fault strategy to mitigate the open-circuit SCR fault in a motor-side converter is to operate this in a single-phase mode. However, if there is only one motor-side converter (6-pulse motor-side configuration), significant oscillations are expected to occur in the DC-link current and torque. Conversely, in case of a 12-pulse (dual-converter) arrangement on the motor side, the fault consequences can be mitigated much more effectively. An interesting approach is described in [60], where the SCRs of the healthy motor-side bridge are fired with suitable delay angles (changing from one phase to the other) in order to minimize the ripple amplitude in the DC-link current and motor torque.

4.6 Diagnostics

Diagnostics and condition monitoring play an important role in large synchronous motor drives because of the high economical value of the equipment, the expensive reparation procedures and the significant production losses and hazards which may result from a drive system trip or fault.

Diagnostic and condition monitoring techniques are well established for large synchronous generators, where the output voltage and current waveforms are very close to the sinusoidal pattern and the general operation, in absence of faults, would be theoretically 'ideal'. For a generator, a deviation from the ideal predicted performance can be relatively easy to identify as the 'signature' or symptom of a fault or anomaly. Conversely, large motor operation under inverter supply, even in healthy conditions, is affected by a large amount of disturbances (like vibrations, voltage and current distortion, etc.) which result from the inverter supply and make it more difficult to recognize the symptoms of a fault. However, many diagnostic methods applied to synchronous generators can be, theoretically, employed also in inverter-fed synchronous motors, and, hence, they will be addressed in the following. In any case, in order to avoid repetitions, the focus will be placed only on those diagnostics technique which are peculiar to large synchronous motor drives, omitting to cover the procedures which are also employed in small-size induction and permanent-magnet motor drives, already covered in Chapter 2. Regarding the converters, the attention will be then focused on the condition monitoring and diagnostic issues peculiar to medium-voltage multi-level converters, which are employed in large synchronous motor drives. As regards the electric motor, attention will be given to the following diagnostic aspects which specifically characterize large wound-field synchronous machines:

- journal-bearing fault detection;
- rotating rectifier fault detection;
- rotor field winding fault detection;

- eccentricity detection;
- medium-voltage and high-voltage stator winding insulation system condition monitoring.

4.6.1 Diagnostics in medium-voltage converters

Detecting a fault in a medium-voltage converter is important either to stop and disconnect the converter before any damage occurs to the converter itself and to the supplied motor or to activate the suitable post-fault operation mode, if available.

Fault detection is, usually, more challenging in VSI-based converters than in CSI-based ones. Hence, the two cases will be separately addressed.

4.6.1.1 Fault detection in VSI-based converters

A comprehensive review of the main fault diagnosis techniques for VSI-based medium-voltage converters is provided in [56] and will be herein summarized taking into account the three main VSI topologies (NPC VSI, FC VSI and SC-HB VSI).

In general, the easiest method to detect the possible fault of a converter switch is to measure the voltage across the switch and compare it with the gating signals [61]: if the voltage remains zero or different from zero, regardless of the gating signals, a short-circuit or an open-circuit fault of the device can be detected, respectively.

Converters based on NPC topology

If the presence of a faulty switch is promptly detected, some of the countermeasures discussed in Section 4.5.1.1 can be implemented, by suitably modifying either the topology of the converter or the control of the healthy switches (or both).

In addition or in alternative to monitoring the voltage across each switch, the fault detection can be performed by measuring the phase current and voltages and comparing them with the expected values, predicted under the assumption of healthy operation. If significant discrepancies are found, this is assumed as a sign of a fault. By processing the error between the measured values and the expected ones, it is sometimes possible to identify the faulty switch.

Converters based on FC topology

As mentioned in Section 4.5.1.2, an FC VSI reacts in a very different way depending on whether a switch fails in an open-circuit or into a short-circuit. In the latter case, a continuity operation can be guaranteed, while in the former case, a trip is generally necessary, unless the inverter topology includes additional switches to short-circuit the faulty devices or has some form of redundancy.

As observed in [56], the occurrence of an open-circuit switch fault can be easily recognized by its apparent consequences in terms of output voltage and current reduction. In the case of a short-circuit fault, the diagnosis is more complicated and is generally based on recognizing an unusually large harmonic component (at the inverter switching frequency) in the input voltage or output current. Suitably processing the phase shift of such harmonic (with respect to the fundamental) can provide information to locate the fault inside the VSI [56].

Converters based on SC HB topology

The identification of a possible fault in a H-bridge cell makes it possible to bypass it, guaranteeing the VSI service continuity as discussed in Section 4.5.1.3. As observed in [56], three main fault detection approaches are proposed in the literature. The first approach compares the measured output voltage of each phase to the theoretical value predicted under the assumption of healthy operation; if a significant discrepancy is found, this is interpreted as the sign of a faulty cell. The identification and location of the fault according to this technique is, however, difficult and possibly uncertain. A second approach uses artificial-intelligent algorithms (especially based on neural networks) capable of recognizing fault-specific patterns in the measured phase-voltage. Finally, the spectral analysis of the output voltage (or current [62]) can be used; the presence of an abnormally large harmonic component at the cell frequency is interpreted as a sign of a faulty cell. The processing of the harmonic phase angle can provide information on the faulty cell location.

4.6.1.2 Fault detection in CSI-based converters

The way how an LCI-based converter reacts to the occurrence of a fault on one of its SCRs depends on whether the switch turns into an open-circuit or a short-circuit.

The short-circuit fault is regarded as the worst: in absence of provisions to increase the converter fault tolerance, it produces a sudden increase in the current flowing through the faulty SCR causing a converter trip for overcurrent. In a fault-tolerant LCI drive, the detection of the overcurrent can be used as a signal to trigger the bypass of an SCR bridge and activate an appropriate remedial control strategy as discussed in Section 4.5.2.1.

If an SCR fails in an open-circuit, the drive behaviour depends on whether the faulty switch is on the grid- or on the motor-side. In the former case, a decrease in the DC-link voltage is observed along with the appearance of a DC component in the input currents. If the SCR is on the motor side, the DC-link current peak increases as shown in Figure 4.64. Such events are relatively easy to detect and, if the drive is endowed with fault-tolerance capability, they can be used to detect the fault and activate the post-fault control methods discussed in Sections 4.5.2.2 and 4.5.2.3.

4.6.2 Diagnostics in large synchronous motors

Next, the main diagnostic issues which specifically characterize large wound-field synchronous motors will be addressed.

4.6.2.1 Journal bearings fault detection

As discussed in Sections 4.1.1.3 and 4.1.2.4, large synchronous motors use journal bearings, which are also known as sleeve bearings, in which there are no rolling elements but sliding cylindrical surfaces separated by a thin oil or grease film. In principle, journal bearings are much more reliable than ball and rolling bearings (used in small electric machines) thanks to their contactless operation which reduces mechanical wear and have theoretically infinite life. However, also journal bearings are subject to possible damage and deterioration, which can result in bearing faults with potentially serious damages to the machine and the coupled mechanical load, particularly to the shaft. The corrective maintenance and repair operations following a bearing fault can be extremely long and costly. Therefore, journal bearings condition needs to be carefully monitored in order to prevent faults and allow predictive maintenance.

The main causes for bearing damages are oil whirl instability, bearing clearance increase [63] and bearing currents associated with shaft voltages [11].

Oil whirl instability and increased bearings clearance

The oil whirl instability occurs when the shaft centre, instead of remaining at a constant stable position inside the bearing, turns along a circular orbit in the rotational direction and 'whirls' together with the surrounding oil at a speed which is usually between 40% and 49% of the shaft rotational speed. This whirling motion may lead to a contact (rub) between the shaft and bearing surfaces, in addition to producing potentially destructive vibrations at certain 'critical' speeds.

Regarding the bearings clearance, it is typically between tens and several hundred micrometres [63]. An excessive clearance (e.g. due to corrosion, mechanical wear due to frequent starts and stops, etc.) changes the lateral dynamics of the shaft line and induces large vibrations with possible intermittent rubs between the shaft and the bearing.

The above-mentioned issues can be effectively detected by monitoring the shaft vibrations by means of accelerometers and/or proximity probes. In particular, a spectral analysis of the measured vibration signal makes it possible to identify the bearing issue. In fact, the oil whirl instability is known to produce an increased harmonic component in the vibration spectrum at a frequency between 40% and 49% of the shaft rotational frequency [63]; on the other hand, an excessive clearance is known to produce vibration harmonic components at frequencies equal to f_{rot}/n where f_{rot} is the shaft rotational frequency and n is a positive integer.

Although the diagnosis of journal bearings is primarily performed through vibration monitoring, very recently some authors have proposed a methodology based on the motor current signature analysis (MCSA) [63]. This is not yet implemented in industrial applications but is promising because it makes it possible to detect journal bearings operation anomalies even in those machines where vibration measurement is not possible, e.g. due to environmental reasons.

Shaft voltages and bearing currents

In inverter-fed rotating machines, an electric voltage V_{shaft} can be induced across the shaft (Figure 4.65) mainly due to two effects: possible asymmetries in the stator and rotor construction and the high-frequency common-mode voltages caused by the supplying inverters [11] and affecting the shaft through the stator-to-rotor capacitance. If no design provisions were adopted, the shaft voltage would drive a current $I_{bearing}$ flowing along the closed path shown in Figure 4.65. As it can be seen, the current would flow through the bearings and cause such damaging phenomena as pitting, frosting, spark tracks and welding.

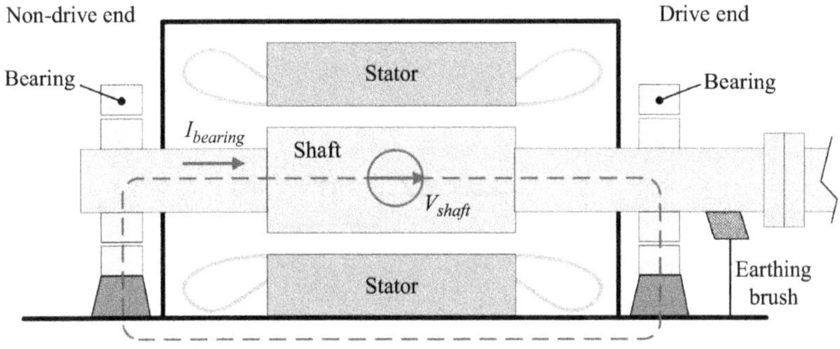

Figure 4.65 Shaft voltage and corresponding bearing current path

In order to prevent or limit the shaft currents (which are considered non-significant or normal, if their amplitude is below 1 A), the electrical configuration shown in Figure 4.21 is adopted: at least the non-drive-end bearing is insulated from the ground and the shaft is grounded. Insulating the non-drive-end bearing prevents bearing currents across the path shown in Figure 4.65, and the shaft grounding provides a low-impedance path in parallel to the drive-end bearing (thus limiting the possible currents flowing through it).

The detection and measurement of shaft currents is usually performed with a Rogowski coil mounted around the shaft on the drive end. As an alternative, a current transformer can be used with an annular ferromagnetic core mounted around the shaft on the drive end; the shaft currents act as a fictitious primary winding of the transformer, while a secondary real winding is wound around the core; the measurements of the current in the secondary winding makes it possible to quantify the amplitude of the shaft current.

The detection of excessively high shaft current indicates the need to re-establish an effective insulation on the non-drive-end bearing in order to avoid serious damages in it.

4.6.2.2 Rotating rectifier fault detection

In large inverter-fed brushless synchronous motors, the rotating rectifier, like the field circuit, is not directly accessible due to the presence of the AC rotating exciter (Section 4.4.2.1).

The rotating rectifier diodes can fail into either open-circuits or short-circuits. In the latter case, the fault causes a large overcurrent through the faulty diode with such a direction that both the field current and voltage are importantly reduced [46]. If the rectifier is equipped with fuses in series to the diodes (Section 4.4.2.1), the overcurrent is blocked by the fuse in series to the faulty diode and the fault is turned into an open-circuit. In absence of fuses, the severe reduction in the field current and voltage following the fault lead the motor voltage regulator system to increase the rotating exciter supply current to such an extent that the exciter supply over-current protection is triggered. Therefore, a short-circuit fault in the rotating rectifier is very easily detected and is not compatible with service continuity in absence of fault-tolerance design provisions (Section 4.4.2.1).

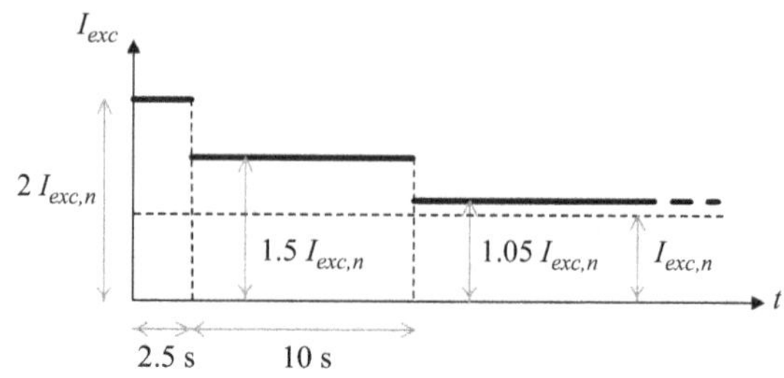

Figure 4.66 Example of thresholds for rotating exciter stator overcurrent protection

If the diode fails in an open-circuit, there is a drop in the average voltage applied to the field and, therefore, the motor voltage regulation system commands an increase in the rotating exciter supply. Such an increase is often not sufficient to trigger the exciter supply overcurrent protection and, in principle, the machine could be able to continue operating at full power even in the presence of the faulty diode [64]. However, this is not advisable because the presence of the faulty diode imposes an overload on the healthy ones as well as on the rotating exciter phases. Hence, it would be desirable that the open-circuit fault was detected. In [64], it is shown how the detection of a faulty diode in a synchronous generator, equipped with a DC-supplied rotating exciter, causes a significant distortion in the exciter supply DC current and, from such a distortion, the fault can be recognized. In large inverter-fed synchronous motors, the detection is generally more challenging due to the rotating exciter having a three-phase AC-supplied stator, with supply currents which are very significantly distorted even during normal operation due to the exciter power-electronics supply system. To mention an example, Figure 4.66 shows a possible thresholds setting in the stator overcurrent protection system of a rotating exciter. In the example, an exciter stator current overload equal to two times the rated value is accepted for a short period (2.5 s) in order to allow for possible fast transient changes in the motor operation (like sudden overloads); a stator current up to 1.5 times the rated value can be tolerated for a maximum time interval of 12.5 s, while in steady-state condition the exciter current is allowed to be at most 1.05 times the rated value without any protection intervention.

When a diode open-circuit fault occurs in the rectifier, this causes the voltage applied to the field to drop by around 30%; as a consequence, the motor voltage regulation system increases the exciter stator current until the rated flux is re-established in the motor. The possibility that such an increase triggers the overcurrent protection mainly depends on how the exciter is sized from an electromagnetic point of view. If it is sufficiently 'oversized', the current increase may not be enough to trigger the overload protection even at steady state. Furthermore, if the motor is not operating at full load or is working in the flux weakening region, the exciter current will be much lower than the rated value before the diode fault and, therefore, it is very likely that after the fault, it does not exceed the overload protection threshold.

Very recently, it has been shown how a detailed knowledge of exciter design and equivalent circuit model can lead to a very accurate estimation of the DC field current [65]. On the other side, a thorough knowledge of the saturated synchronous motor model equivalent circuits makes it possible to predict the field current in any steady-state operating condition. From all these detailed information, a diagnostic method to detect on open diode in the rectifier could be obviously obtained by comparison of the estimated field current from the exciter model and the one required by the synchronous motor according to the motor model. It is however stressed that the method implies an accurate knowledge of both the exciter and the synchronous motor parameters (possibly obtained from tests rather than from design predictions).

4.6.2.3 Field winding fault detection

The field winding of a large synchronous motor is composed of several (typically hundreds) series-connected turns (Section 4.1.2.1). The rated field voltage is usually in the order of few hundreds volts and the dielectric stresses in normal operation are usually low. However, there are cases (like pole-slip events, voltage sags in the supply, temporary asynchronous operation during starting, etc.) in which a transient overvoltage occurs in the field circuit as discussed in Section 4.4.2.1. The use of suitable voltage clamping devices (Section 4.4.2.1) generally limits the overvoltage within the voltage withstand capability of the field winding, which however may suffer a significant occasional dielectric stress. Furthermore, a source of repetitive (non-occasional) dielectric stresses acting on the field winding can be the overvoltage due to the rectifier diode reverse-recovery: in practice, each time the rectifier diode commutates, a voltage spike is generated due to the reverse recovery phenomenon. These spikes, although not dangerous in terms of amplitude, may pose some dv/dt stress on the field insulation. Finally, an important source of stress for the field insulation is due to possible local overheating ('hot spots') where the cooling air flow is not sufficient to effectively drain the heat due to Joule losses.

The mentioned causes may produce the insulation between two adjacent turns to deteriorate resulting in a possible turn-to-turn fault. The presence of a single shorted turn or few shorted turns in the field winding is very difficult to detect and is compatible with an almost normal machine operation [66]. The presence of an induced fault current inside the shorted turn or turns is not a thermal issue for the faulty turns because the fault current is theoretically zero due to the rotor revolving in synchronism with the air-gap rotating field. Conversely, if the fault involves several shorted turns, the machine performance is negatively affected mainly because the field current must be increased to maintain the motor flux to compensate for the shorted turns, which produce no useful magneto-motive force. This can give thermal problems and also trigger an exciter overcurrent protection. Additionally, the pole with the faulted turns produces less magneto-motive force than the other poles, which causes an UMP and consequent vibrations. For these reasons, it is desirable to promptly detect the presence of field shorted turns.

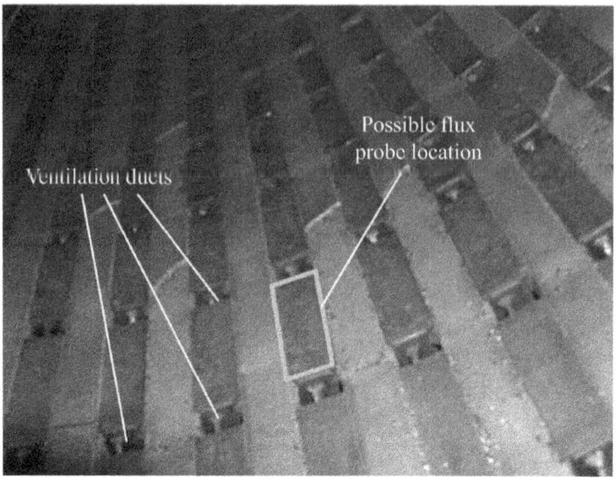

Figure 4.67 Example of location for the air-gap flux probe (consisting in a search coil)

Detection techniques have been mainly developed and implemented for synchronous generators. However, in principle, they can be applied to inverter-fed synchronous motors as well, with the obvious complication due to the harmonic disturbances introduced by inverter commutations in the signals used for the detection. The main approaches proposed in the literature are reviewed next.

Measurement of air-gap flux

This method can be applied to round-rotor synchronous machines, with either solid or laminated rotor (Section 4.1.2.1) where the field winding for each pole is composed of a series of multi-turn concentric coils (Figures 4.11 and 4.13). If a turn-to-turn short-circuit occurs in a certain coil, the slots including the faulty coil will have less current than the other slots and, consequently, a different leakage flux distribution. This difference in the leakage flux distribution slightly affects the air-gap flux. The turn-to-turn detection method under discussion employs flux probe placed inside the stator bore, for example, around a tooth between two successive radial cooling ducts as shown in Figure 4.67. The flux probe can simply consist of a set of series-connected turns across which the induced voltage (proportional to the derivative of the flux linked by the turns) is measured.

The flux measured by the probe includes some oscillations due to the rotor slotting effect. Each time the faulty coil passes near the flux probe, a lower oscillation is measured due to the reduced leakage flux [5,66]. This anomaly makes it possible to detect the presence of a faulty coil and also to locate it.

Measurement of stray magnetic fields

Some authors have shown that the presence of a faulty turn in the field of a salient-pole synchronous machine produces some variation in the stray magnetic field which can be measured in the surrounding of the machine (outside the frame) with a simple search coil [67]. The method is proposed for synchronous generators.

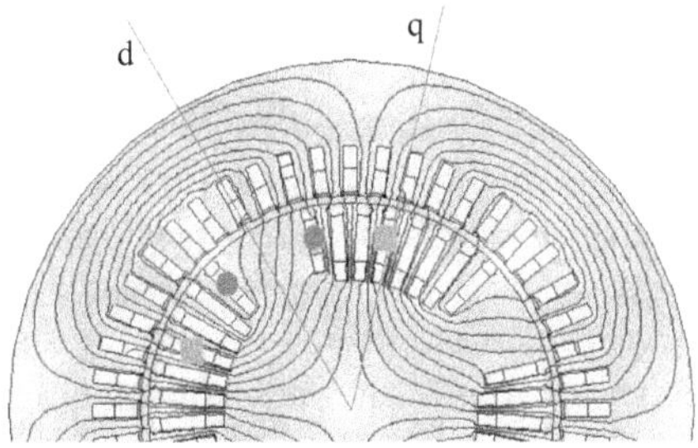

Figure 4.68 Identification of two possible field coils including shorted turns

In case of inverter-fed motors, the detection is likely to become more difficult due to the electromagnetic field harmonics produced by current and voltage distortion.

Measurements of vibrations

This method has been proposed with regard to both round-rotor [5] and salient-pole [67] synchronous machines. Vibrations are usually measured continuously during large electric motor operation by means of accelerometers placed on the bearing housings and/or through proximity probes placed near the shaft. An increase in the measured vibration can be a sign of various issues and, in particular, may indicate the presence of short-circuited turns in the field winding [5,67].

In the case of round-rotor machines, this can be given the following physical interpretation: the rotor coil including the shorted turns carries less current than the other coils and, therefore, has lower temperature. The consequent non-uniform temperature distribution along the rotor periphery causes different temperature-related deformations in different parts of the rotor resulting in an overall particular 'bending' of the rotor. Such bending is recognized as the cause of additional vibration component at the shaft rotational frequency [5] which can be measured when some field turns are shorted.

It may be worth noticing that the impact of the field turn-to-turn short-circuit on vibrations of a round-rotor synchronous machine is recognized to change depending on the location of the faulty coils [5]. In fact, if the shorted turns are included in a coil with wide pitch (i.e. in a coil having its coil sides near the q axis, like that marked with a square in Figure 4.68), the rotor deformation and the consequent vibration increase is much smaller than in the case when the shorted turns are located in the centre of the pole (like that marked with a round in Figure 4.68).

In the case of salient-pole machines, the origin of the additional vibrations is explained with the asymmetry of the air-gap magnetic field which characterizes a machine with shorted field turns (in fact, the pole including the faulty turns produce a weaker flux than the other poles). For salient-pole machines, some particular harmonics in the vibration spectrum are identified as characteristic of the phenomenon [67].

Air-gap width monitoring around the stator circumference

This method specifically applies to salient-pole synchronous machines. An air-gap monitoring of the air-gap width in different positions can be accomplished using special capacitive sensors as discussed in [5]. Each sensor consists of a thin metallic little-conductive plate glued to the inner surface of the stator bore. This plate on one side and the rotor outer surface on the other constitute a capacitor, whose capacitance value depends on the air-gap width. A high-frequency alternate voltage is applied to the conductive plate, while the rotor is grounded. Measuring the current which flows through the plate to the ground, being the impressed voltage known, makes it possible to estimate the probe capacitance and, therefore (after suitable calibrations), the air-gap width. Several probes are applied to various points of the stator bore and the described measurements are performed at a very high rate.

The detection of a turn-to-turn fault in a given pole is possible because this pole produces a weaker flux and is therefore less attracted towards the stator than the pole placed at 180° apart. The air-gap width corresponding in the faulty pole region is therefore larger than that in the region displaced by 180°. Through the measurement of such air-gap width discrepancy (eccentricity), it is then possible to detect the presence of faulty turns in a rotor pole [5].

4.6.2.4 Rotor eccentricity detection

Rotor eccentricity is an abnormal condition in which the centre of the rotor does not coincide with the centre of the stator. A static eccentricity occurs when the rotor centre does not coincide with the stator one and is located in a fixed position; a dynamic eccentricity occurs when the rotor centre does not coincide with the stator centre and moves along a circular or elliptical orbit as the rotor revolves. Both static and dynamic eccentricities can result from various reasons such as shaft misalignment, improper balancing of the rotor, bearing deterioration, abnormal displacements of the stator core and/or the frame, rotor bending due to thermal issues (e.g. resulting from a field turn-to-turn fault). Rotor eccentricity causes an asymmetry in the air-gap field because the flux density tends to increase where the air gap is small and to decrease where the air gap is large. The asymmetric air-gap field distribution produces an UMP due to the fact that the attraction force between the rotor and the stator at a given point of the air gap is roughly proportional to the square of the flux density at that point. The UMP therefore tends to worsen the eccentricity and gives rise to increased vibrations, abnormal bearing stresses and, in the worst case, stator-rotor rub. It is therefore important to detect the eccentricity in order to correct it before significant damages are caused.

An increased vibration, measured by accelerometers on the bearing housing or by proximity probes on the shaft, is certainly a consequence of rotor eccentricity but is generally insufficient to diagnose it. Possible eccentricity detection methods proposed in the literature for large synchronous machines are described next.

Use of air-gap sensors and mechanical measurement

As already discussed in Section 4.6.2.3, the air-gap width distribution along the stator bore circumference can be monitored by means of capacitive sensors applied to the stator bore surface in salient-pole synchronous machines [5]. Of course, the air-gap width can be also checked offline, while the rotor is at standstill by a common mechanical measurement.

Current measurement in stator phase parallel branches

This technique has been proposed and used for synchronous generators [68,69] but no specific issues can be envisioned preventing its application to synchronous motors. The method relies on the fact the eccentricity causes an asymmetric air-gap field distribution compared to that of a healthy machine. If the stator winding is provided with multiple parallel paths (or branches) per phase, such parallel branches will link different amounts of air-gap flux due to the mentioned asymmetry. As a consequence, circulation currents arise inside the parallel paths. The application of current sensors on the end connections makes it possible to measure and compare such circulating currents. A suitable processing of the measurement gives enough information to detect the presence of a static or dynamic eccentricity (distinguishing between the two cases) and to quantify the severity of the eccentricity fault [68,69]. A limitation of this approach is that results are affected by the load level and by magnetic saturation. Furthermore, the technique can be applied only on condition that multiple parallel branches are included in each stator phases.

No-load voltage and field-current measurement

This method uses the fact that air-gap field asymmetries induced by the eccentricity in the air-gap flux gives rise to abnormal harmonic components in the no load synchronous machine voltage and field current [70]. It is shown that the additional harmonics which appear in the no-load voltage spectrum as a consequence of the eccentricity vary from machine to machine as they depend on the winding design. Conversely, it is also shown in [70] that static eccentricity always produces a no-load field current ripple with $2f$ frequency (being f the supply frequency), while a dynamic eccentricity causes the appearance of an additional ripple having f/p frequency (being p the number of pole pairs). It should be noted the use of the no-load voltage spectrum for eccentricity detection is not directly applicable to an inverter-fed motor because it implies the operation of the machine as a driven generator at no load. Also the measurement of the field current is not easily applicable because inverter-fed synchronous motors are usually brushless machines and the field terminals are not accessible.

Sustained short-circuit current measurement

This method requires the synchronous machine to be driven as a generator with the short-circuited stator terminals. The air-gap field asymmetry due to the eccentricity causes some harmonics to appear in the short-circuit currents, and these are used as an eccentricity fault signature [71]. The method is capable of distinguishing the static and dynamic eccentricities and to monitor them separately. Authors suggest a possible way to apply their proposed approach to an inverter-fed synchronous motor: this can be done by bringing the machine at its rated speed at no load by means of the inverter; the inverter is then controlled so as to impose a temporary short-circuit to the stator terminals, while the field current is regulated to give a short-circuit current equal to the rated current. The rotor inertia is expected to be sufficient for the machine to remain in steady-state conditions for the time required to take the necessary short-circuit current measurements [71].

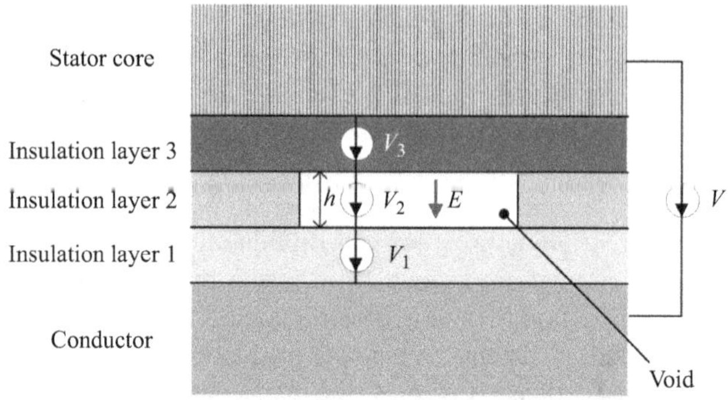

Figure 4.69 Schematic of a void in the insulation system

4.6.2.5 Stator winding insulation condition monitoring

As described in Sections 4.1.1.2 and 4.1.2.2, large synchronous motors used in high-power drives are medium-voltage or high-voltage machines equipped with wound-formed coils. The insulation system is probably the most critical part in a medium-voltage or high-voltage electric motor. In fact, the life time of the whole machine typically coincides with the life time of its insulation system as the most severe faults (those determining the end of the electric motor life) are associated with insulation breakdown events.

There are multiple factors which determine the progressive deterioration of motor insulation over time, some of which are common to low-voltage motors. For example, electrical stresses imposed by the PWM inverter due to the spikes and fast rise-time voltage pulses produced by silicon switch commutations are equally responsible for insulation aging regardless of motor size. Also thermal stresses, mechanical stresses and environmental contaminations are important in determining the insulation system life time in small motors as well as in large ones.

The main aspect which significantly characterizes medium-voltage and high-voltage motors is the occurrence of partial discharges, which are of key importance because it can be estimated that roughly one half of the insulation failures result from them [5]. To understand the partial discharge phenomenon, let us suppose that there are three insulation layers between a motor conductor and the stator core as depicted in Figure 4.69, with the second (intermediate) layer containing a defect constituted by a void (filled with air) of height h. The conductor is supposed to be at a voltage V (depending on motor rated voltage) with respect to the stator core, which is grounded. Since insulation layers (as well as the possible voids inside them) behave like series connected capacitors between the ground (stator core) and the conductor, the voltage V will split into the three voltages V_1, V_2, V_3 such that $V_1 + V_2 + V_3 = V$. The individual values of the three voltages depend on the geometry of the system and on the dielectric constant of its constituting parts.

Inside the void, an alternating electric field of amplitude $E = V_2/h$ arises. If E exceeds 3 kV/mm, it causes a discharge (known as 'partial discharge') in the air between the two sides (upper and lower ones) of the void, along electric field lines.

Partial discharges, which change direction every half period of the supply voltage due to electric field reversal, are dangerous because they deteriorate the void surface and progressively penetrate into the insulation layers. This progression of partial discharges into the surrounding insulation is known as 'electrical treeing'. After a certain amount of time, the electrical treeing expansion leads to the insulation system failure, with a discharge between the conductor and the grounded core (ground fault).

The same phenomenon can obviously occur if the stator core is replaced by another conductor (belonging to a different turn or phase) in Figure 4.69. In this case, the insulation breakdown results in a turn-to-turn or in a phase-to-phase fault.

The described process is typical of medium- or high-voltage machines. In fact, in low-voltage motors, even if a void is present somewhere in the insulation, the electric field inside is lower than 3 kV/mm and is therefore insufficient to produce partial discharges.

Partial discharges may occur not only in voids inside the insulation, but, for the same reasons, also in possible voids between the insulation and the conductor or between the insulation and the stator core. Furthermore, special partial discharges (called surface or corona partial discharges) can also occur on the external surface of the insulation, for example, at the end of the semi-conductive coating as represented in Figure 4.18.

The presence of voids in the insulation system is due to defective impregnation (Section 4.1.2.2) but can also be the consequence of the so-called delamination. A delamination for example takes place when the conductor undergoes severe overheating events (e.g. during overloads): since the copper has a higher thermal expansion coefficient than the insulation around it, the volume of the copper increases more than the insulation causing insulation layers to separate and giving rise to voids.

All this justifies the importance of an accurate periodic condition monitoring of the insulation system in high-voltage and medium-voltage electric motors at least for the following purposes: estimating the remaining life of the machine, helping to schedule maintenance effectively, deciding whether a motor needs to be replaced or rewound.

The insulation system condition monitoring involves a number of tests and activities, some of which can be performed on-line and other off-line (during maintenance or commissioning). The most significant condition monitoring techniques are described next.

Insulation resistance and polarization index

A simple off-line method to assess the insulation system condition is measuring the insulation resistance (IR), which is the total DC resistance (typically in the order of hundreds of megaohms) which can be measured between a motor conductor and the motor ground. For modern medium-and high-voltage form-wound motors, IR values above 100 MΩ are considered acceptable [5]. Low IR values can indicate an insulation failure but may also indicate that the insulation is soaked with moisture, so that normal IR values can be restored after the winding has dried. Actually, the interpretation of IR measurements requires a noticeable experience.

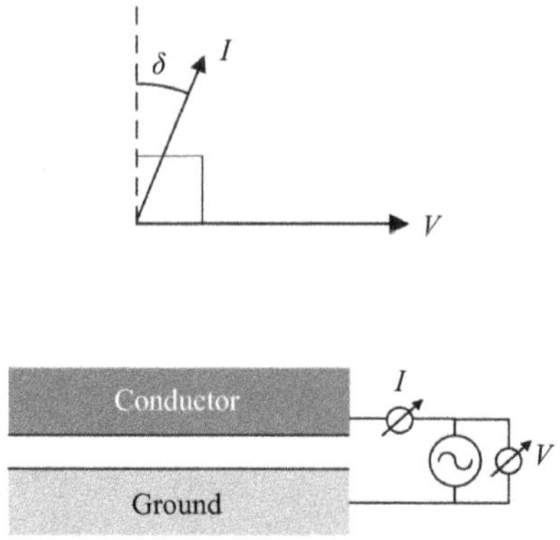

Figure 4.70 Definition of the angle δ

A further useful indicator is the polarization index (PI), defined as the ratio R_{10}/R_1 between the IR measured respectively after 10 min (R_{10}) and after 1 min (R_1) from the instant at which the test voltage is applied. Using this ratio removes the temperature-dependency of IR measurement and quantifies the extent to which the IR increases over time after the test voltage is applied. The increase is due to the progressive extinction of 'absorption' currents which normally flow through the insulation system after voltage application without indicating any damage. The larger the increase in the IR (i.e. the higher the PI), the better is the insulation system condition. In general, PI values around 1 indicate a bad condition for the insulation system, while values above 2 indicate a good condition. Actually, the bad condition may be due to a fault but also to the winding being contaminated or soaked with water.

Insulation capacitance and tanδ measurement

The insulation capacitance C is the capacitance measured between a motor conductor and the motor ground (core). For its measurement, an AC voltage V is applied between the conductor and the grounded core and the resulting current I is measured (Figure 4.70). The angle δ is defined so that the phasors associated with the current I and the voltage V are shifted by $90° - \delta$ degrees (Figure 4.70).

In other words, we can write tanδ as $R/(2\pi fC)$ where R and C are, respectively, the AC resistance and the capacitance between the ground and the conductor and f is the test frequency.

Since the materials used for the motor insulation have a larger dielectric constant than the air and the water, the presence of air voids or moisture in the insulation tends to decrease the capacitance C. On the other side, the presence of water and/or conductive particles (contaminants) reduces the IR.

A single measurement of C and tanδ is usually regarded to be of little significance. More reliable information on the possible insulation aging and degradation can be obtained monitoring the trends of these two indicators over time. If the insulation does not undergo important aging, C and tanδ remain practically

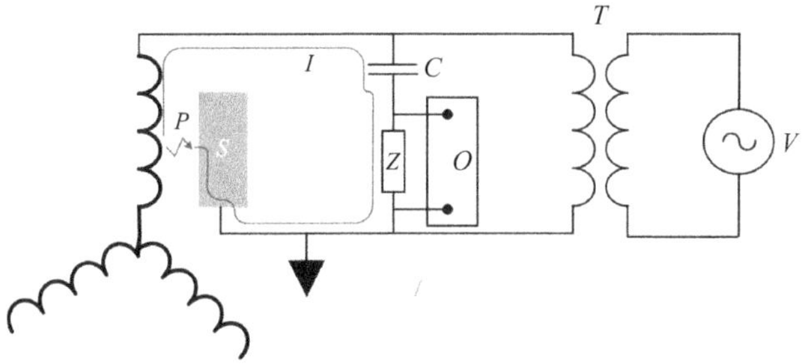

Figure 4.71 Circuit for partial discharge measurement

constant. Conversely, even slight increases or decreases (in the order of ±1% over one year) in the two parameters can indicate an issue in the insulation system. For example, a decrease in C accompanied by an increase in tanδ is a typical symptom of degradation due to thermal stresses (overheating) and consequent delamination; a simultaneous increase in both C and tanδ, instead, indicates that the insulation is contaminated or has absorbed moisture [5].

Partial discharge measurement

Partial discharges are due to a flow of electrons moving from one side of a void (filled with air) to the other side as a consequence of an electric field (larger than 3 kV) inside the void. Each discharge has a very short duration (few nanoseconds) and gives rise to a current pulse which partly travels through motor conductors. These pulses can be detected offline by means of a circuit like that shown in Figure 4.71: the motor winding is energized with an AC (usually 50 or 60 Hz) voltage V through a transformer T such that the insulation system is subject to roughly the same electrical stress as in normal operation. S is the grounded stator core, Z is a measuring resistive or RL impedance, O is an oscilloscope and C is a detection capacitor. When a partial discharge occurs at a point P in the winding insulation, the associated current pulse i arises. Due to its extremely fast rise time, the current pulse views the supply transformer as an open-circuit and the capacitor C as a short-circuit. Hence, the pulse travels along the path indicated in Figure 4.71, producing a voltage pulse across the measuring impedance Z. Such voltage pulse can be detected by the oscilloscope O. Being the capacitor C, small enough to act as an open-circuit at 50 or 60 Hz, the oscilloscope is capable of recording only the high-frequency current pulses due to partial discharges.

As an example, Figure 4.72 shows one period of the supply voltage and, on the same time scale, the recorded voltage pulses related to partial discharges. It can be seen that partial discharge pulses occur when the supply voltage amplitude exceeds a given threshold and reverse every half period.

As a general rule, high partial discharge pulse amplitudes indicate the presence of large voids in the insulation. A high number of pulses in a half period, instead, indicates that there are many voids, i.e. that the deterioration is widespread throughout the winding insulation.

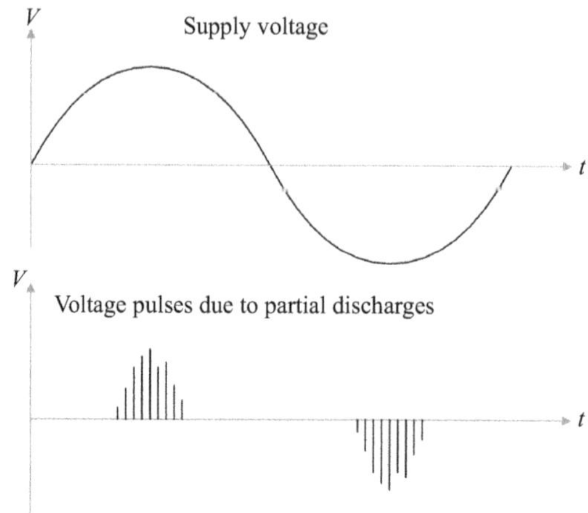

Figure 4.72 Example of partial discharge measurement

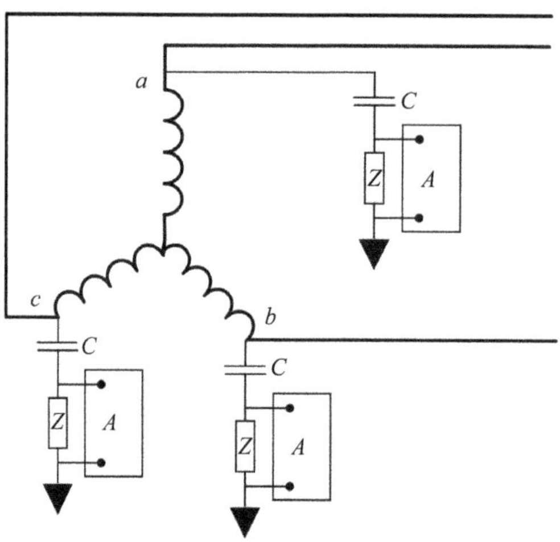

Figure 4.73 Example of circuit for online partial discharge measurement

However, a single partial discharge measurement is usually little significant. What makes sense is to compare the partial discharges (in amplitude and number per period) measured in different phases of the same motor, in different motors with identical design and construction, or at different times on the same machine. In particular, the latter measurement makes it possible to monitor the deterioration process of the winding insulation over time.

Partial discharges can be monitored on line, too. One possibility for online measurement is shown in Figure 4.73: a detection capacitor C and an acquisition system A (with its relevant measurement impedance Z) is installed between each of the motor terminals a, b, c and the ground. The capacitance C is chosen so that the capacitor acts as an open-circuit for the supply voltage and as a short-circuit for the high-frequency partial discharge pulses.

Alternative methods employ special sensors embedded inside the machine frame (e.g. in the slots). Such sensors act as antennas capable of capturing the electromagnetic waves associated with the current pulses produced by partial discharges.

Also for the online monitoring, a single measurement is usually little significant, while the attention is mainly placed on the trend of measurement results in terms of partial discharge pulse amplitude and count. In fact, the experience shows that in the first years of operation, a certain slow increase in the partial discharge activity can be normally experienced, followed by a substantial stabilization. After the stabilization has occurred, a new significant increase in the partial discharge activity is an indicator of insulation system deterioration.

Acknowledgment

This work was partially supported by the European Regional Development Fund (ERDF) through the Operational Programme for Competitiveness and Internationalization (COMPETE 2020), under Project POCI-01-0145-FEDER-029494, and by National Funds through the FCT – Portuguese Foundation for Science and Technology, under Projects PTDC/EEI-EEE/29494/2017 and UID/EEA/04131/2013.

References

[1] A. G. Sarigiannidis, A. G. Kladas, A. Mountaneas, *et al.*, "Design of surface PM motors for pod application utilizing a 3D hydrodynamic model," *2016 XXII International Conference on Electrical Machines (ICEM)*, Lausanne, pp. 2934–2940, 2016.

[2] T. Mazzuca, M. Torre, "The FREMM architecture: a first step towards innovation," *2008 International Symposium on Power Electronics, Electrical Drives, Automation and Motion*, Ischia, pp. 574–579, 2008.

[3] D. M. Saban, C. Bailey, D. Gonzalez-Lopez, L. Luca, "Experimental evaluation of a high-speed permanent-magnet machine," *2008 55th IEEE Petroleum and Chemical Industry Technical Conference*, Cincinnati, OH, pp. 1–9, 2008.

[4] A. Tessarolo, "A survey of state-of-the-art methods to compute rotor eddy-current losses in synchronous permanent magnet machines," *2017 IEEE Workshop on Electrical Machines Design, Control and Diagnosis (WEMDCD)*, Nottingham, pp. 12–19, 2017.

[5] G. C. Stone, E. A. Boulter, I. Cubert, H. Dhirani, *Electrical Insulation for Rotating Machines*, 2014, Piscataway, NJ: Wiley-IEEE Press.

[6] F. Luise, A. Tessarolo, F. Agnolet, *et al.*, "Design Optimization and Testing of High-Performance Motors: Evaluating a Compromise Between Quality Design Development and Production Costs of a Halbach-Array PM Slotless Motor," *IEEE Industry Applications Magazine*, vol. 22, no. 6, pp. 19–32, 2016.

[7] G. Sulligoi, A. Tessarolo, V. Benucci, A. M. Trapani, M. Baret, F. Luise, "Shipboard Power Generation: Design and Development of a Medium-Voltage dc Generation System," *IEEE Industry Applications Magazine*, vol. 19, no. 4, pp. 47–55, 2013.

[8] M. Qiao, C. Jiang, Y. Zhu, G. Li, "Research on Design Method and Electromagnetic Vibration of Six-Phase Fractional-Slot Concentrated-Winding PM Motor Suitable for Ship Propulsion," *IEEE Access*, vol. 4, pp. 8535–8543, 2016.

[9] A. Tessarolo, F. Luise, S. Pieri, A. Benedetti, M. Bortolozzi, M. De Martin, "Design for Manufacturability of an Off-Shore Direct-Drive Wind Generator: An Insight Into Additional Loss Prediction and Mitigation," *IEEE Transactions on Industry Applications*, vol. 53, no. 5, pp. 4831–4842, 2017.

[10] A. Tessarolo, G. Zocco, C. Tonello, "Design and Testing of a 45-MW 100-Hz Quadruple-Star Synchronous Motor for a Liquefied Natural Gas Turbo-Compressor Drive," *IEEE Transactions on Industry Applications*, vol. 47, no. 3, pp. 1210–1219, 2011.

[11] S. Schroder, P. Tenca, T. Geyer, *et al.*, "Modular High-Power Shunt-Interleaved Drive System: A Realization up to 35 MW for Oil and Gas Applications," *IEEE Transactions on Industry Applications*, vol. 46, no. 2, pp. 821–830, 2010.

[12] D. Busse, J. Erdman, R. J. Kerkman, D. Schlegel, G. Skibinski, "Bearing Currents and Their Relationship to PWM Drives," *IEEE Transactions on Power Electronics*, vol. 12, no. 2, pp. 243–252, 1997.

[13] Lataire, "White Paper on the New ABB Medium Voltage Drive System, using IGCT Power Semiconductors and Direct Torque Control," *EPE Journal*, vol. 7, no. 3, pp. 40–45, 1998.

[14] G. Buja, R. Menis, "Dependability and Functional Safety: Applications in Industrial Electronics Systems," *IEEE Industrial Electronics Magazine*, vol. 6, no. 3, pp. 4–12, Sep. 2012.

[15] C. Gerada, K. J. Bradley, "Integrated PM Machine Design for an Aircraft EMA," *IEEE Transactions on Industrial Electronics*, vol. 55, no. 9, pp. 3300–3306, 2008.

[16] S. Kanerva, J.-F. Hansen, "State of the art in electric propulsion – viewpoint on redundancy," *2009 IEEE Electric Ship Technologies Symposium*, 2009. ESTS, Baltimore, MD, pp. 499–504, 2009.

[17] A. Tessarolo, F. Luise, "Design for improved fault tolerance in large synchronous machines," *2015 IEEE Workshop on Electrical Machines Design, Control and Diagnosis (WEMDCD)*, Torino, pp. 45–52, 2015.

[18] B. C. Mecrow, A. G. Jack, D. J. Atkinson, *et al.*, "Design and Testing of a Four-Phase Fault-Tolerant PM Machine for an Engine Fuel Pump," *IEEE Transactions on Energy Conversion*, vol. 19, no. 4, pp. 671–678, 2004.

[19] E. Levi, "Multiphase Electric Machines for Variable-Speed Applications," *IEEE Transactions on Industrial Electronics*, vol. 55, no. 5, pp. 1893–1909, 2008.

[20] Q. Chen, G. Liu, W. Zhao, L. Sun, M. Shao, Z. Liu, "Design and Comparison of Two Fault-Tolerant Interior-Permanent-Magnet Motors," *IEEE Transactions on Industrial Electronics*, vol. 61, no. 12, pp. 6615–6623, 2014.

[21] L. Parsa, H. A. Toliyat, "Five-phase permanent magnet motor drives for ship propulsion applications," *2005 IEEE Electric Ship Technologies Symposium*, Philadelphia, PA, pp. 371–378, 2005.

[22] H.-M. Ryu, J.-H. Kim, S.-K. Sul, "Analysis of Multiphase Space Vector Pulse-Width Modulation based on Multiple *d–q* Spaces Concept," *IEEE Transactions on Power Electronics*, vol. 20, no. 6, pp. 1364–1371, 2005.

[23] S. Dwari, L. Parsa, "An Optimal Control Technique for Multiphase PM Machines Under Open-Circuit Faults," *IEEE Transactions on Industrial Electronics*, vol. 55, no. 5, pp. 1988–1995, 2008.

[24] A. Tessarolo, "Analysis of split-phase electric machines with unequally-loaded stator windings and distorted phase currents," *The XIX International Conference on Electrical Machines – ICEM 2010*, Rome, pp. 1–7, 2010.

[25] J. Dai, S. W. Nam, M. Pande, G. Esmaeili, "Medium-Voltage Current-Source Converter Drives for Marine Propulsion System Using a Dual-Winding Synchronous Machine," *IEEE Transactions on Industry Applications*, vol. 50, no. 6, pp. 3971–3976, 2014.

[26] A. Tessarolo, C. Bassi, "Stator Harmonic Currents in VSI-Fed Synchronous Motors With Multiple Three-Phase Armature Windings," *IEEE Transactions on Energy Conversion*, vol. 25, no. 4, pp. 974–982, 2010.

[27] C. Bassi, A. Tessarolo, "Time-stepping finite-element analysis of a dual three-phase salient-pole synchronous motor under voltage-source supply," *2010 IEEE International Symposium on Industrial Electronics, ISIE*, Bari, pp. 2184–2189, 2010.

[28] Y. Hu, Z.-Q. Zhu, K. Liu, "Current Control for Dual Three-Phase Permanent Magnet Synchronous Motors Accounting for Current Unbalance and Harmonics," *IEEE Journal of Emerging and Selected Topics in Power Electronics*, vol. 2, no. 2, pp. 272–284, 2014.

[29] E. Levi, R. Bojoi, F. Profumo, H. A. Tolyat, S. Williamson, "Multiphase Induction Motor Drives – A Technology Status Review," *IET Electric Power Applications*, vol. 1, no. 4, pp. 489–516, 2007.

[30] J. Birk, B. Andresen, "Parallel-connected converters for optimum reliability and grid performance in the Gamesa G10X 4.5 MW wind turbine," *European Wind Energy Conference*, Brussels, 2008.

[31] C. Bruzzese, A. Tessarolo, M. Mezzarobba, *et al.*, "Study of faulty scenarios for a fault-tolerant multi-inverter-fed linear permanent magnet motor with coil short-circuit or inverter trip," *2014 International Conference on Electrical Machines, ICEM*, Berlin, pp. 2387–2393, 2014.

[32] F. Luise, S. Pieri, M. Mezzarobba, A. Tessarolo, "Regenerative Testing of a Concentrated-Winding Permanent-Magnet Synchronous Machine for Offshore Wind Generation – Part I: Test Concept and Analysis," *IEEE Transactions on Industry Applications*, vol. 48, no. 6, pp. 1779–1790, 2012.

[33] F. Luise, S. Pieri, M. Mezzarobba, A. Tessarolo, "Regenerative Testing of a Concentrated-Winding Permanent-Magnet Synchronous Machine for Offshore Wind Generation – Part II: Test Implementation and Results," *IEEE Transactions on Industry Applications*, vol. 48, no. 6, pp. 1791–1796, 2012.

[34] J. A. Haylock, B. C. Mecrow, A. G. Jack, D. J. Atkinson, "Operation of Fault Tolerant Machines with Winding Failures," *IEEE Transactions on Energy Conversion*, vol. 14, no. 4, pp. 1490–1495, 1999.

[35] B. C. Mecrow, A. G. Jack, J. A. Haylock, J. Coles, "Fault-Tolerant PM Machine Drives," *IEE Proceedings – Electric Power Applications*, vol. 143, no. 6, pp. 437–442, 1996.

[36] A. M. EL-Refaie, "Fault-tolerant PM machines: a review," *2009 IEEE International Electric Machines and Drives Conference IEMDC*, Miami, FL, pp. 1700–1709, 2009.

[37] A. M. El-Refaie, "Fractional-Slot Concentrated-Windings Synchronous PM Machines: Opportunities and Challenges," *IEEE Transactions on Industrial Electronics*, vol. 57, no. 1, pp. 107–121, 2010.

[38] T. Raminosoa, C. Gerada, N. Othman, L. D. Lillo, "Rotor Losses in Fault-Tolerant Permanent Magnet Synchronous Machines," *IET Electric Power Applications*, vol. 5, no. 1, pp. 75–88, 2011.

[39] C. Tong, F. Wu, P. Zheng, B. Yu, Y. Sui, L. Cheng, "Investigation of Magnetically Isolated Multiphase Modular Permanent-Magnet Synchronous Machinery Series for Wheel-Driving Electric Vehicles," *IEEE Transactions on Magnetics*, vol. 50, no. 11, pp. 1–4, 2014.

[40] A. Kock, M. Groninger, A. Mertens, "Modeling and control of fault tolerant drive topologies for electric vehicle applications," *2014 International Conference on Electr. Machines (ICEM)*, Berlin, pp. 2373–2379, 2014.

[41] A. M. El-Refaie, M. R. Shah, K.-K. Huh, "High-Power-Density Fault-Tolerant PM Generator for Safety-Critical Applications," *IEEE Transactions on Industry Applications*, vol. 50, no. 3, pp. 1717–1728, 2014.

[42] M. Michon, R. C. Holehouse, K. Atallah, G. Johnstone, "Effect of Rotor Eccentricity in Large Synchronous Machines," *IEEE Transactions on Magnetics*, vol. 50, no. 11, pp. 1–4, 2014.

[43] A. Di Gerlando, G. M. Foglia, M. F. Iacchetti, R. Perini, "Effects of Manufacturing Imperfections in Concentrated Coil Axial Flux PM Machines: Evaluation and Tests," *IEEE Transactions on Industrial Electronics*, vol. 61, no. 9, pp. 5012–5024, 2014.

[44] W. Oliveira, M. Uemori, J. Rocha, R. Carlson, "Reduction of unbalanced magnetic pull (UMP) due to equipotential connections among parallel circuits of the stator winding," *2009 IEEE International Electric Machines and Drives Conference, IEMDC*, Miami, FL, pp. 771–777, 2009.

[45] A. Tessarolo, "A numeric simulation approach to field protection design in wound-rotor brushless synchronous machines," *2008 IET Conference on Power Electronics, Machines and Drives, PEMD*, York, pp. 139–143, 2008.

[46] T. Zouaghi, M. Poloujadoff, "Modeling of Polyphase Brushless Exciter Behavior for Failing Diode Operation," *IEEE Transactions on Energy Conversion*, vol. 13, no. 3, pp. 214–220, 1998.

[47] J. Sottile, F. C. Trutt, A. W. Leedy, "Condition Monitoring of Brushless Three-Phase Synchronous Generators With Stator Winding or Rotor Circuit Deterioration," *IEEE Transactions on Industry Applications*, vol. 42, no. 5, pp. 1209–1215, 2006.

[48] F. C. Trutt, J. Sottile, J. L. Kohler, "Detection of AC Machine Winding Deterioration using Electrically Excited Vibrations," *IEEE Transactions on Industry Applications*, vol. 37, no. 1, pp. 10–14, 2001.

[49] J. Lutz, "Fast recovery diodes – reverse recovery behaviour and dynamic avalanche," *2004 24th International Conference on Microelectronics (IEEE Cat. No. 04TH8716)*, Nis, Serbia, pp. 11–16, 2004.

[50] X. Jia, Q. Li, J.-S. Lai, B.-M. Song, "Analysis of Polyphase Brushless Exciter," *IEEE Transactions on Industry Applications*, vol. 37, no. 6, pp. 1720–1726, 2001.

[51] G. Stanic, S. Bonato, M. Groppo, A. Tessarolo, "Hybrid synchronous motor-alternator with dual AC/DC excitation system for shipboard generation and propulsion applications," *2014 International Conference on Electrical Machines, ICEM*, Berlin, pp. 2362–2367, 2014.

[52] Q. Chen, G. Liu, W. Zhao, L. Sun, M. Shao, Z. Liu, "Design and Comparison of Two Fault-Tolerant Interior-Permanent-Magnet Motors," *IEEE Transactions on Industrial Electronics*, vol. 61, no. 12, pp. 6615–6623, 2014.

[53] W. Chlebosz, G. Ombach, "Demagnetization Properties of IPM and SPM Motors Used in the High Demanding Automotive Application," *COMPEL*, vol. 32, pp. 72–85, 2013.

[54] H. Chen, R. Qu, J. Li, B. Zhao, "Comparison of interior and surface permanent magnet machines with fractional slot concentrated windings for direct-drive wind generators," *2014 17th International Conference on Electrical Machines and Systems (ICEMS)*, Hangzhou, pp. 2612–2617, 2014.

[55] B. Mirafzal, "Survey of Fault-Tolerance Techniques for Three-Phase Voltage Source Inverters," *IEEE Transactions on Industrial Electronics*, vol. 61, no. 10, pp. 5192–5202, 2014.

[56] P. Lezana, J. Pou, T. A. Meynard, J. Rodriguez, S. Ceballos, F. Richardeau, "Survey on Fault Operation on Multilevel Inverters," *IEEE Transactions on Industrial Electronics*, vol. 57, no. 7, pp. 2207–2218, 2010.

[57] P. Lezana, G. Ortiz, J. Rodríguez, "Operation of regenerative cascade multicell converter under fault condition," *Proc. 11th Workshop COMPEL*, Zurich, pp. 1–6, 2008.

[58] P. Himmelmann, M. Hiller, D. Krug, M. Beuermann, "A new modular multilevel converter for medium voltage high power oil & gas motor drive applications," *2016 18th European Conference on Power Electronics and Applications (EPE'16 ECCE Europe)*, Karlsruhe, pp. 1–11, 2016.

[59] S. Mohamadian, S. Castellan, A. Tessarolo, G. Ferrari, A. Shoulaie, "An Algebraic Algorithm for Motor Voltage Waveform Prediction in Dual-LCI Drives With Interconnected DC-Links," *IEEE Transactions on Energy Conversion*, vol. 31, no. 2, pp. 506–519, 2016.

[60] A. N. Alcaso and A. J. M. Cardoso, "Remedial Operating Strategies for a 12-Pulse LCI Drive System," *IEEE Transactions on Industrial Electronics*, vol. 55, no. 5, pp. 2133–2139, 2008.

[61] S. Li and L. Xu, "Strategies of Fault Tolerant Operation for Three-Level PWM Inverters," *IEEE Transactions on Power Electronics*, vol. 21, no. 4, pp. 933–940, 2006.

[62] K. Thantirige, A. K. Rathore, S. K. Panda, S. Mukherjee, M. A. Zagrodnik, A. K. Gupta, "An open-switch fault detection method for cascaded H-bridge multilevel inverter fed industrial drives," *IECON 2016 – 42nd Annual Conference of the IEEE Industrial Electronics Society*, Florence, pp. 2159–2165, 2016.

[63] J. Jung, Y. Park, S. B. Lee, C. Cho, K. Kim, E. Wiedenburg, M. Teska, "Monitoring Journal-Bearing Faults: Making Use of Motor Current Signature Analysis for Induction Motors," *IEEE Industry Applications Magazine*, vol. 23, no. 4, pp. 12–21, 2017.

[64] M. G. McArdle, D. J. Morrow, "Noninvasive Detection of Brushless Exciter Rotating Diode Failure," *IEEE Transactions on Energy Conversion*, vol. 19, no. 2, pp. 378–383, 2004.

[65] N. Jiao, W. Liu, Z. Zhang, T. Meng, J. Peng, Y. Jiang, "Field Current Estimation for Wound-Rotor Synchronous Starter-Generator with Asynchronous Brushless Exciters," *IEEE Transactions on Energy Conversion*, vol. 32, no. 4, pp. 1554–1561, 2017.

[66] R. Fišer, H. Lavrič, M. Bugeza, D. Makuc, "Computations of Magnetic Field Anomalies in Synchronous Generator Due to Rotor Excitation Coil Faults," *IEEE Transactions on Magnetics*, vol. 49, no. 5, pp. 2303–2306, 2013.

[67] M. Cuevas, R. Romary, J. P. Lecointe, T. Jacq, "Non-Invasive Detection of Rotor Short-Circuit Fault in Synchronous Machines by Analysis of Stray Magnetic Field and Frame Vibrations," *IEEE Transactions on Magnetics*, vol. 52, no. 7, pp. 1–4, 2016.

[68] C. Bruzzese, "Diagnosis of Eccentric Rotor in Synchronous Machines by Analysis of Split-Phase Currents – Part I: Theoretical Analysis," *IEEE Transactions on Industrial Electronics*, vol. 61, no. 8, pp. 4193–4205, 2014.

[69] C. Bruzzese, "Diagnosis of Eccentric Rotor in Synchronous Machines by Analysis of Split-Phase Currents – Part II: Experimental Analysis," *IEEE Transactions on Industrial Electronics*, vol. 61, no. 8, pp. 4206–4216, 2014.

[70] C. Bruzzese and G. Joksimovic, "Harmonic Signatures of Static Eccentricities in the Stator Voltages and in the Rotor Current of No-Load Salient-Pole Synchronous Generators," *IEEE Transactions on Industrial Electronics*, vol. 58, no. 5, pp. 1606–1624, 2011.

[71] T. Ilamparithi, S. Nandi, J. Subramanian, "A Disassembly-Free Offline Detection and Condition Monitoring Technique for Eccentricity Faults in Salient-Pole Synchronous Machines," *IEEE Transactions on Industry Applications*, vol. 51, no. 2, pp. 1505–1515, 2015.

Chapter 5
Capacitors

Acácio M. R. Amaral[1,2] *and M. Sahraoui*[1,3]

In order to minimize the harmful effects of greenhouse gas, the use of clean energy has become vital. When produced from renewable resources, electricity can be considered a clean energy; unfortunately, its storage presents some problems, because it is not possible to store large amounts of electrical energy quickly.

There are mainly two different electrical energy storage technologies: capacitors and batteries. Capacitors present high power density and low energy density, unlike batteries whose energy density is much higher and the power density is lower. These differences are fundamentally due to the fact that capacitors actually store electrons, rather than batteries that use chemical energy for energy storage. For these reasons capacitors are used to transfer energy in short periods of time unlike batteries which are used as primary energy sources.

When the energy is transferred from the primary source to the load, some energy can be lost, other is stored and finally the reminiscent part will be transformed into the form of energy required by the load. If the energy in question is electric, almost all losses can be represented as heat (Joule effect), the stored energy can be preserved both as potential energy (capacitors) or kinetic energy (inductors) and the final form of energy transferred depends on the load requirements. The electronic circuits that perform the described tasks are denominated by power converters and are fundamentally composed of switches, lossless energy storage elements, and magnetic transformers. Figure 5.1 shows a basic electronic system [1].

Capacitors can be found in the DC-link of power converters, namely, in AC–DC–DC, DC–DC–AC, AC–DC–AC, AC–DC and DC–AC power converters (Figure 5.1). The previous configurations cover a huge variety of applications such as wind energy conversion systems, photovoltaic systems, motor drives, electrical vehicles, lighting systems, among others [2,3].

Therefore, capacitors are one of the fundamental elements of power converters but, unfortunately, as it will be shown later, they are one of the most vulnerable elements of these systems [2–7], which make the topic of capacitors

[1]CISE – Electromechatronic Systems Research Centre, Universidade da Beira Interior, Portugal
[2]Polytechnic Institute of Coimbra, Portugal
[3]University of Biskra, Algeria

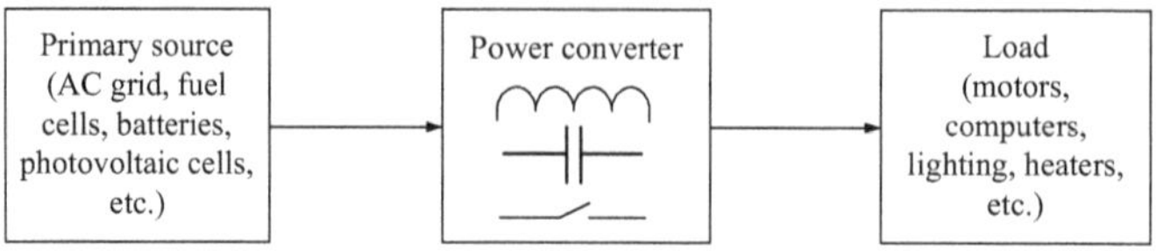

Figure 5.1 Basic electronic system

fault diagnosis a prominent subject in the context of this book. In some applications, capacitors present even the highest failure rate among all the elements of the power section [4–7].

In [4,5], a study was carried out on the distribution of faults in two different converters using the American military specification MIL-HDBK 217F.

The research carried out in [4] used a zero-current-switched secondary resonant half-wave DC–DC forward-type power supply operating at three different temperatures. Four different elements were considered in the study: metal oxide semiconductor field effect transistor (MOSFET), diodes, electrolytic capacitors and polypropylene capacitors. The following conclusions were drawn:

- For an ambient temperature of 25 °C, electrolytic capacitors exhibited a failure rate of approximately 75%.
- For an ambient temperature of 40 °C, electrolytic capacitors failure rate increased to almost 80%.
- For an ambient temperature of 70 °C, electrolytic capacitors failure rate increased to almost 95%.

In [5], a similar study was carried out for a half-bridge DC–DC forward-type power supply operating at an ambient temperature of 25 °C. In this study, the following components were considered: output filter capacitors, MOSFET, diodes, and inductive elements; and it was shown that capacitors had a failure rate of 60%.

Both studies allow us to obtain the following two conclusions: capacitors are the most vulnerable element in the power section of both power supplies, and their failure rate increases significantly with temperature. This information is particularly important since many power converters operate with temperatures ranging from 30 °C to 110 °C or even more, during their lifetime [6,7].

In [7], a questionnaire survey on the reliability of power electronic converters was carried out with the participation of aerospace, automation and motor drive industries, among others. The authors concluded that both power semiconductors and capacitors represent the most fragile components of these systems [7].

The results of another survey on power converters reliability [6], based on 200 products from 80 companies, revealed that capacitors represent the weakest element in power converters, with a failure rate of 30% in the following universe: semiconductors, connectors, solders, printed circuit board (PCB) circuits, and capacitors, among other components.

Power converters, and therefore capacitors, are used in several critical applications, such as medical equipment [8], aerospace industry [9], uninterruptible power supplies (UPS) used in nuclear power plants [10], suspension control systems and braking systems of vehicles, traction systems of hybrid/electric vehicles [6,7,9] and military equipment [7]. Therefore, several industries, such as automotive and aerospace, have introduced thigh reliability constraints on power converters [9].

In this context, the development of monitoring techniques that are able to evaluate, in real time, the capacitor health status is vital, so that it would be possible to schedule maintenance before serious deterioration or breakdowns can occur. In this way, it would be possible to avoid malfunction of the converter or even its stoppage, whose consequences could entail very high costs, or even worse, jeopardizing human lives.

The above-mentioned monitoring techniques allow not only the identification of the failure cause but also, more importantly, if the capacitor life prediction model is known, the prediction of the capacitor health status at a given point in the future [4,5].

In order to develop the above-mentioned techniques, it is essential to know in detail the capacitors composition, equivalent circuit and failure mechanisms. Hence, the following sections will present the main capacitors technologies, giving a particular emphasis on capacitors commonly used in power converters.

The main capacitors diagnostic techniques developed to date are then reviewed in the following sections.

Finally, a summary, challenges and future research directions will be presented.

5.1 Capacitor technologies

In order to understand some of the fundamental aspects related to capacitors diagnostic techniques, it is essential to analyse these components in detail.

A capacitor is composed of two conductive materials, namely, plates, separated by an insulating material designated by dielectric, which prevents the conduction between the two plates. This structure allows the capacitor to store energy in the form of an electric field. When the capacitor is charged, (Figure 5.2), an

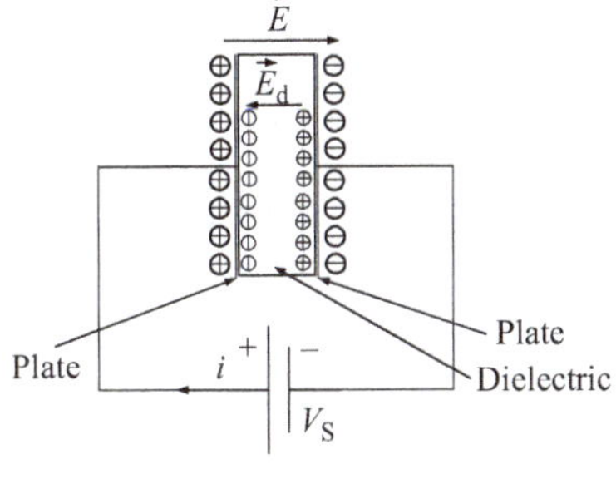

Figure 5.2 Charging a capacitor

electric field ($\vec{E}$) is created according to the polarity of the voltage source, V_S. However, the dielectric creates an electric field ($\vec{E}_d$) that opposes to $\vec{E}$; thus, it is necessary to store more charge in the plates so that the voltage, in the capacitor, equals the voltage of the source. The higher the dielectric constant of the dielectric (ε_r) is, the greater is $\vec{E}_d$, so more charge must be stored in the capacitor.

At this point, it is possible to define a fundamental quantity associated with capacitors: the capacitance (C), which represents the amount of charge (Q) that a capacitor can store:

$$C = \frac{Q}{V} \tag{5.1}$$

where V represents the voltage at capacitors terminals.

The maximum voltage value that the dielectric can withstand is designated by breakdown voltage (BV) and it represents the minimum voltage applied to the capacitor terminals, which makes conductive a small part of the dielectric.

The capacitor capacitance increases with the rise of the surface plate (S), since the capacitor has more space to store charge. On the other hand, if the thickness (d) of the dielectric decreases, the electric field $\vec{E}_d$ increases. Thus, it becomes necessary to store more charge in the plates so that the voltage, in the capacitor, equals the voltage of the source, which means that capacitance increases.

Therefore, the capacitance of the capacitor can also be defined as follows:

$$C = 8.855 \times 10^{-8} \times \varepsilon_r \times \frac{S}{d} \tag{5.2}$$

Capacitors block the flow of a direct current. However, during charge and discharge stages, there is an alternate current flow (i_c) that can be defined as:

$$i_c = C \times \frac{dv}{dt} \tag{5.3}$$

where $\left(\frac{dv}{dt}\right)$ represents the derivative of the capacitor voltage at its terminals.

Capacitors can be grouped in three main categories: electrolytic capacitors, film capacitors and ceramic capacitors. In turn, each category can be classified according to the dielectric, construction, terminal connection method, usage, coating and electrolyte [11]. The most common classification uses the dielectric type, since the amount of stored charges depends fundamentally on the dielectric characteristics, namely, the ε_r and BV.

Table 5.1 shows some dielectric proprieties of the above categories [11,12].

As previously verified, the dielectric properties define the capacitors behaviour. This is true for almost all capacitors; however, for very high-power capacitors, the conductivity of the plates, tabs and terminals can also define the capacitor peak current.

Table 5.1 clearly shows that different capacitors categories have different characteristics, so the capacitor selection depends on the characteristics of the

Table 5.1 Dielectric constants and minimum thickness for different capacitors categories [11,12]

Capacitors categories	Dielectric	Dielectric constant (ε_r)	Dielectric thickness (d)
Electrolyte	Aluminium oxide	[8, 10]	[0.03, 0.7] μm
	Tantalum oxide	[23, 27]	[0.04, 0.5] μm
Film	Polyester film	≅3.2	[0.5, 2] μm
Ceramic	Barium titanate	$[0.5, 20] \times 10^3$	[2, 3] μm
	Titanium oxide	[15, 250]	[2, 3] μm

circuit where it will operate. Some of the most important aspects to consider during the selection process are [11]:

- rated voltage, capacitance value (tolerance and stability), cost, volumetric efficiency;
- reliability and lifetime, energy density;
- ripple current rating and peak current;
- leakage current and insulation resistance, temperature range;
- resonance frequency and capacitor performance, namely, the capacitance dependency with frequency and temperature, as well as, capacitor internal resistance.

5.1.1 Electrolytic capacitors

Electrolytic capacitors can be divided in two main technologies, according to their dielectric type: aluminium electrolytic capacitors (Al-Caps) and tantalum electrolytic capacitors (Ta-Caps) [11,13].

Ta-Caps present very large temperature ratings [−55 °C, 175 °C], better temperature and frequency characteristics and a smaller internal resistance, when compared to Al-Caps, and offer the highest capacitance per unit volume among all types of capacitors, except supercapacitors [13,14]. On the other hand, Ta-Caps have smaller capacitance [47 nF to 2.2 mF] and voltage [2 V to 50 V] ratings are more expensive and the recoverability of the dielectric is not as good as in Al-Caps. This last characteristic can lead to failures, like short-circuits, since the dielectric is prone to defects [11,13–15]. In turn, these short-circuits can, in some cases, start a fire, becoming a critical issue in applications where a bank of capacitors is needed [15]. Ta-Caps are commonly used in audio-visual equipment, such as camcorders and cameras; in informatics equipment, such as laptop computers, tablets, wireless cards and universal serial bus (USB) port decoupling; in telecommunication equipment, such as smartphones and in some applications related to automotive and avionics industries, since their operating temperature can reach 175 °C [14,16]. However, their range of applications in power electronics is limited by the fact that their maximum operating voltage is less than 50 V.

Al-Caps have very high energy density, low cost per Joule, low bias dependency, very large capacitance range [0.1 μF, 3 F], large voltage ratings [5 V, 700 V] and large temperature ratings [−40 °C, 105 °C] [11,17]. However, this technology has high internal resistance and inductance, low current ripple, presents polarity, has relatively high leakage current, the temperature and frequency characteristics are poor, has weak solder heat resistance and some reliability problems [2,11–13,15,17,18]. These capacitors present wear problems that limit their life, leading in most cases, to open-circuit failures. On the other hand, there are reports that indicate that faults in Al-Caps, used in high-voltage banks of 10 or more units, triggered explosions [17]. Al-Caps can be found in the DC-link of power converters used in motor drives that control the motor speed in several applications, such as pumps, washing machines, central heating and air-conditioning systems, in industrial machinery, in electrical propulsion systems (trains and electrical/hybrid vehicles), but also in solar photovoltaic applications (solar inverters), UPS, in advanced technology extended (ATX) power supplies and motherboards, or in pulsed power applications, such as welding equipment, X-rays, and high-frequency plasma torches [2,3,19]. Both Al-Caps and Ta-Caps technologies have toxic materials, which can cause a disposal concern [15].

5.1.2 Film capacitors

Film capacitors (MK) are commonly classified on the basis of their dielectric type and construction. The most common dielectrics are polypropylene (PP), polyethylene terephthalate or polyester (PET), polyethylene naphthalate (PEN) and polyethylene sulphide (PPS) [20–23].

Both PEN and PPS technologies have the same dielectric constant at 1 kHz (3.0). However, PEN has a much higher dissipation factor, which translates in a higher internal resistance, when compared to PPS [21,22]. Despite its good dissipation factor (0.05% at 1 kHz), PPS material is scarce and expensive [21,24]. Both dielectrics are used in special applications where high temperatures are required (PEN up to 125 °C and PPS up to 150 °C) [20], although they do not meet the general requirements of power electronic applications, such as high self-healing proprieties and, simultaneously, low and stable dissipation factor [21,22].

The most used dielectrics in commercial MK capacitors are PET and PP technologies [20,25]. PET material has a high dielectric constant (3.3) and high dielectric strength (400 V/μm) [22], which provides the best power density (50 W/cm^3) of all MK capacitors [20–22,25], good stability with respect to temperature, and the lowest cost among MK capacitors [15,25]. Unfortunately, polyester capacitors (PET) are not the best option for power electronics applications, since their dissipation factor is high (0.5% at 1 kHz) and the self-healing proprieties are modest [15,21,24]. This capacitor is commonly used in applications with small bias DC voltages and/or small AC voltages at low frequencies [22].

Of the four dielectrics, PP capacitors are the most suitable for power electronic applications due to their very low dissipation factor (0.02%), excellent self-healing proprieties, high dielectric strength (600 V/μm) [15,21,22,24] and very good long-term stability [25]. This capacitor is commonly used in the DC-link, snubber circuits, power factor correction (PFC) and LC filters of power converters used in several systems such as motor drives, UPS, photovoltaic (PV) systems, switch mode power supplies (SMPS), electrical welding equipment, electronic ballasts, light-emitting diode (LED) systems, among others. PP capacitors are also used in smart energy meters due to their long-term stability, high insulation resistance and low dielectric absorption, which make them suitable for precision circuits and relatively immune to harsh climate conditions [26].

On the other hand, MK capacitors can be also classified according to their construction. One of the most important aspects of their construction is related to the capacitor's electrode system. MK capacitors present three basic electrode systems: metallized capacitors, film-foil capacitors and hybrid capacitors [24]. The first one presents the highest energy density, low inductance, low losses and self-healing proprieties. Nevertheless, the high pulse withstanding capability is lower when compared with the other two systems [24,27]. Film-foil electrodes ensure high insulation resistance, very good capacitance stability, low losses and excellent high pulse withstanding capabilities; however they do not present self-healing proprieties. Therefore, in order to improve their reliability, the dielectric requires higher thicknesses, which reduces capacitors volumetric efficiency and increases the cost [24,27]. The hybrid capacitors present the advantages of the above-mentioned technologies and are often classified as metallized capacitors [20,25,27].

5.1.3 Ceramic capacitors

Ceramic capacitors are the most used capacitors in electronics and, in this category, the multilayer ceramic capacitor (MLCC-Caps) is the dominant technology. MLCC-Caps can be found in many electronic devices such as cell phones and computers. Commercial available dielectrics can be subdivided into three classes: Class I, Class II and Class III [28].

Class I dielectrics, also denominated by temperature compensation dielectrics, use titanium oxide as dielectric, which has the lower dielectric constant of ceramic technology (Table 5.1). Thus, these capacitors are manufactured with small capacitance values (<100 nF) and with low volumetric efficiency. However, the capacitance is quite stable with voltage, temperature and time, and the dissipation factor is low [28–31], which makes them quite suitable in some power electronics circuits such as snubber circuits and soft-start circuits [29].

Class II dielectrics, also denominated by high dielectric constant's dielectrics, use barium titanate as dielectric which has a high dielectric constant (Table 5.1). Thus, these capacitors are manufactured with large capacitance values and small sizes [29,31], although the capacitance is unstable with voltage, temperature and time, besides the larger dissipation factor [28–30]. The specification of Class II dielectrics is based on a three-character alphanumeric code, in which the last digit defines the maximum capacitance deviation over a temperature range. For example, X5R, X7R and X8R are recommended for power electronics circuits since the capacitance has a maximum capacitance variation of $\pm 15\%$ with the temperature [28]. Both classes have higher dielectric resistance and lower dissipation factor when compared with Al-Caps. However, when compared with MK capacitors, both features are worse [23].

Class III dielectrics are the basis for barrier layer capacitors and present an extremely high capacitance and the best volumetric efficiency of the three classes [31]. Nevertheless, they have a very high dependence with temperature, voltage and frequency. Besides, their operating voltage is quite low (< 25 V) [28]. The Class III ceramic capacitors are commonly used in bypass coupling, where dielectric losses, high insulation resistance and stability are not required [31].

Capacitors with higher dielectric constant, such as class II and III, tend to have lower breakdown fields, which makes them more suitable for low-voltage applications, namely, when large capacitance is required. Another relevant aspect regarding ceramic capacitors is their failure during dielectric breakdown, which leads to the device useless [32].

Recently, a new ceramic capacitor technology has been developed by EPCOS CeraLink^{TM}, which can be well suited for power electronics applications. Recent studies show that this technology can be particularly useful in the DC-link of voltage source inverters due to its promising properties, such as low losses, increasing capacitance with the applied voltage, low series inductance and high capacitance density [33,34].

In summary:

- The best volumetric efficiency is found in Al-Caps, due to the chemically etched electrodes, that creates a very large surface, and to the thin dielectric films. In this way, it is possible to manufacture compact and cheap capacitors with large capacitance range. However, Al-Caps present very high dissipation factor, poor electrical characteristics and high leakage current. This last feature limits the maximum operating voltage to values lower than 1 kV [32].
- Metalized polypropylene film capacitors (MPPF-Caps) have the smallest dielectric constant beyond the three technologies addressed in this section, which makes them bulky and quite expensive for high capacitance values (above 1 μF). However, they present excellent electrical characteristics and a graceful failure. The self-healing characteristic is essential in high power capacitors, because the alternative is fire or explosion [32].
- MLCC-Caps have a wide range of ceramic formulations. Class I is stable with operating temperature, in turn, it has small capacitance. The other classes (II and III) present very high dielectric constant, so these capacitors have very high volumetric efficiency, however, large capacitances are only achieved for low voltage ratings. Furthermore, they present large instability regarding the operating frequency, temperature and voltage. These technical features limit their application in the power converters DC-link.

Currently, the most commonly used capacitors in the DC-link of power converters are Al-Caps and MPPF-Caps. The new technology proposed by EPCOS CeraLink^{TM} will not be addressed in this chapter, since it is quite recent and there are no studies related to its ageing process.

In the following sections, particular emphasis will be placed on the first two technologies.

5.2 Aluminium electrolytic capacitors

Al-Caps are composed of a wound element, impregnated with electrolyte, which is connected to the terminals and inserted into a can (Figure 5.3(a)). The element is composed of two aluminium foils (the anode foil and cathode foil) which are physically separated by means of paper separators saturated in electrolyte (Figure 5.3(b)) [11,12,17,18,35].

The foils are high-purity aluminium and are etched with billions of microscopic tunnels which increase extremely the surface of the foils. The dielectric is electrochemically formed in the surface of the anode foil forming a thin oxide layer (Al_2O_3) capable of withstanding the Al-Cap ratted voltage ([1.1, 1.5] nm/V) and showing a high insulation resistance ([10^8, 10^9] Ω/m) [35]. The cathode foil is also submitted to an etching process. However, the oxide layer (Al_2O_3) is formed spontaneously, which gives a withstand voltage of approximately 0.5 V [35].

The electrolyte, the real cathode, is an ion-conductive liquid that contacts with the dielectric and strongly influences the capacitor characteristics (temperature and frequency response, but also the service life) [12]. Its main functions are to adhere as closely as possible to the surface of both foils in order to extract the maximum possible capacitance and to repair the defects on the anode foil [35]. The electrolyte has a small portion of water that maintains the integrity of Al_2O_3. When the leakage current flows, the water is broken into hydrogen and oxygen by hydrolysis. The oxygen joins the anode foil forming Al_2O_3 in the defect regions, where the thickness of dielectric is too thin [17]. In this way, it corrects the defect.

The paper separators maintain the electrolyte distribution uniform and, simultaneously, keep the distance between both foils, avoiding a short-circuit between them [11,12,17,18,35].

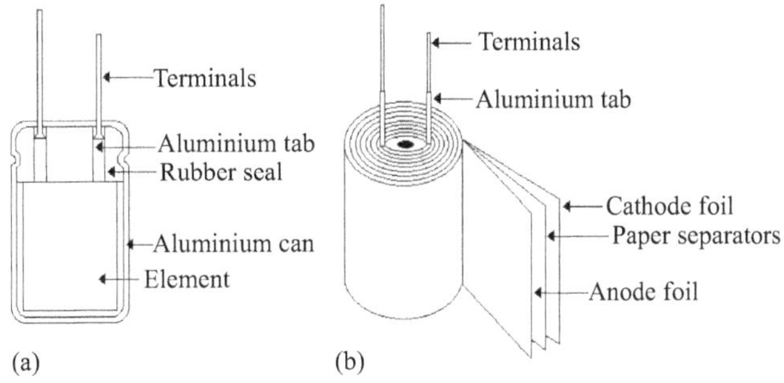

Figure 5.3 Basic structure of an Al-Cap: (a) the all structure and (b) inside the can [17,35]

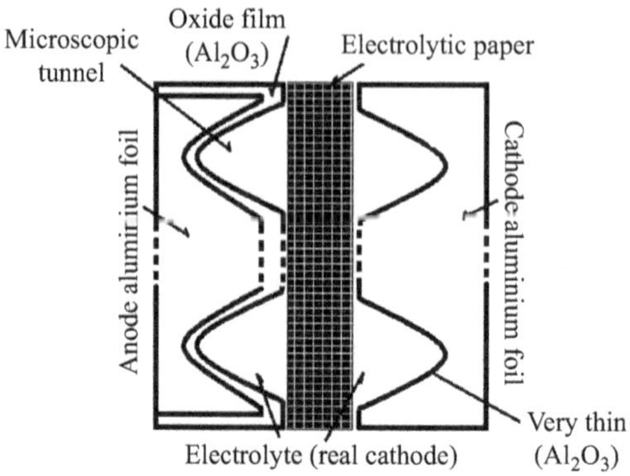

Figure 5.4 Basic composition of an Al-Cap

Figure 5.4 shows the basic composition of an Al-Cap. Actually, the Al-Cap is made of two internal capacitors:

- The anode-foil capacitor (C_A) which consists of an anode foil (the positive plate), the dielectric (capable of withstanding the rated voltage) and the electrolyte (the negative plate).
- The cathode-foil capacitor (C_C) which is composed of the electrolyte (the positive plate), the dielectric (much thinner than C_A dielectric and capable of withstanding a voltage lower than 0.5 V [35]) and the cathode foil (the negative plate).

Both capacitors are in series, however, since C_A has a denser dielectric and, so, lower capacitance, the Al-Cap capacitance (C) is nearly C_A:

$$C = \frac{C_A \times C_C}{C_A + C_C} \tag{5.4}$$

Al-Caps have a self-healing mechanism that is capable of repairing small defects in the dielectric. During dielectric formation, small gaps may arise, as well as regions where the dielectric thickness is narrow. Such regions are not able to withstand the rated voltage, allowing the flow of a current, the leakage current. This current is essential for the self-healing mechanism, because it transports the oxygen ions to the defective regions. Thus, the oxygen combines with the aluminium of the foil in order to form the dielectric (Al_2O_3) and repair the defects. During this process some hydrogen is released and it accumulates on the cathode side. Under normal operating conditions this process is useful, although it leads to the gradual accumulation of hydrogen on the cathode. However, if the capacitor operates under extreme conditions (overvoltage, very high-current ripple or very high temperatures), the damage in the dielectric film will increase as well as the leakage current, which accelerates the self-healing process, leading to the destruction of the capacitor [11].

If the opposite polarity is applied to the capacitor, a formation process will develop on the cathode foil, which increases its Al_2O_3 thickness and, so, a significant reduction of C_C will occur, which causes a significant reduction of C (5.4). This process will generate heat, as well as large amounts of gas at the expense of the electrolyte, which will lead to the destruction of the capacitor.

The gas generated in the processes described in previous paragraphs increases the internal pressure inside the can. In order to avoid explosions, the can is equipped with a vent mechanism which helps release the excess of pressure inside the capacitor.

5.2.1 Al-Caps equivalent circuit

Al-Caps present some parasitic elements that influence their behaviour. Figure 5.5 presents Al-Caps equivalent circuit during normal, overvoltage and reverse-voltage operation [17].

The series resistance, R_S, represents the terminal resistance, tab resistance, foil resistance, paper-electrolyte resistance and tunnel-electrolyte resistance, and decreases with the increase of temperature, as it will be demonstrated in the next paragraph [17,36,37]. R_S is inversely proportional to the capacitor capacitance for a given rated voltage [17].

The resistance of the terminals, tabs, and foils increases slightly with temperature, since the material is almost aluminium. However, it is practically unaffected by the operating frequency (<100 kHz). For higher frequencies, this resistance exhibits the skin effect, therefore it increases with frequency [37]. The paper-electrolyte resistance depends on the distance between both foils, the electrolyte resistance and a paper factor. This resistance decreases with temperature, due to the variations in ionic mobility in the electrolyte, which is affected by viscosity of the electrolyte; still, it is practically unaffected by the operating frequency. Most paper-electrolyte systems drop in resistance by 40% to 80%, as the temperature increases from 25 °C to 85 °C [37]. Thus, the conjugation of both factors allows to conclude that R_S decreases with temperature and it is practically unaffected by the frequency (<100 kHz). The effect of the temperature on the series resistance can be represented analytically as follows:

$$R_{\text{p-e}} = R_{\text{p-e base}} \times e^{\frac{T_{\text{base}} - T_{\text{core}}}{E}} \tag{5.5}$$

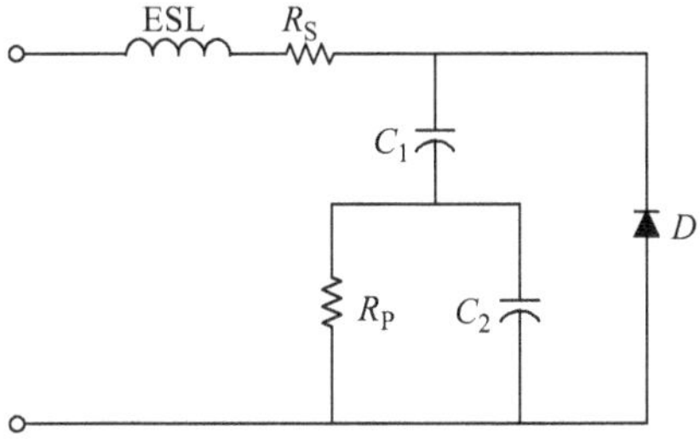

Figure 5.5 Al-Caps equivalent circuit during normal, overvoltage and reverse-voltage operation [17]

where $R_{\text{p-e}}$, $R_{\text{p-e base}}$, T_{core} and E represent the actual paper-electrolyte resistance, the paper-electrolyte resistance at base temperature (T_{base}), the core temperature and a temperature-sensitive factor, respectively [36].

The parallel resistance, R_{P}, accounts for the leakage current in the capacitor. The dielectric has a very high resistance that prevents the flow of direct current. Nevertheless, some regions present defects, which allow the circulation of a small leakage current when the capacitor is subjected to a high DC voltage. The capacitance C_2 represents the dielectric loss capacitance and the parallel combination of R_{P} and C_2 represents the dielectric resistance, R_{d} [36]:

$$R_{\text{d}} = \frac{R_{\text{P}}}{1 + (2 \times \pi \times f \times C_2 \times R_{\text{P}})^2} \tag{5.6}$$

R_{d} describes the energy losses in the alignment of dipoles in the dielectric, and the time it takes for the dipoles to become oriented. Its value increases in high rated capacitors since the dielectric thickness is greater [36].

The effective capacitance (C_1) increases with the rise in temperature and it decreases with increasing frequency [17,36]. The increase in temperature expands the electrolyte, so more surface of the dielectric is covered by the electrolyte (plate). Therefore, the capacitance increases. On the other hand, the dielectric constant of Al_2O_3 decreases with frequency. When the capacitor charges and discharges rapidly, the electrical field changes so quickly that the dielectric dipoles have some difficulty in following it in its fullness. This effect leads to a reduction in the capacitance [38].

The equivalent series inductance, ESL, comes from the tab loop configuration (the loop area from the terminals and tabs outside of the active winding) [36] and it is almost independent of frequency and temperature. Its value can vary between 10 nH (in radial-leaded capacitors) and a maximum of 200 nH (in axial-leaded capacitors) [17].

The Zener diode, D, models the overvoltage and reverse voltage behaviour of the capacitor [17].

Usually, a simpler equivalent circuit is used to represent the Al-Cap (Figure 5.6).

The equivalent series resistance (ESR) represents both the contribution of R_{S} and R_{d}, and it decreases with the increase in temperature and with the increase in frequency. Contrarily, it increases with the rated voltage and, typically, it can vary from 10 mΩ to 1 Ω [17].

5.2.2 Al-Caps failure modes

Al-Caps present two types of failures: catastrophic and parametric failures. The catastrophic failures lead to the destruction of the component through a short or an

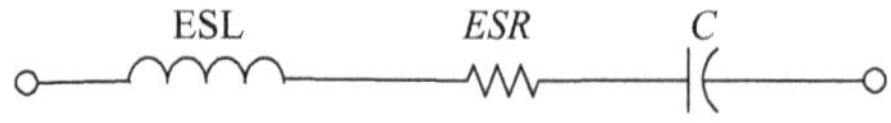

Figure 5.6 Simplified equivalent circuit of an Al-Cap

open-circuit. Thus, the capacitor loses completely its function. In the parametric failure, the capacitor does not lose completely its function; however, its electrical characteristics deteriorate (*ESR* increases and capacitance decreases). Depending on the power electronics application, it is considered a parametric failure when the change of *C* and *ESR* reaches a specific limit. This limit is defined by the required performance of the application. Usually capacitors manufacturers define the end-of-life limit of Al-Caps when the *ESR* doubles its value or the capacitance reduces in 20% when compared with its initial value [2,17]. Those requirements are in line with the Electronics Industries Alliance Interim Standard 479 [17].

Al-Cap failure rate follows the bathtub curve (Figure 5.7).

The period I represents the early failure period, being failures mainly due to deficient manufacturing processes, bad design or inadequate operating conditions, such as very high operating temperatures, very high current ripple, overvoltage or reverse voltage. The most common failure in period I is a short-circuit, due to the weakness in Al_2O_3 [17,35].

In the random failure period, period II in Figure 5.7, the failure rate is very small and Al-Caps show a smaller failure rate in this period than Ta-Caps or semiconductors [17,35]. The time duration of this period is directly related to the operating conditions.

In period III (Figure 5.7), the wear-out failure period, the failure rate increases significantly. During normal operation, the electrolyte will gradually evaporate due to the self-healing mechanism, which causes a reduction of *C* and an increase of the *ESR*. The electrolyte loss causes a reduction of water, so the electrolyte conductivity reduces, which means that *ESR* increases. On other hand, the electrolyte loss leads to a reduction of the surface area of the plates of C_A and C_C, therefore, reducing the overall capacitance.

Al-Caps present essentially six types of failure modes: open-circuit, short-circuit, opened vent, electrolyte leaking, increase of leakage current and, simultaneously, an increase of *ESR* and decrease of *C* [11,12,35].

The open-circuit may occur when one of the following conditions is met: mechanical damage to the lead conditions, corrosion due to the infiltration of a corrosive material, extreme operating conditions and the final stage of a degradation failure. The mechanical damage (lead wire and tab distorted or twisted) results

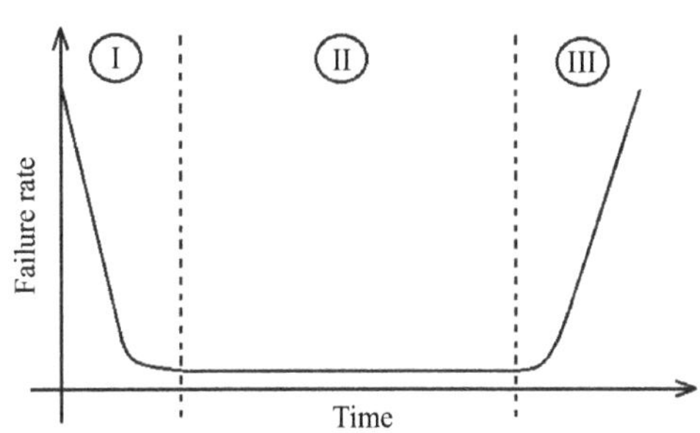

Figure 5.7 Al-Caps failure rate – bathtub curve

from bad connection during manufacture, an extreme force applied to the capacitor during mounting or an excessive stress applied during operation (vibration or impact). The corrosion may result from the cleaning action, in which corrosive products are used, which leads to the erosion of the tabs and foils, ultimately leading to an open-circuit failure. The extreme operating conditions such as reverse voltage, very high operating temperature, very high current ripple lead to a rapid evaporation of the electrolyte which, in the end, leads to an open-circuit. In the final stage of a degradation failure, the Al-Cap completely lose its electrolyte and, so, the two capacitors (C_A and C_C) are no longer electrically connected [11,12,35].

The short-circuit failure is rare; however, it may occur due to a short-circuit between the foils or to the insulation breakdown of the dielectric. The electrical contact between both foils is the result of some defects during manufacture, the application of an overvoltage during operation or to very high mechanical stress. The most common defects during manufacture are the presence of metal or other conductive particles on the electrode foils or weak points in the electrolyte-paper system, which lead to short-circuit during capacitor operation. If high mechanical stress is applied to the capacitor during its operation, the paper separators may damage, interfering with their functions (separation of both foils), and, consequently, leading to a short-circuit between anode and cathode foil. The insulation breakdown of Al_2O_3 is the result of defective oxide layer during manufacture [11,12,35].

The opened-vent failure will occur when the vent mechanism opens abruptly due to the enormous pressure inside the capacitor which, in turn, is due to the very high quantity of gas generated. This occurrence may arise when the capacitor is subject to extreme operating conditions, such as reverse voltage, overvoltage, very high current ripple, very high operating temperature or the use of an AC voltage. This failure mode may be the final result of a degradation failure, because during normal operation the capacitor gradually releases some gas at the cathode side [11,12,35].

The electrolyte leaking may result from defects in the vent mechanism during manufacture, deterioration of the vent mechanism during operation, extreme operating conditions, or the result of the final stage of a degradation failure [11,12,35].

The increase of leakage current is the result of deterioration of the oxide layer which may result from extreme operating conditions or degradation failure. The corrosion due to the infiltration of a corrosive material from the cleaning action may also contribute to the increase of the leakage current [11,12,35].

The simultaneous increase of *ESR* and decrease of *C* is the result of a degradation failure (parametric failure) and it represents the most common failure mode. This failure mode results from the natural ageing process of the capacitor and it does not represent a catastrophic failure as in the case of the first four failure modes previously discussed. In this case, the capacitor can keep operating; however, the probability of developing a catastrophic failure increases significantly. On the other hand, it should be noted that the electrical characteristics of the capacitor changed at this stage, which may affect the correct operation of the power electronics application where it is being used [11,12,35].

5.3 Metalized polypropylene film capacitors

MPPF-Caps are composed of two overlapping long polypropylene strips wound on an insulating mandrel. The polypropylene strips represent the low-loss dielectric formed by pure polypropylene film, which is vacuum metallized, usually on one side, with a thin self-healing mixture of zinc and aluminium that represents the capacitors plates. The aluminium is used because of its good bonding to polypropylene, its resistance to atmospheric corrosion and the ease in which it is applied in vacuum coated in thin layers. The thickness of the electrode layer is very small, and one of the layers extends out the top of the winding leaving a clear margin on the other side, while the other layer extends out to the bottom of the winding leaving a clear margin on the other side. The ends of the capacitor windings are contacted by spraying a metal contact layer, the sprayed end, which assures a reduced resistance and inductance. Once completed, a tab is soldered to the sprayed end (Figure 5.8). This construction has a very high current-carrying capability [39–43].

The electrical-field stress in MPPF-Caps may be much larger than that in film foil MK capacitors, due to the capability of electrodes to self-heal. The metallic film has a very thin thickness, thus, if a breakdown occurs in the dielectric, the electrodes in that region are destroyed due to the heavy transient current. In this way, the defect region is isolated from the rest of the capacitor. This process is called the self-healing propriety [20,41]. In order to reliably achieve the self-healing mechanism, the thickness of the metallized electrodes must be very small, which makes them susceptible to corrosion. On the other hand, a thick metallized layer, in case of a breakdown, will produce greater energy dissipation because the layer resistance is lower, leading to a greater damage [20]. In this way, there is a minimum thickness limit that must not be exceeded, which limits the self-healing property [42].

In order to overcome the problems described in the previous paragraph, the segmented metallization was introduced. This solution limits the size of the damage, and in addition, manages to take full advantage of the self-healing

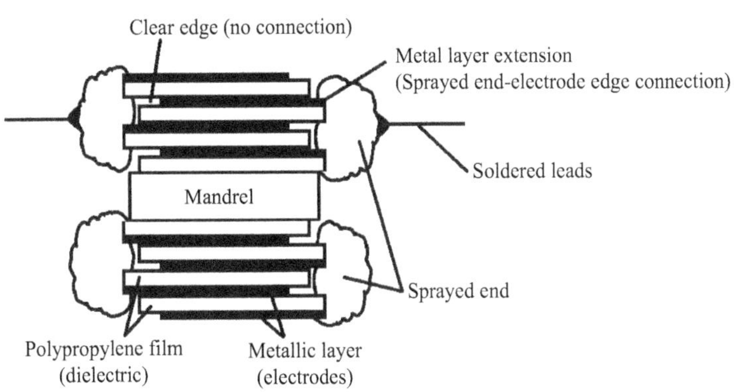

Figure 5.8 Cross-sectional diagram of a cylindrical MPPF-Cap

property. The segmentation increases slightly the capacitor impedance but, in return, improves the self-healing properties. In T-segmentation (Figure 5.9), each individual segment of the electrode is interconnected by narrow current gates that serve as fuses in the case of a breakdown in one of the segments. In this way, it is possible to isolate the failed segment or segments from the rest of the electrode. This structure can be compared to a thousand or millions of little capacitors connected in parallel. In this way, when a breakdown occurs, the segment is isolated and, as a consequence, the capacitor capacitance decreases.

The segmentation reduces the possibility of a short-circuit and simultaneously reduces the amount of heat generated during a breakdown, making the capacitor more reliable. In this way, it is possible to construct MPPF-Caps with higher rated voltages [39–43].

The gaps in the winding are impregnated with degradable plant oil in order to prevent glow discharges and electrode corrosion. This procedure improves the capacitor lifetime and improves the capacitance stability [39].

5.3.1 MPPF-Caps equivalent circuit

MPPF-Caps present some parasitic elements that must be included in their equivalent circuit (Figure 5.10).

ESL represents the equivalent series inductance and results from the magnetic field created by the current that passes through the electrodes, sprayed end, tabs and leads. Thus, ESL depends on the winding structure, geometric design and lengths, and thickness of the contact paths. Its typical value is generally very small, presenting a maximum value of 1 nH per mm of lead length and capacitor length [43,44].

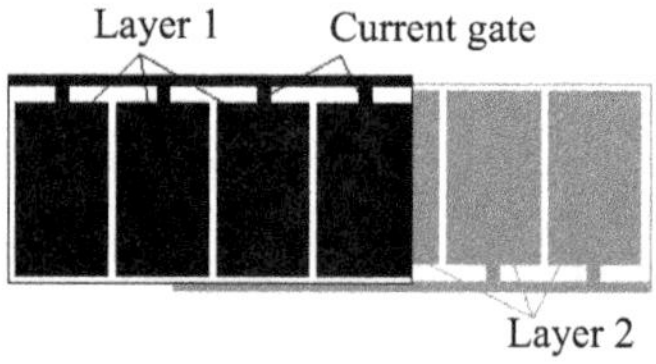

Figure 5.9 T-segmentation of the electrodes in a MPPF-Cap

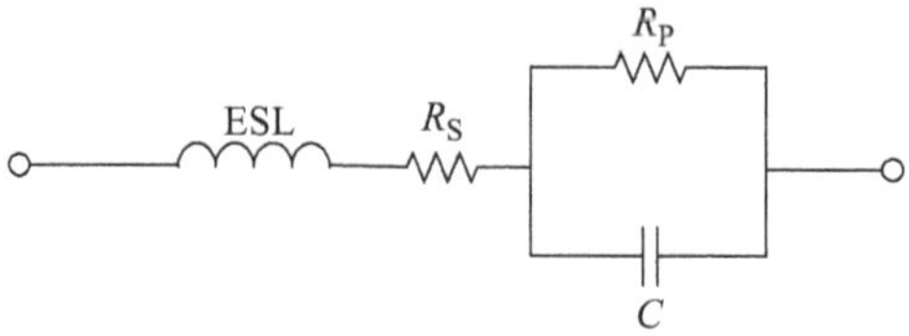

Figure 5.10 Equivalent circuit of MPPF-Caps [43,44]

The capacitance, C, represents the amount of charge that the capacitor can store per unit of voltage. The value of C may vary depending on several factors, namely, temperature, humidity and frequency.

The typical variations of MPPF-Caps capacitance with temperature, humidity, and frequency are shown below:

- MPPF-Caps capacitance can vary between +2% and −2% for a temperature range between −60 °C and 100 °C [44].
- MPPF-Caps capacitance can vary between −0.25% and +0.25% for a relative humidity range between 10% and 85% [44].
- MPPF-Caps capacitance can vary between 0% and 1% for frequency range between 1 kHz and 1 MHz [44].

MPPF-Caps capacitance does not vary significantly with temperature, humidity and frequency for the range of values previously presented, unlike PET and PEN technologies [44].

It should be mentioned that, with usage, MPPF-Caps capacitance decreases, due to the self-healing property.

R_S and R_P represent the series resistance and parallel resistance, respectively. R_S results from the combination of the dielectric resistance, contact resistance between the sprayed end and the electrode layer, the leads resistance, the metal layer resistance and sprayed end resistance, while R_P is due to the insulation resistance [43,44].

The effect of both resistances (R_P and R_S) can be modelled by a single resistance: *ESR*. Thus, the model shown in Figure 5.6 can also be applied to MPPF-Caps. It will be this simplified model that will be used from now on.

MK capacitor *ESR* can vary with temperature, humidity and frequency, as can be seen below [44]:

- In MPPF-Caps, the *ESR* value does not vary significantly with temperature, unlike PET capacitors.
- *ESR* value increases with humidity.
- *ESR* value changes with frequency in the following manner:
  - At low frequencies, the *ESR* is inversely proportional to the frequency $\left(\frac{1}{f}\right)$. The effect of the dielectric losses is predominant.
  - At medium frequencies, the *ESR* is approximately constant. The effect of the conductor losses is prominent.
  - And finally, at very high frequencies, the skin effect arises, so that *ESR* increases by $\sqrt{f}$.

Another important characteristic of this capacitor is the dissipation factor, *DF* or tangδ, which represents the ratio between the *ESR* and the capacitive reactance, X_C:

$$DF = \text{tang}\delta = \frac{ESR}{X_C} = ESR \times 2 \times \pi \times f \times C \tag{5.7}$$

In the previous definition, ESL was neglected. Thus, for frequencies lower than the capacitor resonant frequency, it is possible to subdivide *DF* into three different components [44]:

- The parallel component, DF_P, which results from the insulation resistance. This component can be neglected due to the enormous value of the insulation resistance.
- The dielectric component, DF_D, which represents a measurement of the losses associated with the dielectric. In other words, it represents the energy lost in polarizing and re-polarizing the dielectric in two opposite directions. The value of DF_D in MPPF-Caps remains constant with frequency and it is approximately equal to 10^{-4} [44].
- The series component, DF_S, results from the sum of the contact resistance, the leads resistance, sprayed end resistance and electrode foils resistance. The value of DF_S increases with frequency becoming dominant at high frequencies; it also increases with the capacitor capacitance.

The *DF* value does not vary significantly with temperature in MPPF-Caps, unlike PET capacitors.

5.3.2 MPPF-Caps failure modes

MPPF-Caps are very reliable capacitors and present a much larger lifetime than Al-Caps. However, they are not immune to failures. Like Al-Caps, these capacitors can present two types of failures: catastrophic and parametric failures [2,25]. Catastrophic failures lead to the destruction of the component and the capacitor loses completely its function and, under certain circumstances, can cause fire or even explosions [45,46]. In the parametric failures, the capacitor electrical characteristics deteriorate and its capacitance decreases, however, it does not lose completely its function. Typically, capacitor manufacturers define the life limit of MPPF-Caps when the capacitance decreases a certain amount when compared with its initial value (usually from 2% to 10%) or *DF* increases a certain amount (typically twice) from the initial value [2,46–49]. The maximum capacitance loss that defines the capacitor life limit differs between manufacturers, but typically lies between 2% and 10% [2,46–49].

MPPF-Caps failure rate follows the bathtub curve (Figure 5.7) [20,25]. During period I, failures are mainly due to bad design or deficient manufacturing processes. In a desirable situation, these failures should not be observed by the end user. However, if they are not detected by the manufacturer, they must be covered by the warranty [20]. In period II, the failure rate is very small and its time duration is directly related to the operating conditions. Finally, in period III, the failure rate increases significantly. Usually, capacitors that manage to overpass periods I and II without failures, manifest a parametric failure in period III, which translates into the gradual reduction of the capacitance, due to the self-healing property. Under this circumstance, the capacitor replacement should be defined by the following reasons:

- The required performance of the application, which is defined by the minimum permissible capacitance for the circuit to operate correctly.
- The maximum admissible capacitance loss. According to some manufacturers, the life limit of MPPF-Caps is reached when 2% to 10% of the initial capacitance is lost [2,46–49]. However, the capacitor can still operate. From this point, the ageing process accelerates through a faster loss of capacitance. Under these circumstances, the capacitor may lose more than 10% of its capacitance, without revealing a catastrophic failure. It should be mentioned that in this last period, in which the capacitance loss is notorious, there will be an increment in hydrocarbon-based gas production and, thus, it is advised the capacitor replacement to avoid any possibility of a very severe consequence, such as a fire or an explosion [46].

MPPF-Caps present essentially four types of failure modes: full short-circuit, resistive short-circuit, open-circuit, and wear out [20,45,50]. The causes of such failures may be due to a bad design, a deficient manufacturing process, inadequate operating conditions or simple wear out [20].

During design phase, some of the following causes can lead to failures: the dielectric is too thin, the metallization layer is too thick or too thin, or the conductor size is inadequate.

If the dielectric is too thin, dielectric breakdowns might occur during capacitor operation, which may cause failures in the foil-electrode-tab structure and, as a consequence, a short-circuit failure might happen. On the other hand, dielectric breakdowns can lead to a continuous self-healing process, which will melt and carbonize the film by discharging energy and, as a result of this process, a resistive short-circuit will arise [50].

The resistive short-circuit can be extremely dangerous, since it can produce chimneys of melted polypropylene in the winding. These chimneys represent a conductive channel to the leakage current, so, when this current passes through these channels, it will produce a significant increase of the internal temperature of the capacitor, causing more film to melt, as well as the increase of the pressure inside the capacitor. This process can become cyclic, leading to an extremely severe condition such as fire or even explosions [20].

If the metallization layer is too thick or too thin, the self-healing process might no longer be effective. In turn, it will damage the dielectric film and generate heat that will be transmitted to the next film layer. As a consequence, the dielectric strength will decrease and dielectric breakdowns may arise. As a result, the same failure modes described in the previous paragraphs might occur [20,45].

As stated in Section 5.1, for very high-power capacitors, the conductivity of the terminals can decisively influence the capacitor performance. So, in the case of an inadequate conductor size, the Joule losses may increase; as a result, the capacitor internal temperature will increase, and so the dielectric strength will decrease. In this way, the same failure modes described in the previous paragraph might occur [20].

During manufacturing production some of the following causes can lead to failures: production in a dirty environment, bad space factor control of the dielectric films during winding operation, bad drying, bad sealing, or uncontrolled soldering process [20,45].

The capacitor must be manufactured in a very clean environment to avoid contamination with ionic species, which might stimulate corrosion of the metal film. On the other hand, electrodes corrosion will contribute to the increase of the *ESR*, since the electrodes thickness decrease. Thus, the internal temperature of the capacitor increases and so the dielectric strength will decrease, which might cause one of the first three failure modes [20,45]. A similar problem may occur in the case of bad space factor control of the dielectric films during winding operation, because this situation cause the electrode destruction by corona discharge [20].

MPPF-Caps are sensitive to moisture/humidity exposure, thus bad drying or bad sealing can affect the capacitor lifetime. The presence of humidity inside the capacitor can lead to the following effects: electrode corrosion, corona effect or reduction of the insulation resistance [20,45]. The effect of corrosion was explained in the previous paragraph. The corona discharges on the electrodes edges will lead to fast reduction in the capacitance, and might separate the electrode from the sprayed end [20]. On the other hand, the reduction of insulation resistance increases the leakage current. Therefore, it is possible to conclude that the presence of moisture in the capacitor can cause one of the three first failure modes [20,45].

If the soldering process is not well controlled, it is possible for corrosive materials/ionic materials to propagate through the electrode, which in turns might corrode the metal film resulting in the same type of failures described in the previous paragraph [45].

Under inadequate operating conditions, some of the following causes can lead to failures: overvoltage and/or high-pulse voltage, overcurrent, high-temperature environment, high-humidity environment and shock or vibration [20,45,46,50].

If the capacitor is subjected to an overvoltage and/or high-pulse voltage, dielectric breakdowns might occur, which may cause a full or resistive short-circuit [50].

Very high currents will lead to self-heating, and so, the dielectric strength will decrease. This same effect will occur if the capacitor is subjected to high temperatures. Both situations described above may lead to one of the three first failure modes [20,50].

As previously mentioned, very high humidity environments cause electrochemical corrosion and, if it occurs near the metal film, it might origin an open-circuit failure [20,45,50].

The combined effect of electrical, thermal and mechanical stresses can cause the detaching of the 'sprayed ends' from the capacitor roll. This failure is commonly found in pulse power applications and manifests, at an early stage, by an increase in the capacitor *ESR*, which ultimately results in an open-circuit [46].

It should be noted that some of the causes previously discussed may indirectly lead to other catastrophic failures [20].

Usually, most of the failure causes presented in the last paragraph give rise to a catastrophic failure in the first two periods of the bathtub curve. However, if the capacitor does not present the above-mentioned problems (bad design, a deficient manufacturing process or an inadequate operating conditions), its lifetime is quite large, which means that the capacitor can reach the wear-out failure period. In this case, the most common ageing process results from multiple self-healing cycles during normal operation. As a result, both leakage current and power losses increase and the capacitor capacitance decreases [46–49,51]. If the capacitor is not replaced in due time, the described phenomena can lead to catastrophic failures as it can be concluded from the previous paragraphs. Under certain circumstances, these catastrophic failures might lead to very severe consequences such as fire or even explosions.

In this way, the development of capacitors fault diagnostic techniques is essential to anticipate the occurrence of these consequences, which can lead to very high costs, or even worse, put human lives in danger.

5.4 Fault diagnostic techniques

As mentioned in the previous sections, the capacitors commonly used in the DC-link of power converters can manifest two types of faults: catastrophic failures or parametric failures.

Catastrophic failures may result from manufacturing defects and/or improper use, or are simply the result of the natural ageing process. Catastrophic failures associated with manufacturing defects usually manifest themselves in the early failure period, should not be perceived by the end user, and must be covered by the warranty. On the other hand, if the end user does not comply with the manufacturer's application guidelines, catastrophic failures may suddenly occur, shortening the random failure period.

Fault diagnostic techniques mainly seek to prevent catastrophic breakdowns resulting from a natural ageing process, which are the result of a parametric failure that was not detected in due time. These parametric failures manifest themselves through electrical and physical modifications, namely, in the internal resistance, capacitance and volume of the capacitor. In this way, it is possible to conclude that those capacitor parameters are fundamental to define the capacitor end-of-life criteria, such as the following ones:

- Al-Caps – loss of 20% of the initial capacitance or doubling the initial *ESR* value [2,17].
- MPPF-Caps – loss of 2% to 10% of the initial capacitance [2,46–49].

The above widely used criteria define a threshold condition, from which the ageing process of the capacitor accelerates. On the other hand, some applications may cease to operate in the most appropriate way when the indicated conditions are reached. Therefore, the above-mentioned criteria permit, at any time, to decide if capacitor should be replaced, the recognition of the capacitor degradation level or to estimate the capacitor remaining useful life [3]. Consequently, the majority of capacitors fault diagnostic techniques are based on the identification of the capacitor *ESR* and *C*.

In some applications, it may be necessary to use other criteria, based on the performance of the converter. Indeed, capacitors' ageing increases the capacitor *ESR*, which in turn, leads to an increase of the voltage ripple in the DC-link. The increase of the ripple can affect the converter efficiency or even damage other components. For instance, in solar photovoltaic inverters, the increase of the DC-link voltage ripple decreases the extracted power and it can simultaneously damage the semiconductor switches due to over-voltage [52]. In these cases, it is necessary to introduce a criterion that protects the converter, such as, a maximum voltage ripple limit.

Capacitors fault diagnostic techniques can be subdivided into three main types: off-line, on-line and quasi-online.

Off-line techniques require the removal of the capacitor from the converter, so that *ESR* and *C* values can be estimated. The major disadvantage of these techniques is the need to interrupt the converter operation; however, this can also be considered as an advantage, because they can be used for any circuit, unlike on-line and quasi-online techniques that are designed for a particular application. Thus, in addition to being universal, they can be used in applications where there are no other fault diagnostic techniques available. Furthermore, it is also possible to enumerate other advantages, such as they are usually not very expensive, nor do require the introduction of sensors in the converter; they are extremely precise, simple to apply and their accuracy is not affected by the converter operating conditions (temperature, frequency, voltage, current, humidity, etc.).

On-line techniques are designed for specific applications (DC–DC converters, adjustable speed drives, UPS, etc.) and do not require the converter shutdown. Therefore, the values of *ESR* and *C* can be determined without the stoppage of the converter, which is fundamental in applications whose operation cannot be interrupted. However, these techniques may exhibit some of the following disadvantages: they are invasive (require the introduction of sensors inside of the converter, which can affect its operation), they are complex (the estimation of *ESR* and *C* values should consider the converter operating conditions) and they are costly. (Some techniques require the use of additional high-speed A/D converters.)

Quasi-online techniques do not require the capacitor removal from the converter, however, the measurements are usually taken during a routine pause in the application. These techniques involve the injection of an external signal and/or impose a special working configuration. Therefore, quasi-online techniques can only be used in applications where such operating conditions are allowed. For example, at night for solar photovoltaic systems, in AC–DC–AC PWM converters during no-load condition or just before the switching *on-off* of the inverter used to control the electrical machine speed [52]. These techniques present some of the disadvantages of the on-line techniques. However, they are not so dependent on the converter operating conditions.

5.5 Off-line measurement techniques

The characterization of the capacitors equivalent circuit (*ESR* and *C*) proves to be essential not only for fault diagnosis but also during power converters design. For example, in the design of switched mode power supplies, exact knowledge of the capacitor equivalent circuit (*ESR* and *C*) is essential since both parameters influence the dynamic behaviour of these circuits. On the other hand, this information (*ESR* and *C*) is also relevant in the selection of capacitors used in the DC-link of variable frequency drives.

In order to estimate the capacitors equivalent circuit it is common to use very expensive equipment, such as LCR meters and impedance gain-phase analysers. As an alternative to that equipment, which is capable of characterizing the equivalent circuit of different electronic components, some considerably cheaper off-line measurement techniques have been proposed.

Indeed, it is possible to find in the literature some off-line techniques that characterize the capacitors equivalent circuit. However, some of these techniques cannot be used for the capacitors under analysis (Al-Caps and MPPF-Caps), due to some of the following factors: those techniques were designed for small capacitance capacitors, the frequency range of the measurement techniques is limited, the measuring circuits require standard components (namely, capacitors, inductors and high precision resistors) or alternatively they require extremely expensive equipment such as network analysers [53].

In this way, it was found necessary to develop off-line measurement techniques for the capacitors under analysis. These techniques are capable of estimating both *ESR* and *C*, with great precision, under specific operating conditions. Since the measurements can be always carried out under the same operating conditions (temperature, frequency, voltage, current, humidity, etc.), it is not necessary to consider those effects (unlike on-line and quasi-online techniques), making the off-line techniques simpler and less prone to errors. Another important feature of these techniques is that they allow the calculation of the frequency and temperature multipliers that are vital for the design of on-line and quasi-online fault diagnosis techniques.

Off-line measurement techniques demand the removal of the capacitor from the original circuit (power converter), so that it can be placed in a test circuit. The test circuits shall generate adequate currents and voltages on the device under test (DUT), in order to be possible to characterize its equivalent circuit.

Test circuits can be subdivided into two types: circuits capable of generating a sinusoidal current waveform, and charging/discharging circuits. In addition to the two different test circuits, different mathematical algorithms can be used to characterize the DUT. These algorithms will be presented according to the used test circuits.

5.5.1 *Off-line measurement techniques based on the injection of a sinusoidal current*

The off-line measurement technique (off-line MT) proposed in [54,55] is able to estimate the *ESR* value of Al-Caps by injecting a sinusoidal current into an LC filter similar to the one used in step-down DC–DC converters (Figure 5.11).

If the frequency of the injected current is much higher than the resonant frequency of the LC filter and, simultaneously, lower than the capacitor resonance frequency, it is possible to estimate the *ESR* value using one of the following equations [54,55]:

$$ESR \cong \frac{\sqrt{\left(\frac{V_{out} \times L \times C \times \omega}{V_{in}}\right)^2 - \frac{1}{\omega^2}}}{C} \quad \text{or}$$

$$ESR \cong \sqrt{\frac{\left(\frac{V_{out}}{V_{in}}\right)^2 \times \left(\omega \times L - \frac{1}{\omega \times C}\right)^2 - \frac{1}{(\omega \times C)^2}}{1 - \left(\frac{V_{out}}{V_{in}}\right)^2}} \tag{5.8}$$

where V_{in}, V_{out}, L, C and ω represent the amplitude of the input voltage, the amplitude of the output voltage, the inductor inductance, the capacitor capacitance and the angular frequency, respectively.

The accuracy of the previous technique is dependent on several factors, namely, the tolerance of L and C values provided by manufacturers, the capacitance loss of the aged capacitors, and the measurement frequencies. Moreover, it does not estimate the capacitor capacitance, and the *ESR* value can only be computed for a limited range of frequencies. For that reason, a new technique was proposed in [56].

The off-line MT proposed in [56] is able to estimate both *ESR* and reactance values of capacitors through the injection of a sinusoidal current, i_c, into an RC filter (Figure 5.12).

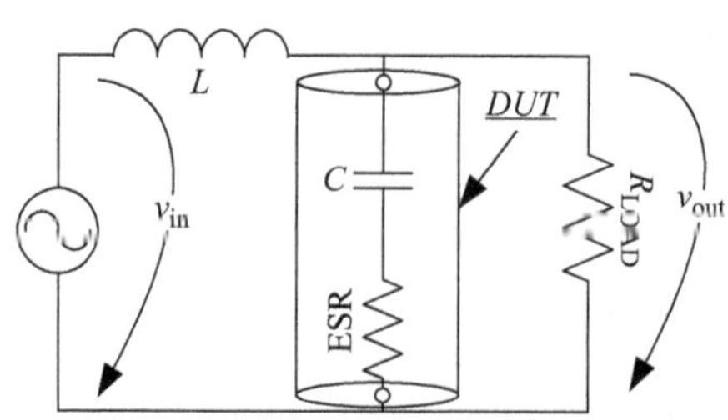

Figure 5.11 LC filter used for ESR estimation [54,55]

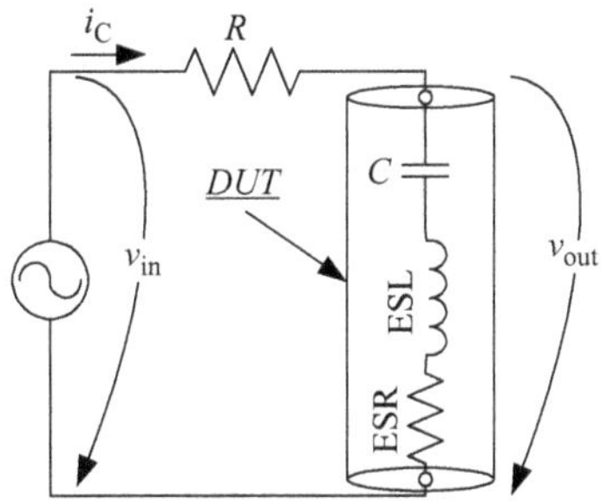

Figure 5.12 RC filter used for ESR and X_{cond} estimation [56]

This technique uses the simplified capacitor model shown in Figure 5.6, so the inductive effect of ESL is taken into account. In this way, the capacitor impedance, Z_{cond}, can be defined as follows:

$$\begin{cases} X_{ESL} = \omega \times L = 2 \times \pi \times f \times L \\ X_C = \dfrac{1}{\omega \times C} = \dfrac{1}{2 \times \pi \times f \times C} \\ Z_{cond} = ESR + jX_{cond} = ESR + j(X_{ESL} - X_C) \end{cases} \tag{5.9}$$

where X_{cond}, X_C, X_{ESL} and f represent the capacitor reactance, the capacitive reactance, the inductive reactance and the frequency, respectively.

Then, it is necessary to estimate both ESR and X_{cond} values. For this purpose, different solutions can be proposed.

The technique proposed in [56] uses the input (v_{in}) and output voltage (v_{out}) waveforms. Through the analysis of the circuit of Figure 5.12, it is possible to write the following relationship:

$$\frac{v_{out}}{v_{in}} = \frac{Z_{cond}}{Z_{cond} + R} = \frac{ESR + jX_{cond}}{R + ESR + jX_{cond}} \tag{5.10}$$

If the value of $|Z_{cond}| \ll R$, then the above equation can be simplified as follows:

$$\frac{v_{out}}{v_{in}} \cong K_X + jK_Y = \left(\frac{ESR}{R}\right) + j\left(\frac{X_{cond}}{R}\right) \tag{5.11}$$

The technique proposed in [56] has a major drawback when the test frequency is small. In these circumstances, the value of Z_{cond} increases significantly due to the effect of X_{cond}, thus, the R value must increase significantly, in order to satisfy the condition $|Z_{cond}| \ll R$. In this case, the method becomes imprecise because the value of v_{out} decreases significantly. Consequently, this technique should not be used for small frequencies.

In order to overcome the aforementioned problem, the previous simplification cannot be used. Thus, after multiplying the numerator and the denominator of (5.10) by the denominator conjugate, it is possible to write the following equation:

$$\frac{v_{\text{out}}}{v_{\text{in}}} = K_X + jK_Y = \frac{ESR \times R + ESR^2 + X_{\text{cond}}^2}{(R + ESR)^2 + X_{\text{cond}}^2} + j\frac{X_{\text{cond}} \times R}{(R + ESR)^2 + X_{\text{cond}}^2} \tag{5.12}$$

Through the analysis of (5.12), it is possible to conclude that *ESR* and X_{cond} must be computed through a system of non-linear equations. In order to solve the mentioned system, it is necessary to use a numerical method. For this purpose, the Newton–Raphson method was used in [57]:

$$\begin{aligned}
ESR_{i+1} &= ESR_i - \frac{u_i \times \dfrac{dv_i}{dX_{\text{cond}}} - v_i \times \dfrac{du_i}{dX_{\text{cond}}}}{\dfrac{du_i}{dESR} \times \dfrac{dv_i}{dX_{\text{cond}}} - \dfrac{du_i}{dX_{\text{cond}}} \times \dfrac{dv_i}{dESR}}, \\
X_{\text{cond}\ i+1} &= X_{\text{cond}\ i} + \frac{u_i \times \dfrac{dv_i}{dESR} - v_i \times \dfrac{du_i}{dESR}}{\dfrac{du_i}{dESR} \times \dfrac{dv_i}{dX_{\text{cond}}} - \dfrac{du_i}{dX_{\text{cond}}} \times \dfrac{dv_i}{dESR}} \\
u_i &= ESR \times R \times (1 - 2 \times K_X) + ESR^2 \times (1 - K_X) \\
&\quad + X_{\text{cond}}^2 \times (1 - K_X) + R^2 \times (-K_X) \\
v_i &= -X_{\text{cond}}^2 \times K_Y + X_{\text{cond}} \times R - ESR^2 \times K_Y \\
&\quad - ESR \times R \times (2 \times K_Y) - R^2 \times K_Y
\end{aligned} \tag{5.13}$$

The application of this method is quite simple, in a first moment, being necessary to obtain the functions $u_i(ESR, X_{\text{cond}})$ and $v_i(ESR, X_{\text{cond}})$, its derivatives $\left(\frac{du_i}{dESR}, \frac{du_i}{X_{\text{cond}}}, \frac{dv_i}{dESR}, \frac{dv_i}{X_{\text{cond}}}\right)$ and the initial guess (ESR_0 and $X_{\text{cond}0}$). The initial guess can be provided by the manufacturer datasheet. Although inaccurate, it does not affect the convergence of this method. Afterwards, an iterative process starts, with $i = 0$ and ends when the following conditions are met: $ESR_{i+1} = ESR_i$ and $X_{\text{cond}\ i+1} = X_{\text{cond}\ i}$, where *i* represents the iteration number.

The technique proposed in [57] is quite accurate for a large range of frequencies. Alternatively, the capacitor current (i_C) and voltage (v_{out}) waveforms can be used; if so, it would be necessary to introduce a current sensor in the circuit [58]. In this case, the computational effort decreases because it is not necessary the use of Newton–Raphson method:

$$\frac{v_{\text{out}}}{i_C} = K_X + jK_Y = ESR + jX_{\text{cond}} = ESR + j\left(\omega \times L - \frac{1}{\omega \times C}\right) \tag{5.14}$$

However, in order to implement the previous off-line MT, it is crucial to extract some data from experimental waveforms (v_{in}, v_{out} or i_C), namely, the values of V_{in}, V_{out}, K_x and K_y. These data can be obtained through manual or automatic processes.

For the methods presented in [54,55], it is essential to compute the amplitude of both input and output voltage. These data can be easily obtained by observing the oscilloscope or, in the case of digital oscilloscopes, it is automatically available.

For the methods proposed in [56–58] to be feasible, it is necessary to extract the values of K_x and K_y from the experimental waveforms. For this purpose, manual and automatic methodologies can be used.

In the following, the manual method will be presented [59]. The experimental waveforms are both sinusoidal (v_{out} and v_{in} or v_{out} and i_C). Thus, to compute K_x and K_y values, it is necessary to determine both amplitudes (A_2 and A_1) and the phase displacement (ϕ):

$$\frac{\text{signal}_2}{\text{signal}_1} = \frac{A_2 \times \sin(\omega \times t + \phi)}{A_1 \times \sin(\omega \times t)}$$
$$\Rightarrow \begin{cases} \dfrac{\text{real}(\text{signal}_2)}{\text{signal}_1} = \dfrac{A_2 \times \cos(\phi)}{A_1} = K_X \\ \dfrac{\text{imag}(\text{signal}_2)}{\text{signal}_1} = \dfrac{A_2 \times \sin(\phi)}{A_1} = K_Y \end{cases} \tag{5.15}$$

The representation of both waveforms in *X-Y* mode is an ellipse (Figure 5.13(a)), wherein the slope of the ellipse represents K_x and the relationship between both ellipse axes represents K_y.

In this way, it is possible to compute the values of K_x and K_y after visual inspection of the graphs (Figure 5.13), as follows:

$$K_X = \phi = \frac{Y_1 + |Y_2|}{X_1 + |X_2|} \quad \text{and} \quad K_Y = \frac{b}{a} \tag{5.16}$$

Figure 5.13(b) can be easily obtained after the removal of the real component of the signal$_2$; in this way, it is possible to represent signal$_{2y}$ as:

$$\text{signal}_{2Y} = \text{signal}_2 - K_X \times \text{signal}_1 \tag{5.17}$$

However, the manual methodology may become inaccurate when the ellipse slope becomes too small, due to user measurement errors. Thus, as an alternative to the previous process, some automatic methods can be used for computing K_x and K_y. In this chapter, two different methodologies will be presented, one based on the discrete Fourier transform (DFT) and the other on sinusoidal fitting.

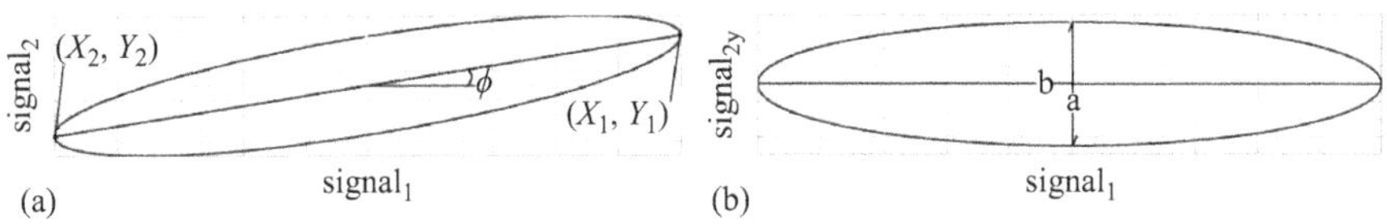

Figure 5.13 Representation of the experimental waveforms in X-Y mode: (a) v_{out} as function of (v_{in} or i_C) and (b) the imaginary component v_{out} as function of (v_{in} or i_C)

Through the use of DFT, it is possible to compute both modulus and phase of the experimental waveforms using the first harmonic, since these signals are essentially sinusoidal. The modulus (*M*) and phase (ϕ) of the experimental waveforms can be computed as follows:

$$M = \sqrt{\left(\frac{2}{NPP} \times \sum_{i=1}^{\mathrm{NPP}} \mathrm{signal}(i) \times \cos(t(i) \times \omega)\right)^2 + \left(\frac{2}{NPP} \times \sum_{i=1}^{\mathrm{NPP}} \mathrm{signal}(i) \times \sin(t(i) \times \omega)\right)^2}$$

$$\phi = \mathrm{arctang}\left(\frac{\sum_{i=1}^{\mathrm{NPP}} \mathrm{signal}(i) \times \sin(t(i) \times \omega)}{\sum_{i=1}^{\mathrm{NPP}} \mathrm{signal}(i) \times \cos(t(i) \times \omega)}\right) \tag{5.18}$$

where *i*, *NPP*, signal, *t* and ω, represent the sampling number, the total number of samplings in a period of the signal, the experimental waveform (v_{in}, v_{out} or i_{C}), the time vector in a period and the angular frequency, respectively.

Finally, it is possible to compute K_x and K_y [60]:

$$\frac{\mathrm{signal}_2}{\mathrm{signal}_1} = \frac{M_2 \times \sin(w \times t + \phi_2)}{M_1 \times \sin(w \times t + \phi_1)} \Rightarrow K_X = \frac{M_2}{M_1} \times \cos(\phi_2 - \phi_1); K_Y = \frac{M_2}{M_1} \times \sin(\phi_2 - \phi_1) \tag{5.19}$$

Another solution based on the least mean square (LMS) algorithm can be used to compute *M* and ϕ [58], where NPP represents the total number of acquired samples:

$$A \times \begin{bmatrix} M_{\mathrm{X}} \\ M_{\mathrm{Y}} \end{bmatrix} = \begin{bmatrix} \sum_{i=1}^{\mathrm{NPP}} \mathrm{signal}(i) \times \cos(\omega \times t(i)) \\ \sum_{i=1}^{\mathrm{NPP}} \mathrm{signal}(i) \times \sin(\omega \times t(i)) \end{bmatrix},$$

$$A = \begin{bmatrix} \sum_{i=1}^{\mathrm{NPP}} (\cos(\omega \times t(i)))^2 & \sum_{i=1}^{\mathrm{NPP}} (\sin(\omega \times t(i)) \times \cos(\omega \times t(i))) \\ \sum_{i=1}^{\mathrm{NPP}} (\sin(\omega \times t(i)) \times \cos(\omega \times t(i))) & \sum_{i=1}^{\mathrm{NPP}} (\sin(\omega \times t(i)))^2 \end{bmatrix}$$

$$M = \sqrt{M_{\mathrm{X}}^2 + M_{\mathrm{Y}}^2} \text{ and } \phi = \mathrm{arctang}\left(\frac{M_{\mathrm{Y}}}{M_{\mathrm{X}}}\right) \tag{5.20}$$

Both methodologies require small computational effort. Finally, to determine K_x and K_y, (5.19) should be used.

In the following, a brief description of the experimental prototype is presented that can be used to implement the proposed off-line MT. For that purpose, a signal generator, a power amplifier, an RC network, an oscilloscope and a microcomputer with numerical computation software are necessary. The latter should be able to manipulate the data from the oscilloscope and, simultaneously, should be capable of implementing the proposed techniques (Figure 5.14). The power amplifier is used to transform the sinusoidal waveform signal, generated by the signal generator, into a sinusoidal waveform with enough power to feed the RC network. It is very important that the DC component of the input voltage is always higher than the amplitude of its AC component, because Al-Caps are polarized. Besides, it should be avoided that the maximum current ripple rating of the capacitor is exceeded, so *R* should be chosen accordingly. On the other hand, *R* must be non-inductive and must have enough power to avoid heating during measurements.

In [56–60], a function generator Topward 8110 was used, a commercial 8 W class AB audio power amplifier, a 10 W non-inductive thick film power resistor that was soldered to the DUT (the capacitor under test), an oscilloscope Tektronix TDS1012 and a microcomputer with MATLAB® software (The MathWorks, Inc., Natick, MA, USA). Still, the oscilloscope could be replaced with a suitable acquisition board.

The proposed off-line techniques are relatively simple and do not require high computational effort, thus, they can be implemented in a low-cost system based on a digital signal processor (DSP).

5.5.2 Off-line measurement techniques based on a charge–discharge circuit

In addition to the previous off-line MT, there is also another off-line technique based on a charging/discharging test circuit [61–63], which will be presented next.

For this new technique to be precise, the capacitor current must be approximately square (Figure 5.15(a)). In this case, it is possible to set the capacitor voltage, v_{Cap}, as follows (Figure 5.15(b)):

$$v_{\mathrm{Cap}} = v_{\mathrm{C}} + ESR \times i_{\mathrm{Cap}} + \mathrm{ESL} \times \frac{di_{\mathrm{Cap}}}{dt} \qquad (5.21)$$

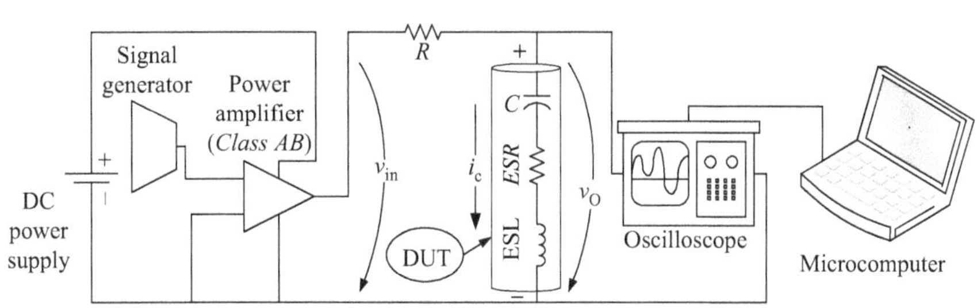

Figure 5.14 Experimental prototype used to implement the techniques proposed in [56–58]

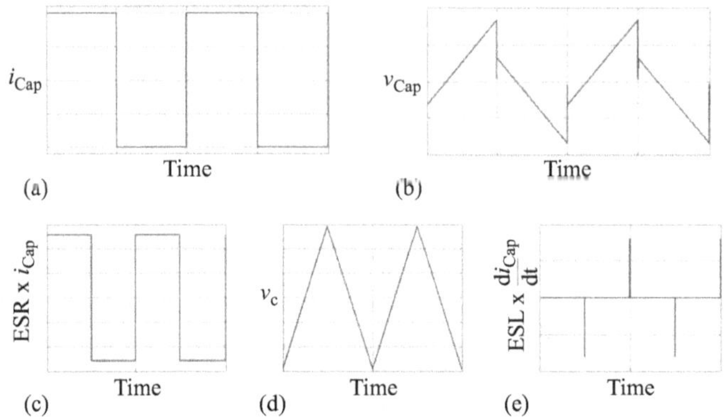

Figure 5.15 Theoretical waveforms of capacitor voltage and current when it is exposed to a square current waveform: (a) capacitor current, (b) capacitor voltage (V_{cap}), (c) the effect of ESR on the V_{Cap}, (d) the effect of C on V_{Cap} and (e) the effect of ESL on V_{Cap}

where i_{Cap}, v_{C} and $\frac{di_{\mathrm{Cap}}}{dt}$ represent the capacitor current, the component of v_{Cap} due to the capacitor capacitance and the capacitor current derivative, respectively.

By analysing the previous expression and considering that the capacitor current is a square waveform (Figure 5.15(a)), it is possible to conclude that:

- The *ESR* effect will manifest itself through a square waveform (Figure 5.15(c)) whose amplitude is directly related to the capacitor *ESR* value $\left(ESR \times i_{\mathrm{Cap}}\right)$.
- The *C* effect will manifest itself through a triangular waveform (Figure 5.15(d)) whose slope is inversely proportional to the capacitor *C* value (v_{C}).
- The ESL effect will manifest itself in the form of small impulses (Figure 5.15(e)) that will occur during the transition between stages (charge and discharge stages). The maximum impulse value is directly proportional to the ESL value $\left(\mathrm{ESL} \times \frac{di_{\mathrm{Cap}}}{dt}\right)$.

The main purpose of this technique is to compute both *ESR* and *C* values, thus the ESL effect will be neglected. Thus, after acquiring the experimental waveforms (v_{Cap} and i_{Cap}) and using the LMS algorithm, it is possible to estimate both values of *C* and *ESR* as follows [61]:

$$X = A^{-1} \times b,$$

$$X = \begin{bmatrix} ESR \\ \dfrac{1}{C} \\ v_c(0) \end{bmatrix}, \quad b = \begin{bmatrix} \sum_{i=1}^{N} \left(v_{\mathrm{Cap}}(i) \times i_{\mathrm{Cap}}(i)\right) \\ \sum_{i=1}^{N} \left(v_{\mathrm{Cap}}(i) \times \int i_{\mathrm{Cap}}(i)\right) \\ \sum_{i=1}^{N} v_{\mathrm{Cap}}(i) \end{bmatrix}$$

$$A = \begin{bmatrix} \sum_{i=1}^{N} \left(i_{\mathrm{Cap}}(i)\right)^2 & \sum_{i=1}^{N} \left(i_{\mathrm{Cap}}(i) \times \int i_{\mathrm{Cap}}(i)\right) & \sum_{i=1}^{N} \left(i_{\mathrm{Cap}}(i)\right) \\ \sum_{i=1}^{N} \left(i_{\mathrm{Cap}}(i) \times \int i_{\mathrm{Cap}}(i)\right) & \sum_{i=1}^{N} \left(i_{\mathrm{Cap}}(i)\right)^2 & \sum_{i=1}^{N} \left(\int i_{\mathrm{Cap}}(i)\right) \\ \sum_{i=1}^{N} \left(i_{\mathrm{Cap}}(i)\right) & \sum_{i=1}^{N} \left(\int i_{\mathrm{Cap}}(i)\right) & N \end{bmatrix} \tag{5.22}$$

where i_{Cap}, v_{Cap}, $\int i_{\mathrm{Cap}}$ and N represent the capacitor current, the capacitor voltage, the integral of capacitor current and the total number of samples acquired, respectively.

In order to implement the prior off-line MT, it is necessary to design a test circuit that imposes a current approximately square to the capacitor (Figure 5.16).

The power stage is supplied by a DC power supply and it is composed of a N channel enhancement mode silicon power field effect transistor, a thick film non-inductive resistor, R_1, soldered to the capacitor (DUT), a variable wire-wound resistance, R, that is used to discharge the capacitor and a power resistor, R_0, that is used to limit the maximum current. The resistor R_1 is used simultaneously to limit the capacitor current ripple and as a current sensor. The control circuit must ensure that the capacitor time constant during both states is high enough, so that the capacitor current is approximately a square waveform.

The data acquisition system is composed of an oscilloscope and a microcomputer with numerical computation software; alternatively, the oscilloscope can be replaced by a suitable data acquisition board.

In the same way as the first off-line MT, the simplicity of this methodology allows its implementation in an embedded system. For that, a low-cost system based on a DSP can be used to implement the proposed algorithm, together with,

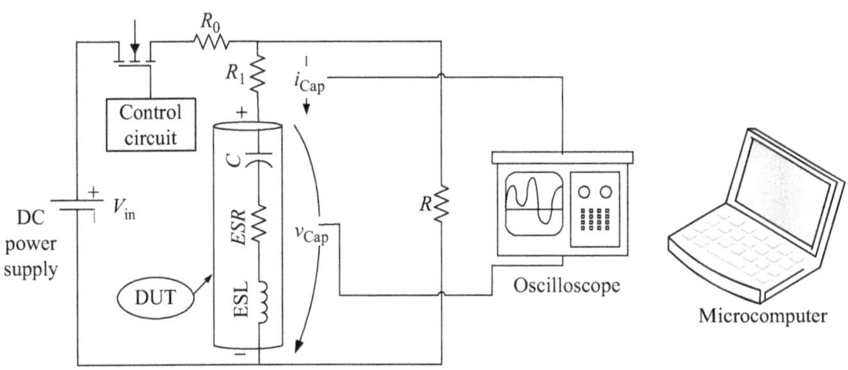

Figure 5.16 Experimental prototype used to implement the technique proposed in [61]

a liquid crystal display (LCD) to display the *ESR* and *C* values. In this way, it is possible to construct a simple LCR meter for the capacitors under analysis.

It should be noted that this last technique (Figure 5.16) computes the DC capacitance, which is slightly higher than the AC capacitance computed using the prototype of Figure 5.14. However, this does not affect their use as off-line fault diagnostic techniques.

The measurement technique based on the prototype of Figure 5.14 involves a more complex experimental prototype than the one based on the prototype of Figure 5.16. However, the implementation of the mathematical algorithms (5.18)–(5.20) requires less computational effort than the algorithm (5.22).

5.5.3 Frequency and temperature multipliers

The capacitors internal resistance and capacitance changes with the operating conditions (temperature and frequency), in particular, for Al-Caps, as it was shown in Section 5.2.1. Therefore, the knowledge of the frequency and temperature multipliers of capacitors is fundamental for design purposes, to assess the reliability of power electronics applications, and for the design of on-line and quasi-online fault diagnostic techniques.

Since different power electronics circuits operate at different operating conditions, designers need to know the capacitors equivalent circuit under these conditions in order to achieve the best design solution. On the other hand, as it was mentioned at the beginning of this chapter, the reliability of these applications is many times dependent on the capacitors lifetime. However, to predict the Al-Caps lifetime it is necessary to exactly know their *ESR* at their operating conditions (temperature and frequency). Still, most manufacturers provide this information for an operating frequency and temperature of 120 Hz and 20 °C, respectively. Therefore, both temperature and frequency multipliers are fundamental for evaluating the capacitors lifetime, and thus, to evaluate the reliability of these systems. Furthermore, the design of on-line (ONDTs) and quasi-on-line fault diagnostic techniques (QONDTs) requires the knowledge of the capacitor temperature and frequency multipliers, since the operating conditions of the circuit change with time.

The computation of frequency and temperature multipliers is particularly important in case of Al-Caps, because both *ESR* and *C* values vary significantly with these factors. For this reason, three algorithms, based on the techniques discussed in the previous section, will be presented, which enable the calculation of the frequency and temperature multipliers for Al-Caps.

In Section 5.2.1, it was shown that the *ESR* of Al-Caps results from the combination of two resistances: the series resistance (R_S) and the dielectric resistance (R_d). The first one (R_S) represents the terminals resistance, tab resistance, foil resistance, paper-electrolyte resistance and tunnel-electrolyte resistance and decreases with the increase of temperature. However, the effect of frequency on R_S is practically negligible. On the other hand, the dielectric resistance (R_d) is practically unaffected by temperature; nevertheless, it is strongly conditioned by the operating frequency, as can be seen from (5.6).

Therefore, it is possible to conclude that the effect of temperature affects preponderantly paper-electrolyte resistance, according to (5.5), while the operating frequency affects primarily the dielectric resistance, according to (5.6). In this way, the mathematical models represented by (5.5) and (5.6) can be used to obtain the temperature and frequency multipliers for the *ESR* of Al-Caps.

In the following, a mathematical model is presented that describes the effect of temperature on the *ESR*, which is accomplished using (5.5):

$$ESR(T) = \alpha + \beta \times e^{-\frac{T}{\delta}} \tag{5.23}$$

where T represents the capacitor core temperature and α, β and δ depend on the capacitor type.

Equation (5.23) can be rewritten as follows:

$$ESR(T) = a \times \left(1 + b \times e^{-c \times T}\right); \quad \alpha = a, \quad \beta = a \times b, \quad \delta = c^{-1} \tag{5.24}$$

In order to estimate the values of a, b and c, it is necessary to use non-linear regression, since (5.24) as a non-linear dependence on their parameters (a, b and c). Therefore, Gauss–Newton algorithm was used to minimize the sum of the squares of residues between the data (experimental results) and the non-linear (5.24), as can be seen below [62]:

$$\begin{cases} a_{j+1} = a_j + \Delta a \\ b_{j+1} = b_j + \Delta b \\ c_{j+1} = c_j + \Delta c \end{cases}, \quad \begin{bmatrix} \Delta a \\ \Delta b \\ \Delta c \end{bmatrix} = \frac{Z_j^T \times D}{Z_j^T \times Z_j},$$
$$D = \begin{bmatrix} y_1 - f(x_1) \\ \vdots \\ y_n - f(x_n) \end{bmatrix}, Z_j = \begin{bmatrix} \frac{df(x_1)}{da} & \frac{df(x_1)}{db} & \frac{df(x_1)}{dc} \\ \vdots & \vdots & \vdots \\ \frac{df(x_n)}{da} & \frac{df(x_n)}{db} & \frac{df(x_n)}{dc} \end{bmatrix} \tag{5.25}$$

Here, $j, j+1, Z_j, D, y_n, f(x_n)$ and n are the initial guess, the prediction, the matrix of partial derivatives of (5.24), the vector of the differences between the experimental data and (5.24), the experimental data, (5.24) values and the number of sampling, respectively.

In order to obtain the experimental data, it is necessary to design an experimental prototype. For this purpose, one of the two circuits shown in Figure 5.14 or 5.16 can be used. However, it is still necessary to modify the capacitor core temperature. In the following, a very simple solution proposed in [62] is presented.

In order to modify the capacitor core temperature, a very simple prototype can be used, consisting of a vitreous glass wire winding resistor (TVEWR), two aluminium winding resistors (ACWR), a 12 V battery and a Tektronix TX3 multimeter. The capacitor temperature can be measured using the multimeter temperature sensor, which should be fixed to the capacitor can. The capacitor must be covered with aluminium foil to help maintain the temperature constant inside the capacitor. Subsequently, it must be inserted into the TVEWR. The temperature inside the TVEWR can be controlled by adjusting the current flowing through it. On the other hand, the current and voltage of the capacitor can be acquired by an oscilloscope and manipulated in a microcomputer with numerical computation software such as MATLAB [62]. Alternatively, a suitably adapted oven can be used.

The thermal model of the capacitor consists of two thermal resistors in series: the core-to-case thermal resistance (R_{thcc}) and the case-to-air thermal resistance (R_{thca}). Since R_{thca} is much larger than R_{thcc}, the latter can be neglected and, therefore, the capacitor core temperature can be approximated to the capacitor case temperature. In this way, the temperature displayed by the multimeter is approximately equal to the capacitor core temperature.

In addition to the *ESR* temperature multipliers, it is also possible to determine the capacitance temperature multipliers. In [53], it was shown, using several experimental data, that the relation between the capacitor reactance (X_{cond}) and the temperature (T) is approximately linear for Al-Caps. This relation can be modelled as follows:

$$X_{cond}(T) = \tau_1 \times T + \tau_2 \tag{5.26}$$

Subsequently, using the LMS algorithm [see (5.27)], it is possible to extract the values of τ_1 and τ_2 from the experimental data, which can be acquired using the prototype presented above together with the circuit of Figure 5.14. In this way, it is possible to estimate the capacitance temperature multipliers, since the ESL value is approximately constant:

$$\begin{bmatrix} \sum_{i=1}^{N} T_i^2 & \sum_{i=1}^{N} T_i \\ \sum_{i=1}^{N} T_i & N \end{bmatrix} \times \begin{bmatrix} \tau_1 \\ \tau_2 \end{bmatrix} = \begin{bmatrix} \sum_{i=1}^{N} (T_i \times X_{condi}) \\ \sum_{i=1}^{N} X_{condi} \end{bmatrix} \tag{5.27}$$

where T_i, X_{condi} and N represent the temperature of sample i, the reactance of sample i and the total number of samples, respectively.

The effect of the frequency on the *ESR* can be modelled through (5.6). However, due to the complexity of (5.6) it will be used the following equation, that is equally valid [63]:

$$ESR = \frac{DF_{OX} + w \times C \times R_S}{w \times C} = \frac{DF_{OX}}{w \times C} + R_S \tag{5.28}$$

where DF_{OX} represents the dissipation factor of the oxide layer (Al_2O_3).

In the following, the mathematical model is presented that describes the effect of frequency on the *ESR*, which is accomplished using (5.28):

$$ESR = \frac{K_1}{f} + K_2; \quad K_1 = \frac{DF_{OX}}{2 \times \pi \times C}, K_2 = R_S \tag{5.29}$$

Finally, using the LMS algorithm it is possible to obtain the frequency multipliers:

$$\begin{bmatrix} \sum_{i=1}^{N} \frac{1}{f_i^2} & \sum_{i=1}^{N} \frac{1}{f_i} \\ \sum_{i=1}^{N} \frac{1}{f_i} & N \end{bmatrix} \times \begin{bmatrix} K_1 \\ K_2 \end{bmatrix} = \begin{bmatrix} \sum_{i=1}^{N} \frac{ESR_i}{f_i} \\ \sum_{i=1}^{N} ESR_i \end{bmatrix} \tag{5.30}$$

where f_i, ESR_i and N represent the frequency of sample i, the ESR of sample i and the total number of samples, respectively [53].

5.5.4 Off-line fault diagnostic techniques

The design of off-line fault diagnostic techniques (OFFDTs) is simpler than off-line measurement techniques. The first ones must guarantee always the same operating conditions, while the second ones should make measurements at different operating conditions, that is, the final prototype should be much more complex, so that it can simulate different operating conditions.

This chapter addresses the topic of fault diagnosis, so the aforementioned off-line measurement techniques can easily be transformed into off-line fault diagnosis techniques, as long as the operating conditions remain unchanged (temperature, frequency, humidity, etc.).

The off-line fault diagnosis technique (OFFDT) based on the LC filter (Figure 5.11) is only feasible under certain conditions, thus, the use of one of the prototypes of Figure 5.14 or Figure 5.16 would be preferable.

Nevertheless, it is possible to use simpler prototypes, such as the one proposed in [64]. This test circuit requires a DC power supply (or a battery in alternative), a low-frequency single-phase transformer, an RC circuit, an oscilloscope and a microcomputer with numerical computation software (Figure 5.17). The RC circuit is composed of a ceramic encased wire-wound resistor, R, connected to the capacitor under test.

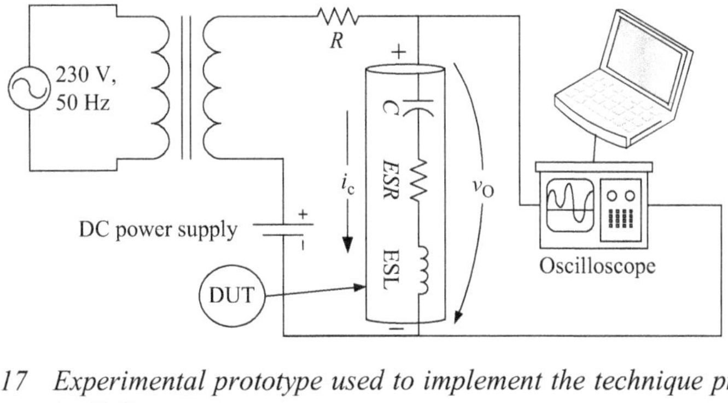

Figure 5.17 Experimental prototype used to implement the technique proposed in [64]

The previous prototype will run under a specific operating frequency (50 or 60 Hz), which is not a problem, since it will be used to implement an OFFDT. One of the algorithms represented by (5.18)–(5.20) can be used to extract both values of *ESR* and X_{cond}.

Another solution also quite simple, which replaces the oscilloscope by a multimeter, was presented in [65]. This new OFFDT assesses the capacitor health status through the estimation of both *ESR* and *C* values, which are not computed for a specific operating frequency. In this case, both *ESR* and *C* are estimated through the capacitor impedance.

If the capacitor operating frequency is lower than its resonance one, it is possible to define that capacitor impedance in the following manner:

$$|Z_{cond}| = \sqrt{ESR^2 + \left(\frac{1}{2 \times \pi \times f \times C}\right)^2} \tag{5.31}$$

Equation (5.31) can be rewritten as follows:

$$Z_{cond}^2 = K_1 + \left(\frac{1}{2 \times \pi \times f}\right)^2 \times K_2; \quad K_1 = ESR^2 \;\; K_2 = C^{-2} \tag{5.32}$$

Using (5.32), the LMS algorithm [see (5.33)] and several values of Z_{cond}, at different operating frequencies, it is possible to estimate both *ESR* and *C*:

$$\begin{bmatrix} N & \sum_{i=1}^{N}\left(\frac{1}{2 \times \pi \times f_i}\right)^2 \\ \sum_{i=1}^{N}\left(\frac{1}{2 \times \pi \times f_i}\right)^2 & \sum_{i=1}^{N}\left(\frac{1}{2 \times \pi \times f_i}\right)^4 \end{bmatrix} \times \begin{bmatrix} K_1 \\ K_2 \end{bmatrix} = \begin{bmatrix} \sum_{i=1}^{N}(Z_{condi}) \\ \sum_{i=1}^{N}\left(Z_{condi} \times \left(\frac{1}{2 \times \pi \times f_i}\right)^2\right) \end{bmatrix} \tag{5.33}$$

where f_i, Z_{condi} and N represent the frequency of sample i, the Z_{cond} of sample i and the total number of samples, respectively, such that f_N must be less than the capacitor resonant frequency.

In order to obtain experimentally the Z_{cond} values, the prototype of Figure 5.14 should be used. However, the oscilloscope will be replaced by a true RMS multimeter. The capacitor impedance can be easily calculated through the ratio of the RMS values between the capacitor voltage (v_{cRMS}) and current (i_{cRMS}):

$$|Z_{cond}| = \frac{v_{cRMS}}{i_{cRMS}} \tag{5.34}$$

The previous OFFDTs use the capacitor voltage and current to estimate both values of *ESR* and *C*.

Alternatively, it is possible to compute *C* using the capacitor discharge time. In [66], it was proposed a very simple prototype based on the time-constant measurement methods, which evaluates the capacitor health status by calculating its capacitance.

By considering a simple RC network with the capacitor fully charged, it is possible to represent the capacitor voltage (v_{cond}) during the discharge state, as follows:

$$C \times \frac{dv_{\text{cond}}}{dt} = -\frac{v_{\text{cond}}}{R} \Rightarrow \log\left(\frac{v_{\text{cond}}(t_{\text{d}})}{v_{\text{cond}}(t_0)}\right) = -\frac{t_{\text{d}} - t_0}{R \times C} \tag{5.35}$$

where v_{cond}, R, t_0 and t_{d} represent the capacitor voltage, the discharge resistor, the initial time and the capacitor discharge time, respectively.

Usually, the capacitors under analysis have high capacitance, so the discharge time is very large. Thus, t_{d} will be replaced by the time the capacitor requires to discharge ¾ of the stored energy, which corresponds to half of the initial voltage. Therefore, the new t_{d} can be computed in the following manner:

$$t_{\text{d}} = R \times C \times \log(2) = R \times C \times 0.6931 \tag{5.36}$$

In this way, if the discharge resistor (R) equals:

$$R = \frac{10{,}000}{\log(2)} \cong \frac{10{,}000}{0.6931} \cong 14{,}430\ \Omega \tag{5.37}$$

the capacitor capacitance can be computed as:

$$C = t_{\text{d}} \times 10^{-4} \tag{5.38}$$

Using the previous relationships, it is possible to design a simple and cheap experimental prototype [66], which is able to compute the capacitor capacitance according to (5.38).

The prototype proposed in [66] can be subdivided in two circuits: the counter circuit and the control circuit.

The first circuit is responsible for counting the number of periods the capacitor needs to reach half of the initial voltage. For this purpose, it uses 4-bit counters connected in cascade, which are synchronized with the same clock signal. The counting ends, each time one of the counters control inputs is activated (enable input). The clock signal can be generated using a 555 CI, or an Attinny85 microcontroller, and must operate at a specific frequency according to the capacitance measurement range. Each 4-bit counter is connected using a BCD to seven-segment driver, to a display of 7 segments, which display the capacitor capacitance according to (5.38).

The control circuit determines when the counter should start and stop counting and, for that, it should generate two control signals, which determine the starting

Table 5.2 Off-line fault diagnostic techniques

Identification	Estimate	Description
$OFFDT_1$	It computes the capacitor *ESR*, for a specific operating frequency.	It uses the prototype of Figure 5.11 and (5.8) to compute the capacitor *ESR* and it is only feasible under certain conditions.
$OFFDT_2$	It computes both *ESR* and X_{cond} for a specific operating frequency.	It uses the prototype of Figure 5.14 and one of the mathematical algorithms described by (5.18)–(5.20) to compute both *ESR* and X_{cond}.
$OFFDT_3$	It computes both *ESR* and *C* for a specific operating frequency.	It uses the prototype of Figure 5.16 and the mathematical algorithm described by (5.22) to compute both *ESR* and *C*.
$OFFDT_4$	It computes both *ESR* and X_{cond} for a specific operating frequency (50 or 60 Hz).	It uses the prototype of Figure 5.17 and one of the mathematical algorithms described by (5.18)–(5.20) to compute both *ESR* and X_{cond}.
$OFFDT_5$	It computes both *ESR* and *C*.	It uses the prototype of Figure 5.14 with a multimeter instead of the oscilloscope. The *ESR* and *C* are computed from the capacitor impedance, using (5.33).
$OFFDT_6$	It computes *C*.	It uses a counter circuit together with a control circuit to show, in four displays of 7 segments, the capacitor capacitance.

point and the ending point of the counting. Thus, two 741 ICs, an RC network and a voltage divide were used. The starting signal is generated when the capacitor is completely charged, while the end point is generated when the capacitor voltage is lower than half of the initial one.

In order to better identify the presented OFFDTs, they are summarized in Table 5.2.

The techniques presented in Table 5.2 estimate the electrical parameters that are commonly used to evaluate the capacitors health status. It is now important to define which criteria should be used to identify the capacitor end-of-life limit.

The most commonly used criteria were presented at the beginning of Section 5.4.

In the case of Al-Caps, it is the loss of 20% of the initial capacitance or doubling the initial *ESR* value. This criterion gives rise to two different methodologies that permit at any time to decide if the capacitor should be replaced:

- Use the capacitance initial value and compare it with the actual one. If the actual capacitance value decreases in more than 20%, the capacitor should be replaced.
- Use the *ESR* initial value and compare it with the actual one. If the actual *ESR* value doubles in comparison with the initial one, the capacitor should be replaced.

Table 5.3 Different criteria for evaluating the state condition of Al-Caps [67]

Criteria	Description
1	Use the maximum dissipation factor (DF_{MAX}) given by the manufacturer at 120 Hz as the reference value (for a sound capacitor) and compare it with the actual one at the same operating conditions. If the actual value is higher than twice DF_{MAX}, than the capacitor should be replaced.
2	Use the first criteria presented above – compare the initial value of *C* (sound capacitor) with the actual one.
3	Use the second criteria presented above – compare the initial value of *ESR* (sound capacitor) with the actual one.
4	Compute the typical *ESR* value (typ_{ESR}) at 1 kHz, which can be obtained from a sampling of ten capacitors of the same type. The value of typ_{ESR} should be used as a reference value. Therefore, if the actual *ESR* value of the capacitor is higher than twice typ_{ESR}, the capacitor should be replaced.
5	Compute the typical *C* value (typ_C) at 120 Hz, which can be obtained from a sampling of ten capacitors of the same type. The value of typ_C should be used as a reference value. Therefore, if the actual *C* value of the capacitor decreases by 20% in relation to typ_C, the capacitor should be replaced.
6	Some manufacturers give the typical *ESR* value at 120 Hz. If that is the case, use this value as a reference value. If the actual *ESR* value of the capacitor is higher than twice the typical *ESR* given by the manufacturer, the capacitor should be replaced.

In order for the above-mentioned criteria to be applied with assuredness, the measurements must be carried out under the same operating conditions and the initial values of *C* and *ESR* must be stored. This process must be repeated for each new capacitor, which means that each new capacitor must be evaluated at its initial stage (sound capacitor) and its initial values of *C* and *ESR* should be stored. The above criteria should be applied in applications where the reliability is of prime importance, so, if one of the parameters (*ESR* or *C*) exceeds the defined limit, the capacitor must be immediately replaced.

In [67], several end-of-life limit criteria for Al-Caps were presented, which are synthesized in Table 5.3.

Each criterion presented in Table 5.3 presents some advantages and disadvantages that will be discussed in the following paragraphs.

Criteria 1 and 6 are simple to implement, because they do not require any measurement in the initial stage. However, they present some drawbacks. In the case of criterion 1, the value of *DF* given by the manufacturer is maximum, which means that the reference value is higher than the real one; therefore, criteria 1 should not be used in applications where the reliability is of prime importance. On the other hand, criterion 6 can only be used if the manufacturer provides the typical *ESR* value at 120 Hz. Both criteria can be applied if $OFFDT_2$ or $OFFDT_3$ is used at 120 Hz. The above techniques can also be applied at frequencies of 1 kHz and 10 kHz, since some manufacturers provide the frequency multipliers for these frequencies.

Criteria 2 and 3 are the most conservative ones; therefore, they should be used in applications where the reliability is of prime importance. The application of these two criteria is more complex, because they require measurements of all new capacitors. Criterion 2 can be applied if $OFFDT_2$ or $OFFDT_3$ or $OFFDT_4$ or $OFFDT_5$ or $OFFDT_6$ is used. The *C* measurement accuracy increases as measurements are performed near low frequencies (50, 60 or 120 Hz), because the ESL effect is negligible. In case of $OFFDT_3$ and $OFFDT_6$, DC capacitance is measured, which is not a problem since the reference value is measured under the same operating conditions. Criterion 3 can be applied if $OFFDT_1$ or $OFFDT_2$ or $OFFDT_3$ or $OFFDT_4$ or $OFFDT_5$ is used. The *ESR* measurement accuracy increases as measurements are performed near the capacitor resonance frequency, thus, a measurement frequency of 1 kHz would be a good choice.

Criteria 4 and 5 require a set of initial measurements to compute the typical *ESR* and *C* values, which can be obtained from a sampling of ten capacitors of the same type. After estimating the *ESR* and *C* values of a sample of ten sound capacitors, the typical values should be computed and, for that, the mean value of the initial measurements (*ESR* and *C*) should be computed. If the *ESR* maximum deviation is higher than 25% or *C* maximum deviation is higher than 5%, the obtained reference values should not be used [67]. The application of these two criteria requires the purchase of ten capacitors of the same type, thus, these two criteria should be applied when several capacitors of the same type are used in the same application. Criterion 4 can be applied if $OFFDT_1$ or $OFFDT_2$ or $OFFDT_3$ or $OFFDT_4$ or $OFFDT_5$ is used, and criterion 5 can be applied if $OFFDT_2$ or $OFFDT_3$ or $OFFDT_4$ or $OFFDT_5$ or $OFFDT_6$ is used.

For MPPF-Caps, the most commonly used criterion is the loss of 2% to 10% of the initial capacitance. Therefore, if the actual capacitance value decreases more than 10%, when compared with the initial one, the capacitor should be replaced. However, in applications where the reliability is of prime importance, the criterion must be more conservative, so, the capacitor should be replaced if the actual capacitance value decreases more than 2% when compared with the initial one. This criterion can be applied if $OFFDT_2$ or $OFFDT_3$ or $OFFDT_4$ or $OFFDT_5$ or $OFFDT_6$ is used.

5.6 On-line fault diagnostic techniques

The development of ONDTs reveals vital in some applications, namely, in applications whose operation cannot be interrupted or in critical systems. In this way, it is crucial to detect, or better pre-detect, the capacitor end-of-life limit, so that it would be possible to schedule the best stopping periods for preventive or predictive maintenance. In addition, ONDTs ensure a greater reliability and safety; furthermore, off-line and quasi-online fault techniques are not appropriate, or even unviable, in some of the previous systems.

However, it should be mentioned that ONDTs require important investments in terms of devices, sensors and control schemes; they are complex; in the majority of cases, they are invasive and are usually specific to a particular application.

ONDTs can be classified according to the health indicator or to the methodology used in the estimation of those indicators.

The end-of-life limit criteria, commonly used in ONDTs, are the ones presented at the beginning of Section 5.4. Therefore, the capacitors C and ESR values are the best health indicators [3]. Temperature should be another factor to consider since the aforementioned indicators are extremely susceptible to temperature, in particular, for Al-Caps.

According to the first classification, ONDTs can be grouped into:

- Techniques that only compute ESR, which are typically used in circuits composed of Al-Caps;
- Techniques that estimate both ESR and C, which can be used in circuits having one or both types of capacitors (Al-Caps and MPPF-Caps);
- Techniques based solely on C calculations, which initially were intended exclusively for MPPF-Caps, but recently, they have also been proposed for Al-Caps.

With regard to the second classification, ONDTs can be grouped into:

- Techniques that use analytical relationships between different quantities, such as capacitor current, capacitor voltage, input current, average power, among others;
- Techniques that use a reference system;
- Techniques that use circuit model-based methods;
- Techniques that use common time and frequency domain-processing approaches, such as digital filters or Fourier transform;
- Techniques based on advanced algorithms, such as genetic algorithms or evolutionary algorithms.

This section will be subdivided into three subsections according to the first criterion (health indicator). It was decided to make this division, because some ONDTs use more than one methodology for calculating their indicators; some ONDTs that use different health indicators use methodologies of the same group; and not all groups of the first criterion (health indicator) use all the methodologies of the second criterion.

5.6.1 On-line fault diagnostic techniques based on ESR estimation

This subsection presents the ONDTs associated to the first group of the first criterion, which are the techniques that are based on the ESR estimation to evaluate the capacitors health condition. They are, therefore, ONDTs commonly used in circuits composed of Al-Caps.

The first ONDTs covered here fall in the first group of the two classifications presented in Section 5.6. These techniques use the capacitor current, capacitor voltage, the input current, and output current waveforms to estimate the capacitors ESR [68–73].

In [68], the authors proposed an on-line fault diagnostic technique (ONDT) for a flyback DC–DC converter (Figure 5.18).

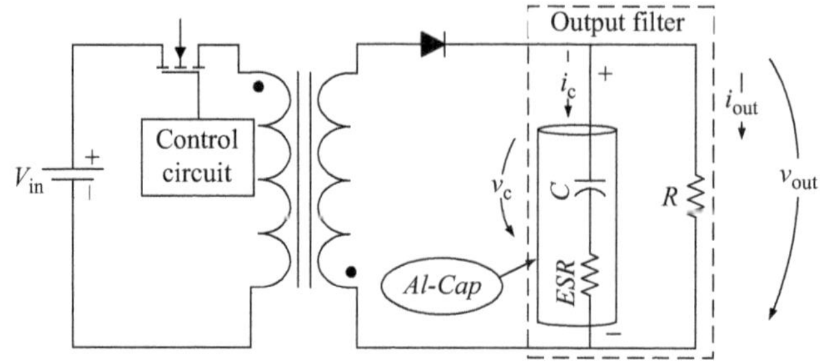

Figure 5.18 Experimental prototype studied in [68] (flyback converter)

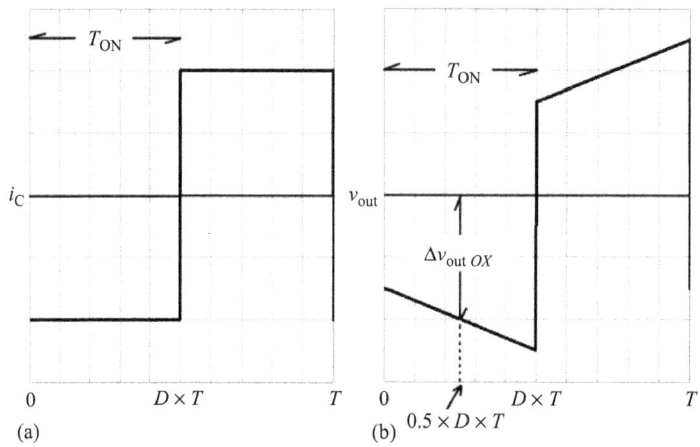

Figure 5.19 Theoretical waveforms of the flyback converter presented in Figure 5.18: (a) capacitor current and (b) capacitor voltage

The authors consider that the capacitor current (i_C) waveform is approximately squared due to the high operating frequency of the converter (Figure 5.19(a)), besides, the ESL effect is neglected. Therefore, the output voltage, v_{out}, can be represented as follows (Figure 5.19(b)):

$$v_{out} = v_C + ESR \times i_C \tag{5.39}$$

Thus, it is possible to represent the theoretical waveforms of i_C and v_{out} as shown in Figure 5.19.

Two different analytical relationships are proposed:

$$ESR \cong \frac{\langle \Delta v_{out} \rangle_{TON}}{\langle i_C \rangle_{TON}} = \frac{\frac{1}{T} \int_0^{D \times T} |\Delta v_{out}| dt}{\frac{1}{T} \int_0^{D \times T} |i_C| dt} \tag{5.40}$$

$$ESR \cong \frac{\Delta v_{\text{out OX}}}{\langle i_{\text{out}} \rangle} = \frac{\Delta v_{\text{out}}\left(\dfrac{D \times T}{2}\right)}{\langle i_{\text{out}} \rangle} \tag{5.41}$$

where $\langle \Delta v_{\text{out}} \rangle_{\text{TON}}$, $\langle i_C \rangle_{\text{TON}}$, $\Delta v_{\text{out OX}}$, $\langle i_{\text{out}} \rangle$, D and T represent the mean value of the output voltage ripple during conduction stage, the mean value of capacitor current during conduction stage, the output voltage ripple halfway the conduction stage, the mean value of output current, and the duty cycle and the switching period, respectively.

To implement (5.40), it is necessary to introduce a current sensor into the capacitor; therefore, it is an invasive technique. In the second case, (5.41), the current sensor can be applied at the output of the converter, and therefore, it can be considered a non-invasive technique.

Later, it was shown that the previous relationships can also be used in some non-isolated converters such as boost and buck-boost converters operating in continuous conduction mode (CCM) and in discontinuous conduction mode (DCM) [53,69,70].

A new ONDT for a boost and buck-boost converters based on an analytical relationship between the output voltage ripple, Δv_o, and the input current, i_{in}, was presented later in [71]. This technique does not need a current sensor inside the converter; in addition, the authors consider the effect of the capacitor temperature.

The *ESR* value can be obtained manually or automatically using the following analytical relationship:

$$ESR \cong \frac{1}{\dfrac{\max(i_{\text{in}})}{\Delta v_o} - \dfrac{I_o}{V_o}} \tag{5.42}$$

where $\max(i_{\text{in}})$, V_o and I_o represent the maximum value of input current, the mean value of output voltage and the mean value of output current, respectively.

The manual process can be implemented through an oscilloscope, being dependent on the operator ability. The automatic process eliminates the human errors. Nevertheless, it requires a prior reduction of the waveforms noise, which is done through linear Lagrange interpolating polynomials. Afterwards, a simple program based on the maximum and minimum search algorithms can be used to compute Δv_o and $\max(i_{\text{in}})$. The feasibility of this technique has been demonstrated in [71], using an experimental prototype, an oscilloscope and a microcomputer with numerical computation software.

The effect of the capacitor temperature was also considered and, for that purpose, two solutions were suggested. The first, simpler, proposes the implementation of the technique at a specific predefined temperature. The second, more complex, requires the prior calculation of the temperature multipliers, which will be used, later, to normalize the *ESR* value [71]. In this case, the temperature multipliers were obtained using experimental results, together with the least mean square algorithm (cubic approximation):

$$ESR(T) = K_1 \times T^3 + K_2 \times T^2 + K_3 \times T + K_4 \tag{5.43}$$

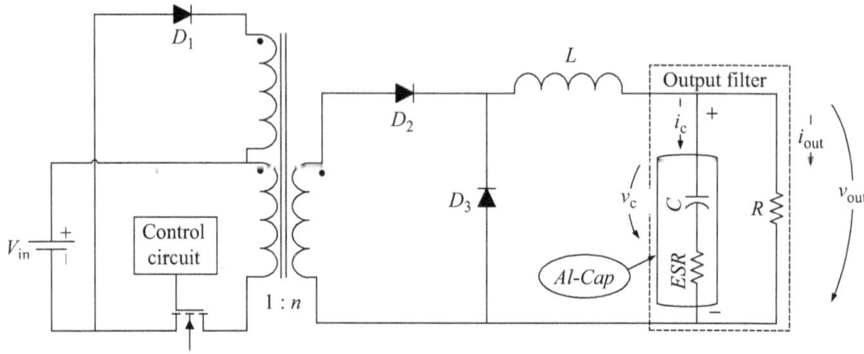

Figure 5.20 Experimental prototype studied in [72] (forward converter)

where $ESR(T)$ represents the mathematical model that describes the evolution of the ESR with the capacitor case temperature T.

In [72], the authors proposed a new deterioration diagnostics method for evaluating the capacitor health status used in the output filter of a forward converter (Figure 5.20).

The following analytical relationship was proposed:

$$ESR \cong \frac{\Delta v_{out}}{\Delta i_C} \tag{5.44}$$

where Δv_{out} and Δi_C represent the output voltage ripple and the capacitor current ripple, respectively.

In order to compute the previous analytical relationship, it is necessary to use a current sensor inside the converter, which makes this technique invasive.

It was later shown that the above relationship can be used in a non-isolated buck-type converter operating in CCM and DCM [73].

However, it should be stated that the previous relationships are only valid under permanent regime, which means that load variations are not considered; beyond that, some authors did not consider the temperature effect.

The following ONDT use a reference system for the evaluation of the capacitor health status.

In [74], the authors present a new fault diagnostics technique for DC–DC converters, namely, for a zero-current switched secondary-resonant half-wave DC–DC forward converter (Figure 5.21).

The power section of the previous circuit is composed of several elements, namely a MOSFET, diodes, polypropylene capacitors (C_{r_1}, C_{r_2} and C_{r_3}) and Al-Caps (C_{o_1}, C_{o_2} and C_{o_3}).

The authors show that the most significant modification due to the wearing out of Al-Caps is an increase in the fundamental component of the output voltage ripple (Δv_{outf}), which is independent of the load variations. If one of the three capacitors of the output filter has reached its life limit (the ESR has doubled), considering that

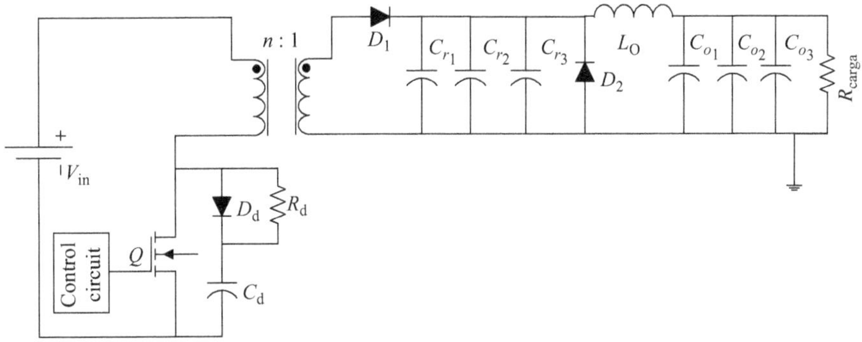

Figure 5.21 Zero-current switched secondary-resonant half-wave DC–DC forward converter [74]

the other two are still new, the Δv_{outf} increases by more than 16%. On the other hand, for the same situation, the increase in the net resistance of the three capacitors (ESR_{filter}) is approximately 20%, which is close to the increase of Δv_{outf}. In this way, the authors propose the use of ESR_{filter} for fault detection, rather than the *ESR* of each individual capacitor.

Then, the authors experimentally obtained the relationship between the percentage increase in the fundamental component of the output voltage ripple (%Δv_{outf}) with respect to the percentage increase of the net resistance of three capacitors (%ESR_{filter}), for an ambient temperature of 25 °C, and used it to evaluate the capacitor health status.

It should be noted that the prior technique requires the computation of an enormous collection of curves, which takes into account not only the effect of load variation or the input and output voltage variations but also the effect of temperature. Thus, its implementation is quite complex and reveals unfeasible in a commercial product. On the other hand, it does not identify which capacitor needs to be replaced.

Later, in [5], the authors proposed a fault diagnostic technique that permits the estimation of the *ESR* value of Al-Caps and, simultaneously, determines their remaining life until failure. This technique is applied to two different DC–DC converters: a half-bridge DC–DC forward-type converter (Figure 5.22) and a zero-current switched secondary-resonant half-wave DC–DC forward converter (Figure 5.21).

In order to implement the previous technique, first, a reference system must be built for a converter with sound capacitors, which should store a set of physical quantities for different operating conditions, namely, the fundamental component of the output voltage ripple (Δv_{outf}), the input voltage (V_{in}), the output current (I_{out}), the ambient temperature (T_a) and the capacitor case temperature (T_c).

Subsequently, during the converter operation, the same quantities (Δv_{outf}, V_{in}, I_{out}, T_a and T_c) must be acquired. The combination of these last values, with the ones of the reference system, makes it possible to assess the output capacitor health status.

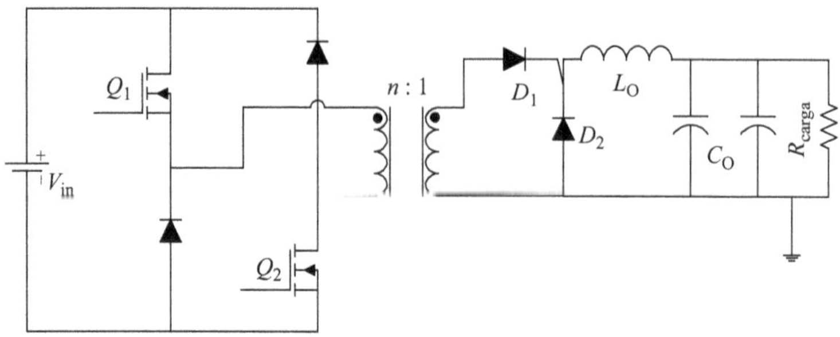

Figure 5.22 Half-bridge DC–DC forward-type converter [5]

In order to compute both *ESR* and the time before failure, the T_c and Δv_{outf} must be measured. The case temperature takes into account T_a and the heating produced by the capacitor current, while Δv_{outf} represents the best image of the output voltage ripple, Δv_{out}. The later represents the only waveform of the converter that modifies with the increase of the capacitor *ESR* value, being almost proportional to the net resistance of filter capacitors (ESR_{filter}) [5,74].

On the other hand, the measured values of T_c and Δv_{outf} depend on V_{in}, V_{out} and T_a. Therefore, the following relationships should be obtained experimentally:

$$\begin{aligned} T_c &\cong f_1(I_{out}, V_{in}, T_a) \\ \Delta v_{outf} &\cong f_2(I_{out}, V_{in}, T_a) \end{aligned} \tag{5.45}$$

Thus, by comparing the present value of Δv_{outf} with the one obtained by the reference system for a sound capacitor, at the same operating conditions (V_{in}, I_{out}, T_a and T_c), it is possible to determine the capacitor *ESR* value.

In order to compute the capacitor remaining life, first, it is necessary to construct the *ESR* prediction model versus time and temperature. For this purpose, the experimental results obtained from the ageing tests carried out on the capacitors in use, as well as Arrhenius's law, were used. Therefore, the authors obtained the following equation:

$$\frac{1}{ESR(t)} = \frac{1}{ESR(0)} \times \left(1 - k \times t \times e^{-\frac{4,700}{T+273}}\right) \tag{5.46}$$

where *ESR(t)*, *T*, *t*, *ESR(0)* and *k* represent the *ESR* at time *t*, the ageing temperature, the ageing time, the *ESR* of a sound capacitor and a constant that depends on the capacitor, respectively. The value of *k* was computed by the least squares method to fit the experimental ageing tests [5].

Finally, using (5.46), *ESR(0)*, the actual *ESR* and T_c, it is possible to compute the capacitor actual operating time (t_1). Thus, the operating time of an aged capacitor (t_2) can be calculated in the same way [5]. The difference between t_2 and t_1 gives the capacitor remaining life.

The technique proposed in [5] was executed through a computer program, which means that all measured quantities need to be converted into a DC value to be processed; for that, a data acquisition board can be used.

Later, the previous methodology was applied with success in the input and output capacitors of two industrial switch-mode power supplies: a forward half-bridge asymmetrical DC–DC converter and a forward half-bridge symmetrical AC–DC converter [75].

In [5,74,75], the characteristics of Al-Caps, as discussed in the previous sections, are further highlighted, as follows:

- Al-Caps are the most vulnerable elements of the entire power section of SMPS.
- The failure rate of this component increases much more than the remaining elements of the power section of SMPS with increasing temperature.
- Al-Caps electrical parameters (ESR and C) change with the temperature.

The previous observations reinforce the importance of the development of fault diagnostic techniques in static converters that take into account the temperature effect.

The circuits analysed in [5,74,75] present several capacitors in the DC-link. This situation is common in several power electronics applications where capacitor banks are often used. In such applications, it is advisable to replace the entire bank when one of the capacitors fails. This procedure is due to the fact that when one capacitor fails, it produces a greater stress on the remaining capacitors, speeding up their ageing process. For this reason, some industrial users define the capacitors end-of-life, based on the maximum permissible ripple voltage [75].

ONDTs based on a reference system are very expensive and complex, depend on the static converter topology and cannot predict the fault of one capacitor if several units are connected in parallel [76], making their use unviable in common commercial circuits.

For this reason, in [76], a smart, economic and simple electronic module, which can be incorporated into the electrolytic capacitor, was presented. This module can signal the moment the capacitor must be replaced. To design the previous circuit, the authors used a simple relationship between capacitor current and voltage, together, with a very simple reference system.

The prior smart circuit estimates the capacitor ESR through the computation of the ratio between the fundamental components of the capacitor voltage ripple, Δv_{outf}, and capacitor current, i_{cf}. Then, the computed ESR value is compared to the ESR of a sound capacitor at the same temperature. If the value exceeds the maximum limit, which typically represents twice the ESR of a sound capacitor, a LED lights up, instructing the operator to replace that specific capacitor.

The circuit proposed in [76] includes three parts:

- Circuit 1 – responsible for computing the ESR value of the capacitor that is operating in the converter (ESR_{use}).
- Circuit 2 – responsible for computing the ESR value of a sound capacitor at the same operating temperature (ESR_{sound}).
- Circuit 3 – compares the previous ESR values, and turns on the LED, when the ESR_{use} exceeds in more than two times the ESR_{sound}.

In [76], it was experimentally verified, through the use of an LCR meter, that the capacitor impedance, Z_{cap}, at the converter operating frequency is nearly equal to the capacitor *ESR*. Consequently, it is possible to write the following equation:

$$ESR_{use} = ESR \cong Z_{cap} = \frac{\Delta v_{outf}}{i_{cf}} \tag{5.47}$$

The above ratio can be computed through circuit 1. For that, the previous circuit measures the capacitor current using a toroidal core and the i_{cf} is obtained thanks to a band-pass filter and an RMS to DC converter. The Δv_{outf} is obtained in the same way as i_{cf}. Finally, (5.47) is executed using the analogue divider (Figure 5.23).

In order to obtain the ESR_{sound} value, first, it is necessary to obtain the mathematical function that describes the relation between the *ESR* of a sound capacitor with the capacitor core temperature, T_c. For this, the authors use an LCR meter to calculate the experimental relationship between the *ESR* and T_c of a sound capacitor. The previous function can be modelled using (5.23) and non-linear regression. Finally, during the converter operation, the capacitor core temperature is measured and used, together with the previous mathematical model, to compute the ESR_{sound} value. The value of T_c is measured using a temperature sensor on the capacitor case together with a signal conditioner. The mathematical model can be represented by and exponential amplifier circuit (Figure 5.23).

The third circuit makes the comparison between ESR_{use} and ESR_{sound}, and for this, a simple circuit with a comparator is used. The maximum *ESR* limit can be defined by selecting the right switch (S_1, S_2 and S_3), as can be seen in Figure 5.23.

The previous circuits were systematized in Figure 5.23.

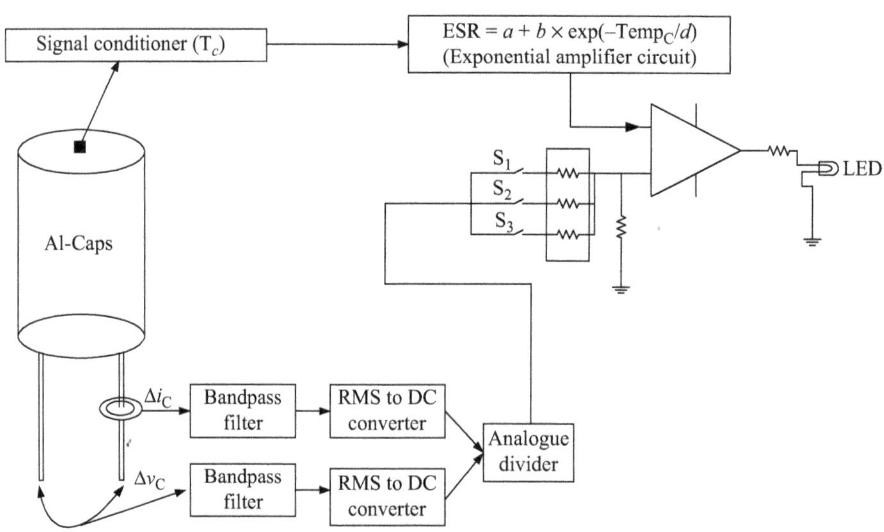

Figure 5.23 The electronic module proposed in [76]

The previous solution requires the use of a high bandwidth filter to avoid both dependence on the converter duty cycle and the capacitor capacitive reactance, namely, in AC–DC converters. The preceding solution leads to an analogue solution of large realization effort. On the other hand, in some drive systems, the switching frequency is relatively small, so, the output signals of the band-pass filters would be very small, which can cause problems to the analogue RSM to DC converters [77].

The concept underlying the previous methodology was later implemented through digital realization in [78–80]: the capacitor impedance (Z_{cap}), which is approximately identical to the *ESR* value near the converter switching frequency, is approximately equal to the ratio between the fundamental components of the capacitor ripple voltage (Δv_{outf}) and capacitor current (i_{cf}). The aforementioned relationship was verified in [76] through (5.47).

In [78], the authors present an ONDT capable of identifying the capacitors and inductors health status, used in the output filter of step-down DC–DC converters. Inductors, like capacitors, can present two failure modes: catastrophic failures (*structural failure*) and degradation failures (*parametric faults*). The most common structural failure in inductors is the open-circuit of the inductor wires due to thermal overstress. In turn, the thermal overstress can be the consequence of short-circuits between adjacent turns, which result from bad insulation of the wires that make up the inductor, the presence of nicks and kinks in the wires, high currents or, simply the result of the natural ageing process. Therefore, these short-circuits can be considered a parametric fault, which manifests itself by the gradual reduction of the inductor inductance [78]. The reduction in the inductance leads to an increase of the inductor current ripple, therefore, the capacitors are subject to higher stress and the output voltage ripple increases. This situation may be particularly critical in applications that have maximum output ripple limits. Beyond that, this failure can also condition the application of ONDTs that are based exclusively on capacitor voltage ripple. It is therefore essential to periodically evaluate this failure.

Figure 5.24 shows the equivalent circuit of an inductor, where R_C represents the core resistance, R_W the wires resistance, L the inductor inductance and C the parasitic capacitance (turn-to-turn and turn-to-core).

The operating frequency of SMPS is considerably lower than the resonance frequency of the inductor; in addition, these inductors have minimum losses. In this way, it is possible to conclude that the inductor impedance near the converter operating frequency is fundamentally due to the inductance [78].

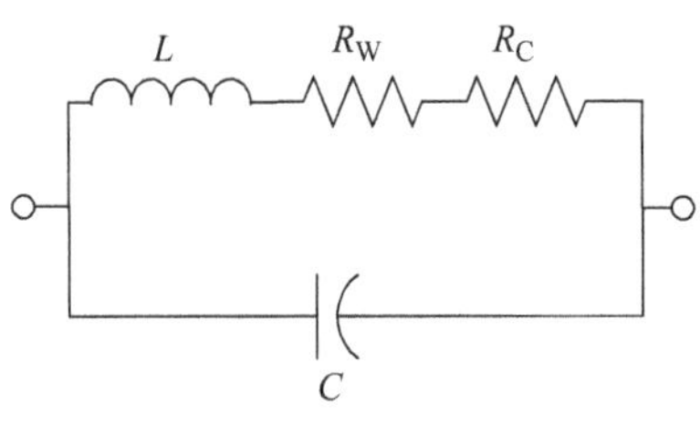

Figure 5.24 Equivalent circuit of an inductor [78]

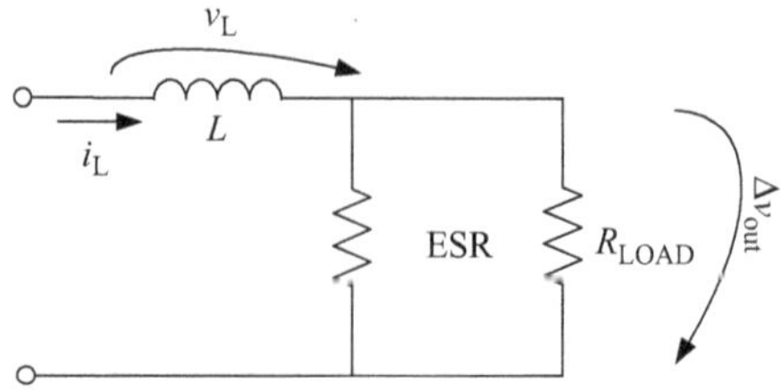

Figure 5.25 AC equivalent circuit of the output filter of a step-down DC–DC converter [78]

Thus, using the previous conclusion, in conjunction with the principle presented in [76], it is possible to obtain a simplified AC model of the output filter for a step-down DC–DC converter [78].

Using the output filter simplified model (Figure 5.25), it is possible to obtain the following equations:

$$L \cong \frac{V_{Lf}}{I_{Lf} \times 2 \times \pi \times f} \tag{5.48}$$

$$ESR \cong \frac{\Delta V_{outf}}{I_{Lf} - \dfrac{\Delta V_{outf}}{R_{LOAD}}} \tag{5.49}$$

where L, ESR, R_{LOAD}, I_{Lf}, V_{Lf} and ΔV_{outf} represent the inductance value, the equivalent series resistance, the load resistance, the amplitude of the first harmonic of inductor current, the amplitude of the first harmonic of inductor voltage and the amplitude of the first harmonic of output voltage ripple, respectively.

Later, in [79], the principle proposed in [76] was also used to design a fault diagnostic technique that is able to evaluate the state condition of Al-Caps used on the primary side of ATX power supplies. These capacitors have a high failure rate, representing one of the weakest components of this equipment. On the other hand, the ageing of Al-Caps may lead to the destruction of other components; in particular, of the transistors that are on the secondary side of ATX power supplies, which reinforces the importance of this ONDT.

The traditional architecture of an ATX power supply consists of two main stages: the primary stage and the secondary stage. The first one, where the mentioned capacitors are, is responsible for converting the AC power into the unregulated DC one. The second one converts the unregulated DC voltage into a regulated one and, for that, it uses a switch mode isolated DC–DC converter, which operates at very high operating frequency.

In this way, it can be concluded that the voltage and current on the capacitors, placed on the primary side of the ATX power supply, have two large harmonics: one at low frequencies (100 Hz or 120 Hz) due to the rectifier bridge and other at the DC–DC converter switching frequency. This second frequency is close to the capacitor resonance frequency; therefore, it is possible to reach the following assumption [79]:

$$ESR = \frac{\Delta v_{cap\_f_2}}{i_{cap\_f_2}} \tag{5.50}$$

where $\Delta v_{cap\_f_2}$ and $i_{cap\_f_2}$ represent the capacitor voltage and current harmonics, at the DC–DC converter switching frequency, respectively.

The effect of the capacitor temperature was also considered in [79] and, for that purpose, the same solution proposed in [71] was used.

The two prior techniques can be implemented through a data acquisition board and PC equipped with an appropriated numerical computation program. The first harmonic, in [78], and the harmonic at the converter switching frequency, in [79], can be obtained from the experimental data using a DFT algorithm.

In [80], the same concept was implemented in a DC–DC boost converter through a DSP.

In this case, and due to the purely digital implementation, the Fast-Fourier-Transform was used to compute both Δv_{outf} and i_{cf}. The experimental results presented in [80] show that the previous values (Δv_{outf} and i_{cf}) are susceptible to noise and to the capacitance reduction, resulting from the capacitor ageing process. Therefore, to account for the previous problems and to implement a threshold point in the system software, the time-average value of both Δv_{outf} and i_{cf} must be computed. This study is performed on the basis of a very severe ageing test.

The following equation shows the time-average value of Δv_{outf} ($\overline{\Delta v_{outf}(n)}$):

$$\overline{\Delta v_{outf}(n)} = \frac{\Delta v_{outf}(n) + (n-1) \times \overline{\Delta v_{outf}(n-1)}}{n} \tag{5.51}$$

where $\Delta v_{outf}(n)$ represents the value of Δv_{outf} at the sampling time n.

The time-average value of i_{cf} ($\overline{i_{cf}(n)}$) can be computed in a similar way, as $\overline{\Delta v_{outf}(n)}$, representing the ratio between the two previous quantities the time-average impedance, $\overline{Z_{cap}(n)}$

$$\overline{ESR(n)} \cong \overline{Z_{cap}(n)} = \frac{\overline{\Delta v_{outf}(n)}}{\overline{i_{cf}(n)}} \tag{5.52}$$

According to the authors, the previous ratio can be used to predict the capacitor health status in the future [80].

The former technique presents some drawbacks, namely, it requires the use of fast analogue-to-digital (A/D) converters with high dynamic resolution for the direct real-time sampling of the capacitor current and voltage [77]; the temperature effect is not considered which may lead to erroneous conclusions; it requires a current sensor inside the power supply, in the same way as the techniques presented in [78] and [79], and the authors do not explain how to predict the capacitor health status in the future.

The use of the extra sensor inside the power supply to acquire the capacitor current brings some shortcomings, namely, the need of space and additional wiring from the capacitor to the output. In turn, extra wiring increases the output voltage ripple and, simultaneously, increases the inductive effect, which is particularly critical in circuits that operate at high frequencies such as SMPS. Thus, some authors have proposed new ONDTs in which other quantities were used instead of the capacitor current.

In [81], the proposed ONDT uses the input current and output voltage ripple, together with a DFT algorithm and LMS algorithm, to predict the capacitor *ESR* of the Al-Cap used in the output filter of step-down DC–DC converters. The *ESR* is computed from a simple analytical relationship between the first harmonic of inductor current, i_{Lf}, and the first harmonic of output voltage ripple, Δv_{outf}:

$$ESR = \frac{\Delta v_{\mathrm{outf}}}{i_{\mathrm{Lf}}} \tag{5.53}$$

In order to avoid the introduction of a current sensor inside the buck converter, the inductor current, i_{L}, is reconstructed using the input current, i_{in}, and an LMS algorithm.

$$\begin{aligned}
i_{\mathrm{L}}(t) &= \begin{cases} m_1 \times t + b_1, t \in [0, D_1 \times T[\\ m_2 \times t + b_2, t \in [D_1 \times T, T[\end{cases} \\
m_1 &= \frac{NPT_1 \times \sum_{i=1}^{NPT_1} (i_{\mathrm{in}}(i) \times t(i)) - \sum_{i=1}^{NPT_1} i_{\mathrm{in}}(i) \times \sum_{i=1}^{NPT_1} t(i)}{NPT_1 \times \sum_{i=1}^{NPT_1} (t(i))^2 - \left(\sum_{i=1}^{NPT_1} (t(i))\right)^2} \\
b_1 &= \frac{\sum_{i=1}^{NPT_1} (t(i))^2 \times \sum_{i=1}^{NPT_1} i_{\mathrm{in}}(i) - \sum_{i=1}^{NPT_1} (i_{\mathrm{in}}(i) \times t(i)) \times \sum_{i=1}^{NPT_1} t(i)}{NPT_1 \times \sum_{i=1}^{NPT_1} (t(i))^2 - \left(\sum_{i=1}^{NPT_1} (t(i))\right)^2} \\
m_2 &= \frac{-2 \times \left(\frac{\langle v_O \rangle}{R} - b_1\right)}{(1 - D) \times T} \quad b_2 = D \times T \times (m_1 - m_2) + b_1
\end{aligned} \tag{5.54}$$

where NPT_1, m_1, m_2, D, t, T and R represent the number of samplings during conduction stage, the slope of i_{in} and i_{L} during conduction stage, the slope of i_{L} during non-conduction stage, the duty cycle, the time vector, the switching period and the load resistance, respectively.

i_{Lf} is computed using DFT and the reconstructed waveform.

However, the previous methodology requires high sampling frequency, which limits its implementation in a low-cost system based on a DSP. In order to overcome the previous drawbacks, in [82] a new technique was proposed, which uses the following analytical relationship to compute the *ESR*:

$$ESR \cong \frac{\frac{dv_{\mathrm{out}}}{dt} \times R}{\frac{di_{\mathrm{L}}}{dt} \times R - \frac{dv_{\mathrm{out}}}{dt}} \tag{5.55}$$

To prevent the introduction of a current sensor inside the buck converter, the slope of the inductor current is calculated from the input current during the conduction stage. In this methodology, the authors consider the temperature effect on the capacitor and, for that, the prior calculation of the temperature multipliers is required. This information will be used, later, to normalize the *ESR* value.

In [83], the authors proposed a new method to detect the rise of the *ESR* of Al-Caps related to the LC filters used in SMPS (Figure 5.25); for this, only the output voltage waveform was used. The preceding technique is based on the following assumptions, some of which have already been mentioned:

- The output voltage ripple is determined by the capacitor's *ESR* and inductor's current ripple.
- Under steady-state regime, the amplitude of inductor current ripple remains unchanged, so, the amplitude of the output voltage ripple is determined by the capacitor's *ESR*.
- Most SMPS have an LC filter to accomplish the output voltage regulation.
- The output voltage ripple is a good indicator of the rise of *ESR*, which, in turn, is an indicator of the capacitor ageing status.

In order to implement this technique, the authors proposed a very simple circuit composed of:

- A band-pass filter, whose function is to allow the closest harmonics to the converter switching frequency to pass, preventing the passage of the remaining harmonics.
- A rectifier and a low-pass filter that converters the previous waveform into a DC one. This voltage is correlated to the *ESR*.
- A hysteretic comparator that compares the previous value with a pre-determined reference voltage, whose output can trigger a warning device.
- A time-delay circuit whose function is to avoid errors due to start-up transients.

However, the effect of temperature is not considered in [83]. Moreover, it has many of the disadvantages of the technique presented in [76], and assumes that *L* remains unchanged.

All methodologies presented so far have been tested on switch mode DC–DC converters, unlike the next two techniques that were applied to PWM adjustable speed drives (ASDs).

The ONDT presented in [84] relies on the fact that, in steady-state regime, the power in the capacitor is fundamentally the result of the losses due to the *ESR*. Therefore, the *ESR* is computed though the ratio between the average power dissipated in the capacitor ($\overline{P}$) to the square of the RMS value of capacitor current (I_{DC}^2):

$$ESR \cong \frac{\overline{P}}{I_{\mathrm{DC}}^2} = \frac{\frac{1}{T}\int_0^t p(t)dt}{\frac{1}{T}\int_0^t i_{\mathrm{DC}}(t)^2 dt} \tag{5.56}$$

where $p(t)$, $i_{\mathrm{DC}}(t)$ and T represent the capacitor instantaneous power, capacitor instantaneous current and the period, respectively.

The method proposed in [84] was implemented in a three-phase 6 kVA/230 V ASD through an analogue-DSP and the temperature effect was also considered.

In [85], the authors report that the failure rate in Al-Caps is roughly equal to twice the failure rate of the power transistors, which reinforces the importance of their early diagnosis. The ONDT presented in [77,85] also uses (5.56), and it was implemented in a low-cost single-chip microprocessor [77,85]. The implementation of the ONDT proposed in [77,85] is quite simple, as can be seen below:

- After acquiring the capacitor voltage ripple and current waveforms, it is possible to obtain both $p(t)$ and $i_{\mathrm{DC}}(t)^2$ by means of multipliers.
- The average values of $p(t)$ and $i_{\mathrm{DC}}(t)^2$ can be derived through low-pass filters.
- The *ESR* value can be computed by a simple division operation.
- The temperature effect is also considered and the temperature multipliers will be used to normalize the *ESR* value.

The authors state that the proposed method does not require specific frequency compensation as in [76] or high-performance DSP as in [80]. Thus, the proposed monitoring unit is directly connected to the Al-Cap and it was realized using a low-cost single-chip microcontroller, implemented in a small PCB, installed between the capacitor screw terminals and the converter bus-bars. The current is sensed through a low resistance (shunt resistor) inserted into the GND current path. The specific position of the monitoring unit, in the vicinity of the electrolytic capacitor, provides a simple way for measuring the capacitor temperature required for evaluating the estimated *ESR*. This can be easily performed since the microcontroller feature an on-chip temperature sensor.

However, since the shunt resistor for the current measurements was inserted in the power wiring of the DC-link, this may reduce the converter's reliability. Nevertheless, the authors claim that this problem can be mitigated with the use, as an alternative to the shunt resistor, of a current-sensing device based on a PCB Rogowski coil sensor (RCS) [85].

In [86], the authors state that the methods already used to determine the capacitor power losses are not accurate because of the capacitor model (Figure 5.6). In the previous model, *ESR* varies with the operating frequency; therefore, in order to compute the total power losses, it is necessary to identify the *ESR* values for all harmonics presented in the capacitor current. For this reason, in [86], a new electrical model for the capacitor is proposed, which considers the variation of the electrolytic capacitor parameters with the temperature and frequency (Figure 5.26).

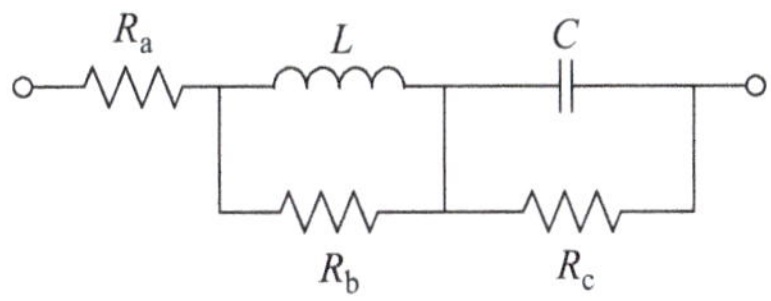

Figure 5.26 Electric model of an Al-Cap [86]

The electrical model parameters (R_a, R_b, R_c, C and L), which are frequency independent, unlike *ESR* and *C* in Figure 5.6, are tuned using a genetic algorithm (GA). GA models with precision the behaviour of the capacitor for large ranges of temperature and frequency. The optimization step was performed by minimizing the error between measurement and algorithm calculation using a cost function given by the following expression [86]:

$$J_{\min} = \frac{1}{2}\sum \left(\log|Z|_{\text{measures}} - \log|Z|_{\text{computation}}\right)^2 \tag{5.57}$$

It was shown that R_a, which is the main component of the capacitor *ESR*, is the best fault indicator since it varies with ageing, unlike R_b, R_c and L, which remain practically constant during Al-Cap wear out.

The proposed Al-Cap model was integrated into a boost-type PFC working under CCM. Two control types were used: PWM and hysteresis controls. The authors revealed that, at steady-state operation of the PFC, the spectral component of the output voltage ripple at the converter switching frequency gives a good image of the resistance R_a. Consequently, the proposed method computes R_a using a reference system and some real-time measurements, such as the converter output voltage ripple, output current and the capacitor case temperature. On the other hand, this method also estimates the remaining lifetime of Al-Caps up to failure.

In [87], a new on-line *ESR* estimation method for solar PV-based DC system has been proposed. The target electrolytic capacitor is connected at the terminals of the solar PV in order to absorb the switching current ripples produced by the converter. It was shown that this method has the ability to work for both CCM and DCM during steady-state regime.

The previous method can be described as follows. First, the PV voltage is used to detect the steady-state regime, after which the operation mode (CCM or DCM) using PV voltage and current is identified. In the following, some coefficients (K_{CCM} and K_{DCM}) are computed, from which *ESR* is subsequently calculated.

ESR can be computed as follows:

- In CCM mode, *ESR* can be calculated as follows:

$$ESR = \frac{L \times K_{\text{CCM}}}{T_s} \tag{5.58}$$

where K_{CCM} is a coefficient reflecting the CCM operation mode, and it is computed using the difference between the solar PV voltages, v_{pv}, sampled at

$t = 0$ and $t = D \times T_S$:

$$K_{\mathrm{CCM}} = \frac{\left\{ v_{\mathrm{PV}}(t)\big|_{t=0} - v_{\mathrm{PV}}(t)\big|_{t=D\times T_s} \right\}}{V_{\mathrm{PV}} \times D} \tag{5.59}$$

The inductor inductance, L, is computed from the inductor current waveform, during initial testing of the converter, D is the duty cycle and T_S the switching period.

- In DCM mode, ESR can be calculated as follows:

$$ESR = \frac{K_{\mathrm{DCM}}}{2} \tag{5.60}$$

where K_{DCM} is a coefficient reflecting the DCM operation mode, and it is computed using the difference between the solar PV voltages sampled at $t = 0$ and $t = 2 \times t_1$:

$$K_{\mathrm{DCM}} = \frac{\left\{ v_{\mathrm{PV}}(t)\big|_{t=0} - v_{\mathrm{PV}}(t)\big|_{t=2\times t_1} \right\}}{I_{\mathrm{PV}}} \tag{5.61}$$

where t_1 represents the time when the capacitor current crosses zero.

$$t_1 = \frac{L \times I_{\mathrm{PV}}}{V_{\mathrm{PV}}} \tag{5.62}$$

The estimated ESR cannot be used directly to monitor the electrolytic capacitor health status because the temperature effect must be taken into account. Therefore, the estimated ESR should be compared to the one of a sound capacitor at the same operating temperature. The proposed technique does not require additional current or voltage sensors. However, the ESR estimation error depends strongly on the accuracy of K_{CCM} and K_{DCM} calculation; moreover, this technique is only feasible in DC–DC converters used to connect the solar PV to the DC system. The authors also present a formula for the capacitance; however, C was not used for the diagnosis and, therefore, was not experimentally validated. For this reason, the prior art has been placed in this section.

In [88], the empirical mode decomposition (EMD) algorithm, combined with Hilbert–Huang Transform (HHT), has been used to detect, in real time, the changes occurred in the ESR value of the Al-Cap presented on the output filter of DC–DC buck converters. It is shown that EMD, which is a signal-processing technique, allows the determination of a number of intrinsic mode functions starting from the output voltage and inductor current. The instantaneous values of the capacitor voltage and current ripples are obtained by applying the EMD method and HHT which, in turn, permits the computation of ESR.

In [89], a new type of RCS has been proposed for the purpose of applying a new ONDT. This method was implemented in non-isolated single-switch DC–DC

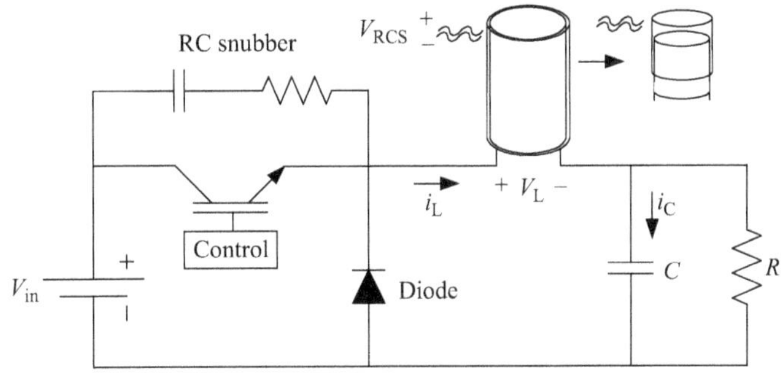

Figure 5.27 The buck converter with RCS [89]

converters operating in different conditions including frequency and temperature variations. The proposed Rogowski coil is like a current transformer, but with no-magnetic core. It has low cost and it is characterized by a linear response for a wide range of frequency (about 2.74 MHz).

It was shown that *ESR* can be successfully estimated using the output voltage ripple (Δv_{out}) and the Rogowski coil output voltage (V_{RCS}) as follows:

$$ESR \approx \frac{\Delta v_{out} - \dfrac{1}{M \times C}\displaystyle\int\left(\int (V_{RCS})dt\right)dt}{\dfrac{1}{M}\displaystyle\int (V_{RCS})dt} \tag{5.63}$$

where M and C represent the mutual inductance of the RCS and the capacitor capacitance, respectively.

This method was tested experimentally using a converter working under continuous conduction mode (Figure 5.27). The effectiveness of this method was studied in different operation conditions including temperature and frequency variations.

5.6.2 *On-line fault diagnostic techniques based on* ESR *and* C *estimation*

In this subsection, some ONDTs will be presented that evaluate the capacitors state condition through the simultaneous identification of the values of *ESR* and *C*. These techniques can be applied to circuits composed of one or both types of capacitors (Al-Caps and MPPF-Caps).

The first ONDT presented in this subsection uses circuit model-based methods; these new methods were not used in the ONDTs presented in the previous section.

In [90,91], the authors use the hybrid model of the circuit under analysis, together with the recursive least mean square (RLMS) algorithm, to estimate both *ESR* and *C* values of the DC-link capacitors.

Therefore, in order to implement the technique proposed in [90], first, it is necessary to obtain the circuit hybrid model. In [90,91], a buck converter was used to validate the proposed technique, whose hybrid model is presented below:

$$\begin{bmatrix} \dot{i}_{\mathrm{L}} \\ \dot{u}_{\mathrm{out}} \end{bmatrix} = A \times \begin{bmatrix} i_{\mathrm{L}} \\ u_{\mathrm{out}} \end{bmatrix} + (s_1 + s_2) \times B \times \begin{bmatrix} i_{\mathrm{L}} \\ u_{\mathrm{out}} \end{bmatrix} + s_1 \times \begin{bmatrix} \dfrac{V_{\mathrm{in}}}{L} \\ \dfrac{ESR \times R \times V_{\mathrm{in}}}{(ESR + R) \times L} \end{bmatrix}$$

$$A = \begin{bmatrix} 0 & 0 \\ 0 & \dfrac{-1}{(R + ESR) \times C} \end{bmatrix}, B = \begin{bmatrix} 0 & \dfrac{-1}{L} \\ \dfrac{R}{(R + ESR) \times C} & \dfrac{-R \times ESR}{(R + ESR) \times L} \end{bmatrix} \tag{5.64}$$

where i_{L}, u_{out}, R, ESR, C, L and V_{in} represent the inductor current, output voltage, load resistance, capacitor equivalent series resistance, capacitor capacitance, the inductor inductance and the input voltage, respectively. s_{i} represents the switch vector: s_1 the MOSFET and s_2 the diode; thus s_1 and s_2 cannot be ON at the same time.

The RLMS method is used to identify the parameters on the hybrid model, from which it is possible to determine the values of the R, ESR, C and L.

The RLMS method is based on linear regression models of the form [90,91]:

$$y(t) = \varphi(t) \,.\, \theta^T \tag{5.65}$$

where $y(t)$, $\varphi(t)$ and θ represent the output vector, the regression matrix and the parameter vector, respectively.

For the buck converter, at instant t, the output vector and the regression matrix can be written in the following manner [90,91]:

$$\begin{aligned} y(t) &= \left[i_{\mathrm{L}}(t), u_{\mathrm{out}}(t)\right]^T \\ \varphi(t) &= \left[i_{\mathrm{L}}(t-1) \cdot I_{2\times2} \;\; u_{\mathrm{out}}(t-1) \cdot I_{2\times2} \;\; s_1(t-1) \cdot I_{2\times2} \right. \\ &\quad \left. (s_1(t-1) + s_2(t-1)) \cdot i_{\mathrm{L}}(t-1) \cdot I_{2\times2} \; (s_1(t-1) + s_2(t-1)) \cdot u_{\mathrm{out}}(t-1) \cdot I_{2\times2}\right] \end{aligned} \tag{5.66}$$

where $I_{2\times2}$ represents the 2×2 unit matrix.

Therefore, after sampling i_{L}, v_{out} and s_1 waveforms during two switching periods, it is possible to get the following sets:

$$\mathbf{\Phi}_N = \begin{bmatrix} \varphi(1) \\ \varphi(2) \\ \vdots \\ \varphi(N) \end{bmatrix}, Y_N = \begin{bmatrix} y(1) \\ y(2) \\ \vdots \\ y(N) \end{bmatrix} \tag{5.67}$$

where N represents the number of samples.

s_1 waveform is equal to 1, when the MOSFET conducts, and zero, when it does not conduct. Therefore, s_1 can be easily obtained from the MOSFET control signal. s_2 waveform is equal to the s_1 complement in CCM, so, it can be obtained from s_1. In DCM, to obtain s_2, it is also necessary to check when the current i_L is zero.

The estimation vector can be obtained in the following manner:

$$\begin{aligned}
&\theta = \left[{\Phi_N}^T \cdot \Phi_N\right]^{-1} \cdot {\Phi_N}^T \cdot Y_N \\
&\theta = \left[\theta_1\ \theta_2\ \theta_3\ \theta_4\ \theta_5\ \theta_6\ \theta_7\ \theta_8\ \theta_9\ \theta_{10}\right] \\
&\theta = \left[1\ 0\ 0\ \left(1 - \frac{T}{(ESR + R) \times C}\right)\ \frac{V_{in} \times T}{L}\ \frac{ESR \times R \times V_{in} \times T}{(ESR + R) \times L}\right. \\
&\qquad \left. 0\ \frac{R \times T}{(ESR + R) \times C}\ -\frac{T}{L}\ -\frac{ESR \times R \times T}{(ESR + R) \times L}\right]
\end{aligned} \tag{5.68}$$

Finally, it is possible to obtain R, ESR, C and L [90]:

$$\begin{aligned}
&L = \frac{V_{in} \times T}{\theta_9};\ R = \frac{(\theta_2 + \theta_6) \times V_{in}}{(1 - \theta_4 - \theta_8) \times V_{in} - \theta_{10}} \\
&C = \frac{T \times R - \theta_{10} \times \dfrac{L}{V_{in}}}{(\theta_2 + \theta_6) \times R};\ ESR = \frac{\theta_{10} \times L}{(\theta_2 + \theta_6) \times V_{in} \times C}
\end{aligned} \tag{5.69}$$

where T represents the sampling time.

In [90], a high speed data acquisition card was used with a maximum sampling frequency of 20 MHz together with an industrial PC.

Later, in [91], a simplification was proposed in the regression model for both buck and boost converters. In this way, it is possible to reduce the computational effort involved in the calculations.

The estimation vector (θ) of both converters (buck and boost) presents several null and unity terms and also some terms that are proportional among them. In this way, it is possible to reduce the order of the regression model without affecting the computation of the desired parameters (R, ESR, C and L). Thus, for the buck and boost converters, after simplification, the output vector ($y(t)$), the regression matrix ($\varphi(t)$) and the estimation vector (θ) can be obtained.

Thus, for the buck converter [91]:

$$\begin{aligned}
&y(t) = \left[i_L(t) - i_L(t-1), u_{out}(t)\right]^T \\
&\varphi(t) = \begin{bmatrix} 0 & 0 & \begin{pmatrix} s_{12}(t-1) \cdot u_{out}(t-1) \\ -V_{in} \cdot s_1(t-1) \end{pmatrix} & 0 \\ u_{out}(t-1) & s_{12}(t-1) \cdot i_L(t-1) & 0 & \begin{pmatrix} s_{12}(t-1) \cdot u_{out}(t-1) \\ -V_{in} \cdot s_1(t-1) \end{pmatrix} \end{bmatrix} \\
&s_{12}(t-1) = s_1(t-1) + s_2(t-1) \\
&\theta = \left[\left(1 - \frac{T}{(ESR+R) \times C}\right)\ \frac{R \times T}{(ESR+R) \times C}\ -\frac{T}{L}\ -\frac{ESR \times R \times T}{(ESR+R) \times L}\right]
\end{aligned} \tag{5.70}$$

And, for the boost converter [91]:

$$
\begin{aligned}
y(t) &= [i_L(t) - i_L(t-1), u_{out}(t)]^T \\
\varphi(t) &= \begin{bmatrix} 0 & 0 & \begin{pmatrix} s_2(t-1)\cdot u_{out}(t-1) \\ -V_{in}\cdot s_{12}(t-1) \end{pmatrix} & 0 \\ u_{out}(t-1) & s_2(t-1)\cdot i_L(t-1) & 0 & \begin{pmatrix} s_2(t-1)\cdot u_{out}(t-1) \\ -V_{in}\cdot s_2(t-1) \end{pmatrix} \end{bmatrix} \\
s_{12}(t-1) &= s_1(t-1) + s_2(t-1) \\
\theta &= \left[\left(1 - \frac{T}{(ESR+R)\times C}\right) \frac{R\times T}{(ESR+R)\times C} - \frac{T}{L} - \frac{ESR\times R\times T}{(ESR+R)\times L} \right]
\end{aligned}
\tag{5.71}
$$

From the estimation vector, it is possible to determine *ESR*, *C*, *L* and *R* values using RLMS method.

By comparing (5.66) and (5.68) with (5.70), one can easily conclude that the method proposed in [91] involves a smaller computational effort than the method proposed in [90].

The previous methodologies, although accurate, require a very high sampling frequency, at least in the order of MHz. This situation is due to the fact that the derivatives are computed through the classical differential approach, whose accuracy depends on the signal sampling frequency. To overcome this inconveniences, in [92], the continuous time model of the buck converter was used, together with the LMS algorithm, taking into account only the converter non-conduction stage.

This new approach requires a much smaller sampling frequency since the derivatives are evaluated through polynomial interpolation. Besides, the computational effort is lower when compared to the previous technique, because the calculus is simpler and smaller.

The main steps for the implementation of this new method are presented below in an orderly fashion [92]:

1. In a first moment, the output voltage, v_{out}, the inductor current, i_L, and the MOSFET gate signal, v_{gate}, are sampled during two switching cycles of the converter under analysis (buck converter).
2. Then, both average values of inductor current, $\langle i_L \rangle$, and output voltage, $\langle v_{out} \rangle$, are obtained, through trapezoidal integration:

$$\langle x \rangle = \frac{1}{N-1} \sum_{i=2}^{N} \frac{x(i) - x(i-1)}{2} \tag{5.72}$$

   where x and N represent the state variables (i_L, v_{out}) and the number of samples in one switching period, respectively.
3. Following, both i_L and v_{gate} waveforms are used to identify the converter conduction stage (CCM or DCM).

4. After the previous step, it is already possible to compute the load resistance value, R, through the converter average state model:

$$\langle i_L \rangle_{CCM} = \frac{\langle v_{out} \rangle_{CCM}}{R} \quad \langle i_L \rangle_{DCM} = \frac{\langle v_{out} \rangle_{DCM}}{R \times (D_1 + D_2)} \tag{5.73}$$

where D_1 and D_2 represent the duty cycle and the period the inductor is being discharged, respectively.

5. ESR is estimated from the following equation, proposed in [72]:

$$ESR = \frac{\Delta v_{out} \times R}{\Delta i_L \times R - \Delta v_{out}} \tag{5.74}$$

where Δv_{out} and Δi_L represent the output voltage ripple and inductor current ripple, respectively.

6. Finally, the values of L and C are evaluated from the converter continuous time model obtained during non-conduction stage:

$$\begin{cases} \dfrac{di_L(t)}{dt} = -\dfrac{v_{out}(t)}{L} \\ \dfrac{dv_{out}(t)}{dt} = \left(\dfrac{R}{(R + ESR) \times C}\right) \\ \qquad \times i_L(t) + \left(\dfrac{-1}{(R + ESR) \times C} - \dfrac{R \times ESR}{(R + ESR) \times L}\right) \times v_{out}(t) \end{cases} \tag{5.75}$$

where the derivatives are obtained through polynomial interpolation and both values of L and C are computed through the application of the LMS algorithm to the previous equations.

To guarantee the reliability of the proposed technique, the authors impose the following condition [92]:

$$D_2 \times \frac{f_{samp}}{f_{SW}} \geq 2 \times (m + 1) \tag{5.76}$$

where f_{samp} and f_{SW} represent the sampling frequency and the converter switching frequency, respectively; and m is used for evaluating the number of samples in the data window for computing the derivative of each sample.

A methodology similar to the previous one was applied to both boost and buck-boost converters in [93]. However, in this case, ESR is computed through (5.41) and both C and L are obtained during conduction stage, by means of the LMS algorithm. Thus, for L:

$$\begin{aligned} \theta &= \frac{V_{in}}{L} = \left[\varphi^T \cdot \varphi\right]^{-1} \cdot \varphi^T \cdot y \\ y &= \left[i_L(2) - i_L(1), i_L(3) - i_L(1), \ldots, i_L(N) - i_L(1)\right]^T \\ \varphi &- \left[t(2) - t(1), t(3) - t(1), \ldots, t(N) - t(1)\right]^T \end{aligned} \tag{5.77}$$

and for C:

$$\theta = \frac{-1}{C \times (ESR + R)} = \left[\varphi^T \cdot \varphi\right]^{-1} \cdot \varphi^T \cdot y$$
$$y = \left[\ln\left(\frac{v_{out}(2)}{v_{out}(1)}\right), \ln\left(\frac{v_{out}(3)}{v_{out}(1)}\right), \ldots, \ln\left(\frac{v_{out}(N)}{v_{out}(1)}\right)\right]^T \quad (5.78)$$
$$\varphi = [t(2) - t(1), t(3) - t(1), \ldots, t(N) - t(1)]^T$$

where N represents the number of samples obtained during the conduction stage.

However, the accuracy of the techniques proposed in [92] and [93] depends on the duty cycle. In the case of the technique proposed in [92], if the duty cycle is too high, the number of samples of interest will be reduced, so that condition (5.76) will not be respected; a similar situation will occur in the technique proposed in [93] but, in this case, if the duty cycle is too low. This could disregard one of the greatest advantages of the techniques proposed in [92] and [93] over the techniques proposed in [90] and [91], which is the less number of samples needed for their application.

In order to overcome the previous drawback, a unified method was proposed in [94] for three non-isolated converters: buck, buck-boost and boost converters. In this new method, the samples of interest can be acquired either in the conduction or non-conduction stage, accordingly to the higher number of samples available. Therefore, the estimation of L and C is performed for the state with the highest number of samples and, for that, the continuous-time model related to that specific state is used. The authors reported that this technique can be quite precise with only 25 samples per converter switching period.

The techniques proposed in [92–94] require a smaller number of samples; in addition, the proposed algorithms are considerably simpler and faster, so, they can be easily implemented in a DSP. On the other hand, the techniques proposed in [90,91], although precise, require a very high sampling frequency, at least five times higher when compared with the techniques proposed in [92–94]. Moreover, the mathematical calculations are heavy, thus, their implementation involves an industrial PC for better performance.

However, it should be mentioned that the techniques presented in [90–94] require the introduction of, at least, three sensors in the converter; in addition, the effect of the temperature is not taken into account.

In [95], the authors proposed an ONDT for capacitors used in the output filter of boost converters connected to an unregulated AC–DC converter. Thus, the ESR is obtained in the following manner:

$$ESR = \frac{V_{fs}}{I_{fs}} \quad (5.79)$$

where V_{fs} and I_{fs} represent the RMS values of the capacitor voltage and current at the converter switching frequency, respectively.

In turn, the capacitive reactance, X_{Cap}, and so, the capacitance, C, can be calculated at the frequency of 120 Hz (twice the line frequency), as follows:

$$X_{Cap} = (2 \times \pi \times 120 \times C)^{-1} = \frac{V_{120\,Hz}}{I_{120\,Hz}} \tag{5.80}$$

where $V_{120\,Hz}$ and $I_{120\,Hz}$ represent the RMS values of the capacitor voltage and current at 120 Hz, respectively.

In order to implement the proposed methodology, the capacitor current and voltage must pass through a band-pass filter, whose central frequency depends on the computed value: in the case of *ESR*, the centre frequency must be close to the converter switching frequency; while in C it must be close to 120 Hz. Then, the obtained values must pass through a RMS calculator. Finally, (5.79) and (5.80) must be performed through an automatic-gain-controller [95].

The authors reported that the proposed technique works in non-stationary systems provided that there is a good matching in the band-pass filter magnitude and phase response for the capacitor voltage and current [95]. However, the previous technique does not consider the temperature effect and it requires the use of a current sensor near the capacitor. Besides, it is conditioned by the fact that the DC–DC converter must be connected to an unregulated AC–DC converter; therefore, its use is conditioned by the topology of the circuit.

In [96], the authors proposed a new ONDT for UPS. This new methodology takes the advantage of the fact that it only uses the resources already existing in the UPS, such as some sensors and powerful computational resources. Therefore, the authors claim that the proposed methodology can work in background task without disturbing the normal operation of the system. The proposed methodology uses the capacitor transfer function, together with a Kalman filter, to extract both values of C and *ESR*.

The capacitor ESL can be neglected because the UPS converter operates far below the resonant frequency of the capacitor. Thus, the capacitor transfer function can be represented as follows:

$$H(s) = \frac{V_C(s)}{I_C(s)} = \frac{ESR \times C \times s + 1}{C \times s} \tag{5.81}$$

However, since the converters are digitally controlled, it is necessary to convert the transfer function using the z-transform [96]:

$$H(z^{-1}) = \frac{b_0 + b_1 \times z^{-1}}{1 - z^{-1}}; b_0 = ESR + \frac{T_S}{2 \times C}; b_1 = \frac{T_S}{2 \times C} - ESR \tag{5.82}$$

where T_S represents the sampling period.

In order to extract both values of b_0 and b_1, the authors use different forms of Kalman Filter.

The converter used was a boost converter, which had a current sensor in the inductor. Therefore, the capacitor current, I_c, was obtained through the following equation [96]:

$$I_c = I_l \times \overline{\mathrm{PWM}} - \langle I_l \times \overline{\mathrm{PWM}} \rangle \tag{5.83}$$

where I_l, PWM and $\langle\rangle$ represent the inductor current, the pulse width modulator and the average value, respectively.

The temperature effect is also considered in the computation of C and ESR.

In addition to the computation of the ESR and C values, the authors also present an algorithm that determines the capacitor time-to-failure, t_{failure}, and, for this purpose, they use the computed values of ESR and C. Thus, the t_{failure} is calculated from the lowest value between the time-to-failure obtained from ESR ($t_{\text{failureESR}}$), or from C (t_{failureC}) [96]:

$$\begin{aligned}
t_{\text{failure}} &= \min(t_{\text{failureESR}}, t_{\text{failureC}}) \\
t_{\text{failureESR}} &= t'_{\text{ESR}} - t_{\text{ESR}};\quad t_{\text{ESR}} = \frac{\log\left(\dfrac{ESR}{ESR_0 + A_1}\right)}{B_1}; \\
t'_{\text{ESR}} &= t_{\text{A ESR}} \times \exp\left(\frac{E_{\text{A ESR}}}{k} \times \frac{T' - T_{\text{A}}}{(T' + 273) \times (T_{\text{A}} + 273)}\right) \\
t_{\text{failureC}} &= t'_{\text{C}} - t_{\text{C}};\quad t_{\text{C}} = \frac{C - E \times C_0}{F}; \\
t'_{\text{C}} &= t_{\text{AC}} \times \exp\left(\frac{E_{\text{AC}}}{k} \times \frac{T' - T_{\text{A}}}{(T' + 273) \times (T_{\text{A}} + 273)}\right)
\end{aligned} \tag{5.84}$$

where:

- T_{A} and T' represent the ambient and the ageing temperature, respectively.
- t_{AESR} and t_{AC} represent the ageing time for the ESR and C limit at T_{A} with ESR and C as ageing indicators, respectively.
- t'_{ESR} and t'_{C} represent the lifetime limit at T' with ESR and C as ageing indicators, respectively.
- k, E_{AESR} and E_{AC} represent the Boltzmann constant and the activation energy of ESR and C as ageing indicators, respectively.
- A_1, B_1, E and F values are obtained experimentally and depend on the capacitor.
- ESR_0 and C_0 represent the values of ESR and C of a sound capacitor, respectively.

In this case, one of the greatest advantages of the ONDT can also be seen as a disadvantage, since many power electronics applications do not have the sensors and/or the computational power of the prototype used in [96].

In [97], the authors presented an ONDT capable of assessing both Al-Caps and MPPF-Caps health status used in railway high-power applications; therefore, the proposed technique is based on the double estimations of C and ESR.

In order to evaluate the effectiveness and accuracy of the proposed technique, a simple boost converter was used, because the technique is intended for the capacitors present in the DC-link after a boost stage. The authors claim that the proposed technique can be extended to other topologies.

In a boost converter, during conduction stage and discontinuous stage, the relationship between capacitor current, i_C, and voltage, v_C, can be modelled by (5.85), because i_C is almost constant:

$$\frac{dv_c}{dt} = \frac{1}{C} \times i_C(t) \tag{5.85}$$

where $\frac{dv_c}{dt}$ represents the first derivative of v_C.

However, during non-conduction stage, the previous relationship changes to (5.86), because i_C varies with time despite being linear.

$$\frac{dv_c}{dt} = ESR \times \frac{di_C(t)}{dt} + \frac{1}{C} \times i_C(t) \tag{5.86}$$

where $\frac{di_c}{dt}$ represents the first derivative of i_C.

The principle of the proposed technique is based on the previous equations. Therefore, first, C is computed using (5.85); afterward, ESR is estimated through (5.86). The above principles represent a good approximation since the ESL effect can be neglected during each stage, as well as, ESR, whose value is negligible when compared to the load resistance, R_{Load}.

However, to implement the proposed technique, the authors suggest the use of the input current, i_{in}, instead of i_C, in order to avoid the problems associated to the introduction of a current sensor in the branch of the capacitor.

In this way, using i_{in}, it is possible to identify each state and, thus, separate the samples of interest relative to i_{in} and v_C. In turn, using the previous data, it is possible to reconstruct i_C as follows:

- For conduction and discontinuous stage:

$$i_C(t) = -i_{out}(t) = -\frac{v_C(t)}{R_{Load}} \tag{5.87}$$

- For non-conduction stage:

$$\frac{dv_c}{dt} = ESR \times \frac{di_C(t)}{dt} + \frac{1}{C} \times i_C(t) \tag{5.88}$$

where i_{out} represents the output current.

Finally, using i_C and v_C, it is possible to estimate the ESR and C values using (5.85) and (5.86), in conjunction with the LMS algorithm. In the computation of the derivatives, a polynomial fitting was used.

In the case of a stationary resistive load, its value (R_{Load}) can be estimated through the converter average state model as it was done in [92–94]. If this is not the case, it is necessary to introduce a current sensor at the output of the converter to measure i_{out}.

This technique requires low sampling frequency and little computational effort; therefore, it can be easily implemented in a DSP, as demonstrated in [97].

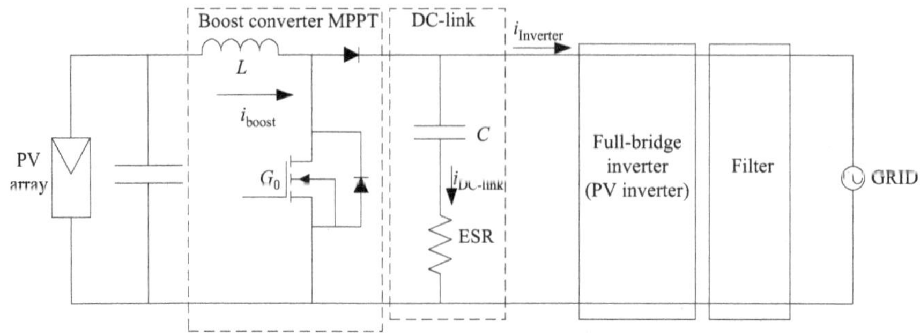

Figure 5.28 Power circuit of grid-connected photovoltaic inverter [98]

On the other hand, it can be used in any application that contains a boost converter, namely, in rail traction drive applications. However, in such applications, it is necessary to introduce some more current sensors as shown in [97]. Additionally, the authors pointed out that this technique can be used in train workshops to make off-line measurements of demounted capacitors of trains under maintenance.

In [98], a non-invasive on-line method for fault detection of Al-Caps, in the DC-link of grid-connected PV inverters, was presented. This method allows the estimation of both C and ESR using the DC-link voltage measurement and an indirect measurement of the DC-link current. The DC-link current can be measured through a current sensor; however, this solution has the drawbacks already mentioned. According to Figure 5.28, the DC-link current can be obtained indirectly through the following expression:

$$i_{\text{DC-link}} = \overline{G_0} \times i_{\text{boost}} - i_{\text{inverter}} \tag{5.89}$$

where G_0 represents the gate signal of the boost semiconductor switch.

The capacitor capacitance can be computed using the values of the DC-link voltage ($V_{\text{DC-link}}$) in two different instants (t_{n-1} and t_n), in which the DC-link current is zero ($I_{\text{DC-link}} = 0$). In this circumstance, the DC-link voltage equals the capacitor voltage (V_{C}). Consequently, the capacitance can be computed using the following expression [98]:

$$C = \frac{\int_{t_{n-1}}^{t_n} I_{\text{DC-link}}\, dt}{\Delta V_{C_{tn}}} \tag{5.90}$$

Afterwards, the obtained C, the measured DC-link voltage and the DC-link current, when its value is non-zero, are used to calculate the ESR as follows:

$$ESR = \frac{V_{\text{DC-link}_{(tm)}} - \left(\frac{1}{C}\int_{t_n}^{t_m} I_{\text{DC-link}}\, dt + V_{C_{tn}}\right)}{I_{\text{DC-link}(tm)}} \tag{5.91}$$

At t_m, the DC-link current is non-zero.

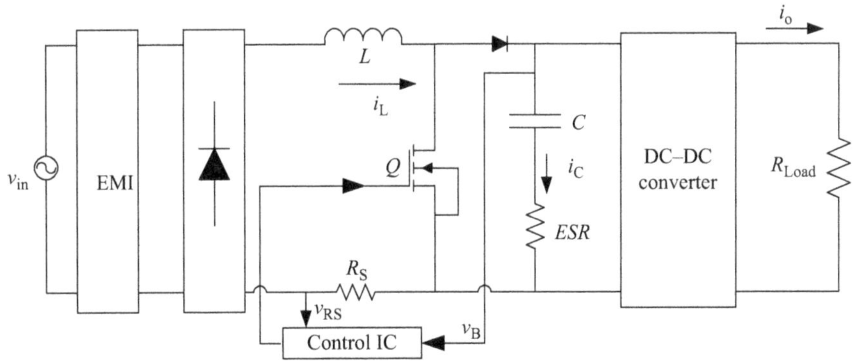

Figure 5.29 Main circuit of the boost PFC converter [99]

It is clear that the DC-link current and voltage ripples affect significantly the *ESR* and *C* values. For this reason, the authors have used the RLMS algorithm to optimize the *ESR* and *C* estimation. The temperature effect was also considered by implementing, with look-up tables, the relationship between the temperature and the estimated *ESR* and *C* values. The proposed technique can be implemented in the photovoltaic inverter central microcontroller, which allows the *C* and *ESR* values to be normalized according to the actual operating temperature.

In [99], another technique was successfully used for on-line estimation of *ESR* and *C* applied in a boost PFC converter (Figure 5.29). The major advantage of this ONDT is that it does not require a current sensor, but instead it uses two values of capacitor voltage, in particular moments within a line cycle, to compute the Al-Caps *ESR* and *C*.

According to [99], *ESR* and *C* can be calculated using the following expressions:

$$ESR = \frac{V_B \times [V_B - v_B(0)]}{P_0} = -\frac{V_B}{P_0} \times \tilde{v}_B(0) \tag{5.92}$$

$$C = \frac{P_0}{2 \times \omega \times V_B \times \left[V_B - v_B\left(\frac{\pi}{4}\right)\right]} = -\frac{P_0}{2 \times \omega \times V_B \times \tilde{v}_B\left(\frac{\pi}{4}\right)} \tag{5.93}$$

where $\tilde{v}_B(0)$ and $\tilde{v}_B\left(\frac{\pi}{4}\right)$ are, respectively, the capacitor voltage ripple at 0 and $\frac{\pi}{4}$ in a half-cycle; V_B is the average capacitor voltage and P_0 is the output power, which is equal to the average value of the input power. This last assumption is true if the PFC's efficiency is supposedly 1.

This technique uses two capacitor voltage values in two particular instants within a half-line cycle. Therefore, its implementation requires an external trigger circuit isolated from the main power circuit. The trigger signals are generated at 0 and $\frac{\pi}{4}$ using the AC input voltage. Hence, the overall system is made by a PFC converter, a DC–DC converter, a trigger circuit, a current-isolated amplifier, a voltage-isolated amplifier, an MCU (microcontroller unit) and an LCD.

Similar on-line technique was also applied for a buck converter operating in CCM at different conditions [100]. For this application, the authors have proposed two expressions to calculate C and ESR:

$$ESR = \frac{2 \times L \times f_s \times \left[\tilde{v}_O(0) + \frac{2 \times (2 \times D - 1)}{(2 - D)} \times \tilde{v}_O\left(\frac{D \times T_s}{2}\right)\right]}{V_O \times (D - 1)} \tag{5.94}$$

$$C = \frac{V_O \times (2 - D) \times (D - 1)}{24 \times L \times f_s^2 \times \tilde{v}_O\left(\frac{D \times T_s}{2}\right)} \tag{5.95}$$

where L, f_s, T_s and D represent, respectively, the inductance, the switching frequency, the switching period and the duty cycle; V_O is the average output voltage and $\tilde{v}_O$ is output voltage ripple.

According to (5.94) and (5.95), ESR and C calculation requires trigger signals at 0 s and $\frac{D \times T_s}{2}$ s, which allow the measure of the output voltage at these specific instants. Therefore, the use of capacitor current is not necessary. The trigger circuit uses the pulse-width-modulation signal from the control circuit (SG3525). This on-line technique is well suitable for DC–DC converters working in CCM, but this mode cannot be guaranteed in all operating conditions.

In [101], the short-time least square Prony's (STLSP) method has been proposed for estimating the ESR and C values of Al-Caps, used in DC–DC boost converters connected to unregulated AC–DC power supplies (Figure 5.30).

It is known that the changes on the electrolytic capacitor parameters strongly affect the ratio between the capacitor voltage ripple and their current ripple. This ratio is dominated by C at low frequencies and by the ESR in the high frequency range. Indeed, C and ESR can be expressed by the following ratios:

$$C = \frac{I_{f_m}}{2 \times \pi \times f_m \times V_{f_m}} \tag{5.96}$$

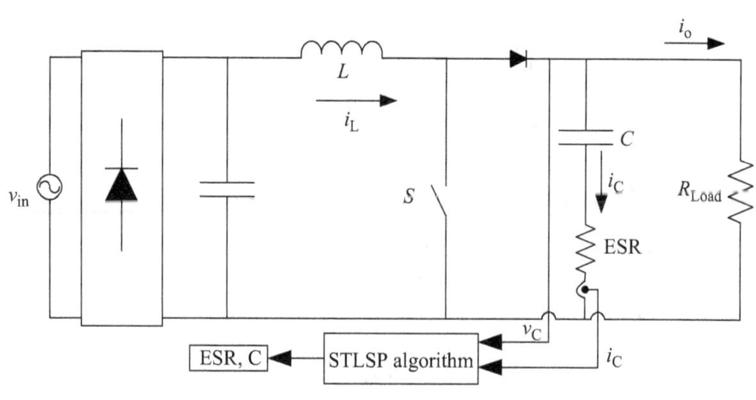

Figure 5.30 The DC–DC boost converter [101]

$$ESR = \frac{V_{\mathrm{f_{sw}}}}{I_{\mathrm{f_{sw}}}} \tag{5.97}$$

where V_{fm} and I_{fm} are the amplitudes of double-mains supply frequency harmonic ($f_{\mathrm{m}} = 100$ Hz), for the capacitor current and voltage ripples; V_{fsw} and I_{fsw} are the amplitudes of switching frequency harmonic, for the capacitor current and voltage ripples, respectively.

The magnitudes of the double-grid frequency component and the switching frequency component, which always exist in the capacitor voltage and current ripples, can be instantaneously tracked. Therefore, the target ratio can be also instantaneously computed, which allows the on-line estimation of the C and ESR values. This operation has been successfully performed, in [101], using the STLSP.

The original Prony's method is a signal-processing technique for the extraction of sinusoidal or exponential signals by solving a set of linear equations. Assuming the signal $x(t)$ and its N complex samples, the Prony's method approximates the sampling data with linear combination of P complex exponential functions [102]:

$$x[n] = \hat{x}[n] + \varepsilon[n] = \sum_{k=1}^{P} h_k z_k^{n-1} + \varepsilon[n] \tag{5.98}$$

with $h_k = A_k e^{j\varphi k}$, $z_k = e^{(\alpha_k + j2\pi f_k)T_S}$ and T_S is the sampling time.

The model parameters A_K, f_k, φ_k and α_k represent, respectively, the unknown amplitude, frequency, phase angle and damping factor of the kth component; and ε represents the approximation error between the original data samples, $x[n]$, and the linear approximation, $\tilde{x}[n]$. This error is assumed to be white and with a Gaussian distribution.

Equation (5.98) represents a difficult non-linear problem which can be solved by using the Prony's method. In fact, Prony's method turns the parameters estimation non-linear problem, into a solution of a linear system and roots calculation of the polynomial. Therefore, the Prony's method constructs a homogenous linear differential equation with constant coefficients (with $a_0 = 1$):

$$\sum_{k=0}^{P} a_k x[n-k] = \varepsilon[n] \tag{5.99}$$

The available N data samples are used to rewrite (5.99) in a matrix form:

$$\begin{bmatrix} x[P] & \cdots & x[1] \\ \vdots & \ddots & \vdots \\ x[N-1] & \cdots & x[N-P] \end{bmatrix} \begin{bmatrix} a_1 \\ \vdots \\ a_P \end{bmatrix} = - \begin{bmatrix} x(P+1) \\ \vdots \\ x(N) \end{bmatrix} \tag{5.100}$$

The vector of the unknown parameters a_k is selected to minimize the linear prediction total squared error. The minimization can be solved by using the least squares method. Then, a characteristic polynomial with roots z_K can be formed using the linear prediction parameters as follows:

$$F(z) = \sum_{k=0}^{P} a_k z^{P-k} \tag{5.101}$$

As a result, the damping factor and the frequency can be deduced directly from the roots z_K of (101):

$$\alpha_k = \frac{\ln|z_k|}{T_s} \text{ and } f_k = \frac{1}{2\pi T_s}\tan^{-1}\left[\frac{\text{Im}(z_k)}{\text{Re}(z_k)}\right]$$

Finally, the roots z_K are used to write the P equations of (5.98) in a matrix form as:

$$Z \times H = C, \text{with } Z = \begin{bmatrix} 1 & \cdots & 1 \\ z_1 & \cdots & z_P \\ \vdots & \cdots & \vdots \\ z_1^{N-1} & \cdots & z_P^{N-1} \end{bmatrix}, \ H = \begin{bmatrix} h_1 \\ \vdots \\ h_P \end{bmatrix}, \ C = \begin{bmatrix} x(1) \\ \vdots \\ x(N) \end{bmatrix} \tag{5.102}$$

The estimation of the complex parameters h_K is switched also to a linear least-squares procedure, and consequently, the exponential amplitudes, A_K, and phase angles, φ_K, can be obtained using the following relationships [102]:

$$A_k = |h_k| \text{ and } \varphi_k = \tan^{-1}\left[\frac{\text{Im}(h_k)}{\text{Re}(h_k)}\right] \tag{5.103}$$

It is worth mentioning that the complexity of the Prony's method depends strongly on the model order, P, and the number of treated data samples, N. In fact, the efforts done for proper selection of P can be considerably reduced if there is a prior knowledge of the number of searched harmonics. It was shown that $P = 3$ is sufficient to obtain good results, while N is chosen according to the swiftness required and the signal quality.

Therefore, the signals of the capacitor voltage and current ripples are divided into short overlapped time windows and each one is analysed by the proposed method (Figure 5.31).

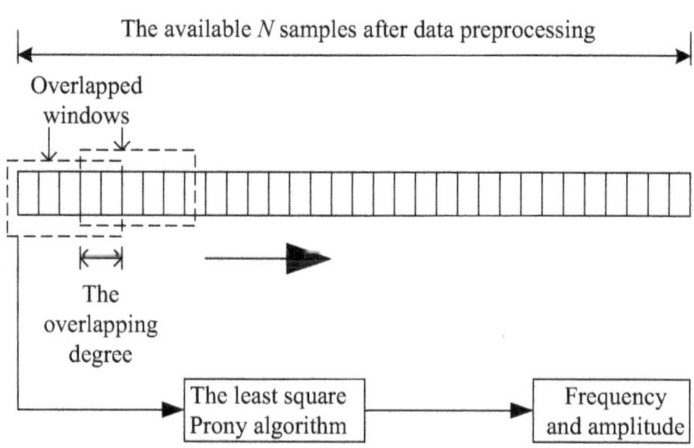

Figure 5.31 Description of the STLSP [101]

It is shown that the STLSP technique is well suited for such application since it has the ability to estimate and track the amplitude and frequency of any spectral component, even for noisy and non-stationary signals, using a small number of data samples. This reduces considerably the calculation time and the storage requirements.

The scheme of the proposed parameter estimation algorithm is presented in Figure 5.32. The capacitor's voltage and current are first pre-processed, and then they are analysed through the application of the STLSP method which permits the estimation and tracking of the magnitudes of both fundamental and switching harmonic components. The capacitor parameters can then be easily deduced by using (5.96) and (5.97).

This technique was experimentally validated and the need for a current sensor in the capacitor represents a major drawback that could be overcome by indirect measurement of capacitor current. This can be done using the converter's input current and the switch's control signal. On the other hand, the capacitor core temperature should also be considered in this technique.

5.6.3 On-line fault diagnostic techniques based on C *estimation*

In this subsection, some ONDTs will be presented that evaluate the capacitors' state condition through the identification of their C value. As discussed at the beginning of this section, these techniques were initially developed for MPPF-Caps.

As stated in Section 5.1, MPPF-Caps can withstand much higher levels of surge voltage, higher RMS currents and offer longer lifetimes when compared with Al-Caps. These features, despite their large size, weight and cost, make them the

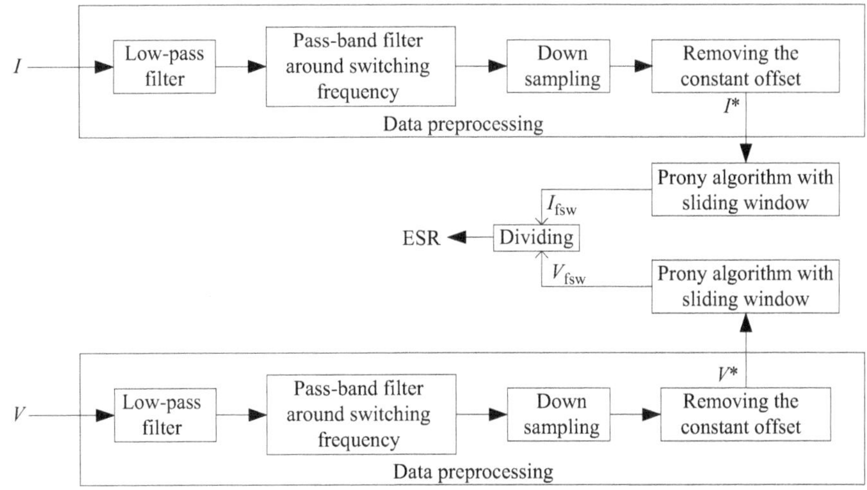

Figure 5.32 Illustrative scheme of the proposed parameter estimation technique [101]

best option for high-power applications, such as railway traction applications. It was also clear from Section 5.4 that the best end-of-life limit criterion for MPPF-Caps is the loss of 2% to 10% of the initial capacitance [2,46–49]. Therefore, in circuits composed of MPPF-Caps, the best failure indicator is C.

In [46], the authors propose an ONDT for the MPPF-Caps, installed in the DC-link of railway power trains, which evaluates their health status through the estimation of the capacitor capacitance by means of curve fitting.

The authors claim that the technique's main advantages are the simplicity, the low sampling frequency required for the signals, and that it does not require the use of digital filters and the introduction of extra sensors. This last advantage is particularly relevant, since it avoids the introduction of a current sensor in series with the DC-link capacitor and the disadvantages that may arise from it. In addition, unlike the quasi-online techniques, it is not necessary to inject an external signal or force a special working configuration.

In order to better understand the technique proposed in [46], it is important to identify the main blocks of a general traction scheme, as well as describe the main operation stages from the moment the traction drive is energized until it is completely de-energized.

The main blocks of a general traction scheme are main circuit breaker, power transformer, charging resistor, single-phase rectifier, DC-link capacitor, 2F frequency filter, braking chopper, three-phase inverter and three-phase motor [46].

When the main circuit breaker is closed, there is a first operation stage (the precharge state), in which the capacitors are smoothly charged through the charging resistor and the diodes of single-phase rectifier. Therefore, the capacitors' high peaks current that might destroy them is avoided.

When the voltage in DC-link reaches the desired value, the second stage starts (charge stage). In this stage, the charging resistor is disconnected and the DC-link voltage reaches its nominal voltage.

Once the DC-link voltage reaches its nominal value, the single-phase rectifier, the inverter and the motor start being operative, and the circuit enters in a new state (working stage), in which the train is running.

When the train stops, the main circuit breaker is opened to de-energize the system, so that the capacitors are discharged. In a first moment, the capacitors are discharged slowly (slow discharge stage) through their respective discharge resistors that are connected in parallel. Finally, during the last stage (fast discharge stage), the braking rheostat whose resistance is much smaller than one of the discharge resistors, is connected to the capacitors [46].

The technique proposed in [46] is implemented during the fast discharge stage. This stage occurs not only when the train stops but also when the train is running, for instance, when there is an AC grid phase change [46]. The DC-link equivalent circuit is shown in Figure 5.33.

The C_{DC} is the DC filtering capacitance, L_{2F} is the 2F filter inductance, C_{2F} is the 2F filter capacitance, R_{rh} is the resistance of the braking rheostat, i_{2F} is the current through the 2F filter and v_{DC} is the DC-link voltage.

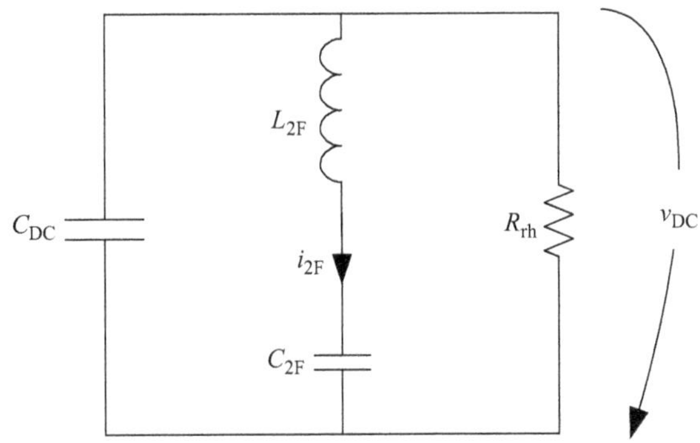

Figure 5.33 Equivalent DC-link circuit during fast discharge stage [46]

This technique estimates not only the DC-link capacitor capacitance, C_{DC}, but also the values of L_{2F}, C_{2F} and R_{rh}. For this purpose, it uses two differential equations that model the previous circuit together with an LMS algorithm.

$$C_{2F} \times \frac{dv_{DC}}{dt} - L_{2F} \times C_{2F} \times \frac{d^2 i_{2F}}{dt^2} = i_{2F} \tag{5.104}$$

$$C_{DC} \times \frac{dv_{DC}}{dt} + \frac{1}{R_{rh}} \times v_{DC} = -i_{2F} \tag{5.105}$$

The first and second derivatives of the state variables (i_{2F}, v_{DC}) are computed using a continuous-time model, by means of polynomial interpolation and LMS algorithm, which reduces the sampling period in comparison with the classic formulas. Therefore, the previous ONDT can be implemented in a simple DSP.

Nevertheless, since the power drive has not a current sensor in the 2F filter, but rather in series with the rheostat, the authors proposed an alternative methodology that will be presented in the following paragraphs [46]. This new methodology is similar to the previous one; however, it uses the current in the rheostat, i_{rh}, instead of i_{2F}.

In a first moment, the value of R_{rh} is computed using the LMS algorithm. Despite its relationship being linear, the LMS algorithm was used to attenuate the noise effect. Therefore, the use of digital filters is avoided.

The DC-link voltage during fast discharge stage consists of an exponential component, due to the two capacitors (C_{DC} and C_{2F}) and an AC contribution due to the L_{2F}. The second contribution of v_{DC} can be filtered using an LMS algorithm. Therefore, the value of C ($C = C_{DC} + C_{2F}$) is estimated through the application of the LMS algorithm to the equation:

$$R_{rh} \times C \times \frac{dv_{DC}}{dt} = v_{DC} \tag{5.106}$$

Finally, the remaining parameters can be estimated by applying the LMS algorithm to the following equations, which were obtained through *Kirchhoff* laws and after some mathematical manipulations [46]:

$$(-C_{2F} - C_{DC}) \times \frac{dv_{DC}}{dt} - \frac{C_{2F} \times L_{2F}}{R_{rh}} \times \frac{d^2 v_{DC}}{dt^2} - C_{2F} \times L_{2F} \times C_{DC} \times \frac{d^3 v_{DC}}{dt^3} - \frac{v_{DC}}{R_{rh}} \tag{5.107}$$

$$(C_{2F} \times L_{2F} \times (C_{2F} - C)) \times \frac{d^3 v_{DC}}{dt^3} - \frac{C_{2F} \times L_{2F}}{R_{rh}} \times \frac{d^2 v_{DC}}{dt^2} = C \times \frac{dv_{DC}}{dt} + \frac{v_{DC}}{R_{rh}} \tag{5.108}$$

The technique proposed in [46] uses the LMS algorithm twice, not only to estimate the four parameters (C_{DC}, C_{2F}, L_{2F} and R_{rh}), but also to estimate the coefficients of the *n*th-order polynomial function, which is used to compute the different derivatives [46].

However, recently the ONDTs based on *C* estimation have also been proposed for Al-Caps, because when these capacitors are subjected to certain operating conditions capacitance degradation is faster than the *ESR*. Detailed studies on the Al-Caps under electrical overstress [103] and thermal overstress [104–106] have reported that capacitance value decreases quickly and achieves its failure limit before the *ESR* value reaches the same condition.

In [107], the authors suggested monitoring the electrolytic capacitor by evaluating its impedance at twice the grid frequency (100 Hz). As previously mentioned at this frequency range, the capacitor impedance is fundamentally due to its capacitance. This technique was applied on a single-phase grid-connected PV system where the grid power fluctuates with twice the fundamental grid frequency. This produces second harmonic ripple in the DC-link voltage.

The available PV and inductor currents were used for estimating capacitor current and so, no additional current sensors were required. These signals are available since they are used for converter control. The capacitor voltage and current signals were processed using a second-order generalized integrator tuned to extract the second harmonic ripple components of double-grid frequency. The ratio of the RMS values of the extracted components gives the impedance of the electrolytic capacitor.

The authors stated that the previous technique can also be applied to MPPF-Caps. However, since the inductor current has significant switching ripple, the selected sampling instants may introduce additional error in the capacitor current estimation and, consequently, on the capacitor impedance estimation, which may affect its accuracy, proving to be particularly critical for MPPF-Caps.

In [107], a study was also presented which shows that inverters are the most vulnerable element in solar photovoltaic systems, accounting for 65% of failures in these systems. In turn, Al-Caps presented in the DC-link of inverters represent the most vulnerable element, being responsible for 30% of their failures. This study reinforces the relevance of the topic of fault diagnosis in capacitors.

5.7 Quasi-online fault diagnostic techniques

The QONDTs, unlike off-line techniques, do not require the removal of the capacitors from the power converter. However, their implementation usually imposes an unusual converter operation, by injecting an external signal, imposing a special working configuration, or both.

QONDTs, as ONDTs, can be subdivided according to the same criteria used in Section 5.6. Since this class is relatively recent, the number of proposed methodologies is small, thus, it was decided not to create a subsection by criterion (health indicator), but rather to present them sequentially.

In [67], the authors proposed a QONDT that can be used in step-down switched mode DC–DC converters. This technique uses the injection of a signal into the converter and, simultaneously, imposes a specific configuration. The injected signal is a sinusoidal voltage with a frequency of 1 kHz. The special configuration is attained by turning the PWM off, i.e. turning the MOSFET off. Afterwards, the capacitor current and voltage are measured and, finally, using the previous waveforms together with equations (5.18–5.20), it is possible to compute the Al-Caps ESR.

The major advantage of this methodology over the off-line techniques is that it is not needed to remove the capacitors from the converter. In addition, it presents some of the main advantages of off-line techniques over on-line ones, such as it is not expensive, it is extremely precise, simple to apply, and the accuracy is not affected by the converter operating conditions. However, the converter must interrupt its normal operation.

In [108], a new quasi-online condition monitoring for both IGBT modules and DC-link electrolytic capacitors for a single-phase DC/AC converter was proposed. This method is based on the introduction of a controlled short-circuit, generated by the converter controller, on one bridge arm of the power converter during a short period of time (2–5 μs). The resulting short-circuit current of the IGBT module will flow through the DC-link capacitor causing a significant change on the capacitor current ripple. The authors presented an expression that estimates the *ESR* value by using the short-circuit current of the IGBT (I_{SC}) and the step voltage of the DC-link capacitor (ΔU_{DC}):

$$ESR = \frac{\Delta U_{DC}}{I_{SC}} \tag{5.109}$$

This methodology uses the same principle proposed in the off-line technique described in [61], which can be seen in Figure 5.15(c). In this methodology, the ESL effect is ignored.

The short-circuit current of the IGBT modules can be measured by clipping the current sensors around each half-bridge module terminal [108], which can increase the system cost. It is reported that this method can also identify the bond wires fatigue. Although authors have stated that short-circuit test has no influence on the key components of the power converter, it seems that repeating the short-circuit test many times, certainly has negative influences on the DC-link capacitor and on the IGBT modules.

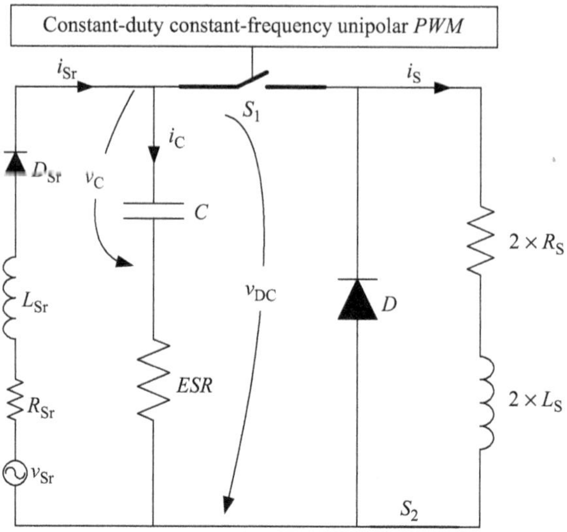

Figure 5.34 Equivalent DC-link circuit during special configuration [109]

In [109], the authors propose a QONDT that can evaluate the capacitors health status used in the DC-link of ASD, through the estimation of both *ESR* and *C* values. The proposed methodology uses the inverter whenever the motor is stopped (before start-up), imposing a special configuration. The previous configuration allows the DC-link capacitor to be discharged through two-phase windings of the motor, which is achieved through the application of a constant-duty constant-frequency unipolar PWM to one of the switches of the inverter (S_1), while the other switch (S_2) is turned ON. In this way, the following equivalent circuit is obtained (Figure 5.34).

v_{Sr}, D_{Sr}, R_{Sr}, L_{Sr}, R_S, L_S and D represent the rectified source voltage, an ideal diode, source resistance, source inductance, per-phase stator resistance, per-phase stator inductance and the freewheel diode, respectively.

During this configuration (Figure 5.34), there are two stages: the charge stage and the discharge stage. In the charge stage, D_{Sr} is ON and i_C is positive, while in the discharge stage D_{Sr} is OFF and i_C is zero or negative depending of S_1 state. In the discharge stage, two situations may occur:

- The first one (S_{D1ON}), S_1 is ON and D is OFF, the capacitor discharges through the motor windings;
- The second one (S_{D1OFF}), S_1 is OFF and D is ON, the motor current freewheels through D [109].

The proposed methodology is applied during S_{D1ON}, in which it is possible to write the following equations:

$$\frac{dv_C}{dt} = \frac{1}{C} \times i_C = -\frac{1}{C} \times i_S \tag{5.110}$$

$$v_{DC} = v_C + ESR \times i_C = v_C - ESR \times i_S \tag{5.111}$$

ESR and *C* can be derived from the following analytical relationships between i_S and v_{DC}, which can be obtained from (5.110) and (5.111):

$$ESR = \frac{\Delta v_{DC,avg}}{I_{S,DC}} \tag{5.112}$$

$$C = \frac{D \times T \times I_{S,DC}}{\Delta v_{C,avg}} \tag{5.113}$$

where D, T, $I_{S,DC}$, $\Delta v_{DC,avg}$ and $\Delta v_{C,avg}$ represent the PWM duty cycle, switching period, DC component of i_S, average variation of v_{DC} and average variation of v_C, during S_{D1ON}.

In [109], the temperature effect over the capacitor is also considered and, in order to compute it, the authors assume that the stator temperature (T_S) is similar to the capacitor one under thermal equilibrium after the system has been shut down for a while. In turn, T_S can be computed from stator resistance (R_S), during discharge stage:

$$R_S = \frac{\left(v_{DC,dc} - 2 \times V_{ce,ON}\right) \times D + \left(-V_{ce,ON} - V_{ak,ON}\right) \times (1 - D)}{2 \times I_{S,DC}} \tag{5.114}$$

$$T_S = T_{S_0} + \frac{R_S - R_{S_0}}{\alpha \times R_{S_0}} \tag{5.115}$$

where:

- $v_{DC,DC}$ and $I_{S,DC}$ represent the DC components of v_{DC} and i_S, respectively, obtained over an integer multiple of electrical cycles;
- $V_{ce,ON}$ and $V_{ak,ON}$ represent the forward voltage drops across the IGBT and diode, respectively;
- R_{S_0} and α represent the stator resistance at reference temperature (T_{S_0}) and the temperature coefficient of resistivity, respectively.

The authors claim that the main advantages of the proposed technique are the simplicity and the accuracy. It is simple since it does not require additional hardware, because the measurements of i_S and v_{DC} are performed for inverter control/protection. It is accurate since it does not depend on the load conditions, and the temperature effect is also considered [109]. However, it exhibits the same problem as the prior technique [67]; it requires the converter to interrupt its normal operation.

In [110], a QONDT based on a biogeography-based optimization (BBO) algorithm was proposed, which is able to estimate *ESR* and *C* of Al-Caps, the inductor inductance, and the sum of the losses in the inductor, MOSFET and PCB trace of DC–DC buck converters. The system under analysis is composed of a PWM-controlled DC–DC converter, a digital controller and the parameter estimation system. The parameters estimation process requires the injection of a pseudorandom binary sequence perturbation which produces voltage and current ripples. Therefore, the proposed technique must be framed in this section, since the normal operation of the converter is affected.

This approach is characterized by a very low sampling rate and it uses the converter average model, where the parameter estimation is formulated as a multivariable optimization problem, solved through the BBO algorithm.

The BBO algorithm can be regarded as a new evolutionary algorithm which can give very good results compared to other optimization algorithms such as GA or particle swarm optimization. Compared with the conventional RLS estimation method, the BBO has provided accurate and more stable solution even under noisy deviation in the experiment [110]. Despite the interesting results obtained by this approach, the high computational load remains its main drawback.

As previously stated, Al-Caps fault diagnostic techniques that use solely C as health indicator have recently emerged, due to the fact that under certain operating conditions the capacitance degradation is faster than ESR [103–106].

Due to the reasons stated in the previous paragraph, and to overcome these limitations, authors in [111] suggest the measurement of low frequency impedance of Al-Caps which is dominated by the capacitance value. This method was applied on a PV-based DC–DC boost converter system (Figure 5.35).

Using the input PV current and voltage, the converter operating mode is determined (CCM or DCM). Then, a low-frequency oscillation ($f_p = 120$ Hz) with small amplitude was generated in the original duty cycle signal. This produces low-frequency ripple in PV current, PV voltage and inductor current. The low-frequency AC components of these three signals were extracted using a filter tuned at f_p. This method does not require additional current sensor because the AC

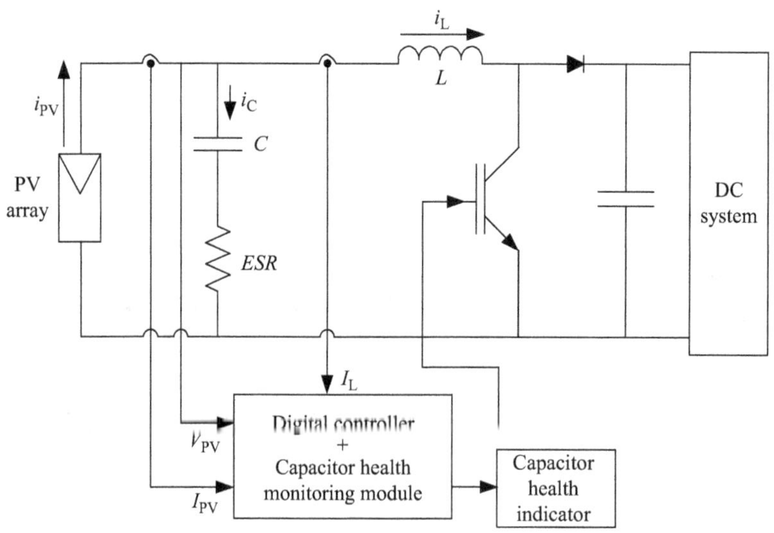

Figure 5.35 PV-fed boost converter with the health monitoring scheme [111]

component of the capacitor current, $i_{C}^{f_p}(t)$, was computed using the AC components of the PV current, $i_{PV}^{f_p}(t)$, and inductor current $i_{L}^{f_p}(t)$ as follows:

$$i_{C}^{f_p}(t) = i_{PV}^{f_p}(t) - i_{L}^{f_p}(t) \tag{5.116}$$

Finally, the obtained capacitor voltage and current were processed to compute their RMS values and divided to calculate the low-frequency impedance of the electrolytic capacitor:

$$Z_C(f_p) = \frac{V_{PV,RMS}^{f_p}}{I_{C,RMS}^{f_p}} \tag{5.117}$$

where $Z_C(f_p)$, $V_{PV,RMS}^{f_p}$ and $I_{C,RMS}^{f_p}$ represent the Al-Cap low-frequency impedance, the RMS value of capacitor voltage and current, respectively.

The inductor current has both high- and low-frequency components, thus the sampling instants must be chosen carefully, in order to accurately represent the low-frequency component. Therefore,

- for CCM, the sampling must be done at midpoint of the on duration/period of the switch [111].

$$t[n] = \frac{D[n]}{2} \times T_S \tag{5.118}$$

  where n represents the number of the switching cycle (T_S) and $D[n]$ the duty cycle in the nth switching cycle.
- for DCM, the sampling must be done at:

$$t[n] = \frac{v_O[n]}{2 \times (v_O[n] - v_{PV}[n])} \times (D[n])^2 \times T_S \tag{5.119}$$

  where $v_{PV}[n]$ and $v_O[n]$ represent the PV input voltage and the output DC microgrid voltage in the nth switching cycle, respectively.

These computations were done in the digital controller used for the MPPT of the PV system [111]. The error in the computation of the capacitor impedance increases in DCM, since the instants of time in which the signals are sampled depend on the measured values of the input and output voltage.

This approach based on the injection of low-frequency perturbation was also proposed in [112], where a regulated AC component was injected in the q-axis stator current of a three-phase induction machine drive system supplied via a PWM-inverter. This causes oscillation in the DC-link voltage and current. It was reported that the injection of the AC component in the d-axis may cause core saturation. The magnitude and frequency of the injected signal are determined according to the allowable torque ripple and capacitor current ripple, as well as the

DC-link voltage ripple. The capacitor current was calculated from the stator currents and switching times of the inverter. Then, the recursive least-squares algorithm together with (5.110) was used to estimate the capacitance from the AC component of the DC-link voltage and current ripple. However, this method can give accurate results only during regenerative operation of the motor. In addition, the injection of the AC component needs to adapt the control algorithm which limits the widespread application of this technique.

Some QONDTs are simpler when compared to ONDTs; however, they have the disadvantage of imposing an abnormal converter operation, either through an imposed special working configuration, or the injection of an external signal, or both.

Although some authors pointed out that abnormal converter operation is carried out in short time frames, it should be investigated the extent to which these operation may or may not affect the functioning of a commercial product.

5.8 Summary

In this concluding section, some key ideas are presented that synthesize the advantages and disadvantages of the fault diagnostic techniques, and some proposals for the future are also discussed.

Capacitors can be found in the DC-link of several power converters. Unfortunately, they are one of the most vulnerable elements of such systems, some of which are presented in critical applications. Therefore, the development of monitoring techniques that are able to evaluate the capacitor's health status is essential to avoid malfunctioning of the converters or even their stoppage, whose consequences could entail very high costs or jeopardizing human lives.

The most commonly used capacitors in the DC-link of power electronic converters are the Al-Caps and the MPPF-Caps, both of which may present two types of failures: catastrophic and parametric failures.

- Catastrophic failures lead to the destruction of the component through a short or an open-circuit, thus, the capacitor loses completely its function. It should be noted that under certain conditions these failures can give rise to fire or even explosions, which can ultimately lead to the complete destruction of the systems where they are inserted. The main purpose of the fault diagnostic techniques is to prevent these faults from arise.
- In the parametric failures, the capacitor does not lose completely its function; however, its electrical characteristics deteriorate. In Al-Caps, these failure manifest itself by an increase of ESR and a decrease of C, while in MPPF-Caps it expresses mainly by the decrease of C. Depending on the application, a parametric failure is considered when the previous electrical parameters reach a specific limit.

The above-mentioned limits are typically defined by capacitors manufacturers. Therefore, the end-of-life limit of Al-Caps is reached when ESR doubles its value or the capacitance reduces in 20% when compared with their initial values.

On the other hand, in MPPF-Caps, that limit is defined as a decrease of the capacitance from 2% to 10% when compared with its initial value (sound capacitor). These are the end-of-life limit criteria commonly used in the design of fault diagnostic techniques for capacitors used in the DC-link of power converters.

Capacitors fault diagnostic technique can be subdivided in three types: off-line, on-line and quasi-online.

- Off-line techniques require the removal of the capacitor from the converter, so that the electrical parameters can be estimated. In this way, the circuit has to be turned off and the capacitor removed. Therefore, this methodology presents all the drawbacks associated with this procedure, still, they are quite simple, quite precise and cheap and can be used in applications where there are no on-line and quasi-online techniques available.
- On-line techniques are designed for a specific application and do not require the converter shutdown. Therefore, the capacitor electrical parameters can be estimated without the stoppage of the converter which is fundamental in critical applications, ensuring a greater safety and reliability. However, these techniques may exhibit some of the following disadvantages: they are invasive, complex, costly, and typically they are not as accurate as off-line techniques.
- In quasi-online techniques, the removal of the capacitor from the converter is not required; however, the measurements are taken during a routine pause in the application. Therefore, the quasi-online techniques can only be applied in applications where such operating condition is permissible. These techniques present some of the disadvantages of the on-line techniques, such as complexity, cost, and some are quite invasive; however, they are not so dependent on the converter operating conditions as the on-line ones, which makes them more accurate.

5.8.1 Off-line fault diagnosis techniques

Despite the aforementioned disadvantages, off-line measurement techniques play a fundamental role in the diagnosis of capacitors' faults, not only due to their simplicity and accuracy, but also because they allow the determination of the capacitor frequency and temperature coefficients, which are vital for the design of on-line and quasi-online fault diagnostic techniques. Moreover, the characterization of the electrical parameters of the capacitors with the frequency and temperature is essential in the design of power converters, which reinforces the importance of these techniques. With regard to the future, some aspects should be addressed, namely, the calculation of ESL, which is particularly critical in the design of some converters, and the use of more complex capacitor models that consider the effects of temperature and frequency.

5.8.2 On-line fault diagnosis techniques

On-line techniques are essentially used in critical applications and can be classified according to the health indicator or to the methodology used in the estimation of those indicators.

According to the first classification, they can be subdivided as:

- ONDTs that only estimate *ESR*, which are applied in circuits composed of Al-Caps;
- ONDTs that estimate simultaneously *ESR* and *C*, which are commonly used in circuits containing Al-Caps and/or MPPF-Caps;
- ONDTs that only compute *C*, which were originally intended for circuits containing MPPF-Caps, and currently, are also used in circuits having Al-Caps, in particular, in applications where Al-Caps are subjected to very high stress.

According to the second classification, the methodologies used in the estimation of the capacitors electrical parameters can be subdivided as:

- ONDTs that use analytical relationships between different quantities. Some of these techniques present some disadvantages, namely, they can only be applied in permanent regime, and they require several sensors, in particular, in the DC-link. The latter is particularly critical because the current sensor requires space and additional wiring. The extra wiring increases the output voltage ripple and simultaneously increases the inductive effect, which might condition the converter operation and/or the precision of the fault diagnostics technique. Another important aspect concerns the methods used on the calculation of the power losses. Some authors reported that these are not accurate, because the model of the capacitor used does not consider the frequency effect. The majority of these techniques are quite simple to apply.
- ONDTs that use a reference system. These techniques are quite accurate, however, they are expensive and complex and depend on the static converter topology. Therefore, they should be used in very critical applications, being impracticable in common commercial circuits.
- ONDTs that use circuit model-based methods. Some of these techniques use the RLMS algorithm, which requires very high sampling rates and high computational effort. Therefore, it is advisable to use other methodologies that do not present the previous shortcomings. The implementation of these techniques typically requires the use of a large number of sensors.
- ONDTs that use frequency domain-processing approaches. The digital ones usually require the use of fast A/D converters with high dynamic resolution, and/or high sampling frequency; therefore, a technique should be chosen that does not present the above disadvantages. In the case of the analogue approach, it requires a large realization effort and, in drive systems where the switching frequency is relatively small, they may not be the best solution if the method is based on the *ESR* estimation. Some of these methodologies require a current sensor in the DC-link, which causes the problems mentioned above. Nevertheless, the majority of these techniques are quite simple to apply.
- ONDTs that use advanced algorithms. Usually, these methodologies have very high computational complexity and longer convergence time.

As it was possible to verify from the above, each of previous methodologies has its own merits and shortcomings. In addition to the above aspects, it should be noted that:

- Some of the presented methodologies do not take into account the temperature effect on the capacitor, although this is vital, in particular, in applications where an Al-Cap is present.
- Some of the mentioned ONDTs have been developed for a specific circuit, which already contains the sensors and the necessary processing capacity; however, that makes inviable their use in other applications that do not have such resources.
- The ONDTs developed for DC/AC converters are usually more difficult to implement because the converter control is more complex when compared to DC–DC converters.

For the future, these techniques should seek to improve some aspects, namely, the reduction of the total number of sensors and complexity and, if possible, they should be non-invasive. Another equally important aspect is the development of on-line diagnostic techniques for capacitor banks that could accurately identify the failed element.

5.8.3 Quasi-online fault diagnosis techniques

The quasi-online techniques require an unusual converter operation, by injecting an external signal, imposing a special working configuration, or both. They can be grouped in the same way as ONDTs, presenting generically the same problems. However, the fact of imposing an abnormal converter operation makes them less dependent on the operating conditions, such as temperature and frequency.

Another equally important issue that is only focused by some authors, but it is transversal to off-line, quasi-online and on-line techniques, is the need to evaluate the capacitor remaining life. This is a central aspect, in particular, when preventive or predictive maintenance strategies are being used.

Acknowledgement

This work was supported by the European Regional Development Fund (ERDF) through the Operational Programme for Competitiveness and Internationalization (COMPETE 2020), under Project POCI-01-0145-FEDER-029494, and by National Funds through the FCT – Portuguese Foundation for Science and Technology, under Projects PTDC/EEI-EEE/29494/2017 and UID/EEA/04131/2013.

References

[1] Krein, P. *Elements of Power Electronics* (Oxford, Oxford University Press 1998).

[2] Wand, H. and Blaabjerg, F., 'Reliability of Capacitors for DC-Link Applications in Power Electronic Converters – An Overview', *IEEE Transactions on Industry Applications*. 2014; vol. 50(5), pp. 3569–3578.

[3] Soliman, H., Wang, H., and Blaabjerg, F., 'A Review of the Condition Monitoring of Capacitors in Power Electronic Converters', *IEEE Transactions on Industry Applications*. 2016; vol. 52(6), pp. 4976–4989.

[4] Venet, P., Surveillance D'Alimentations a Decoupage. Application a la Maintenance Predictive, [PhD Thesis], Lyon, France, L'Universite Claude Bernard; 1993.

[5] Lahyani, A., Venet, P., Grellet, G., and Viverge, P., 'Failure Prediction of Electrolytic Capacitors During Operation of a Switchmode Power Supply', *IEEE Transactions on Power Electronics*. 1998; vol. 13(6), pp. 1199–1207.

[6] Yang, S., Xiang, D., Bryant, A., Mawby, P., Ran, L., and Tavner, P., 'Condition Monitoring for Device Reliability in Power Electronic Converters: A Review', *IEEE Transactions on Power Electronics*. 2010; vol. 25(11), pp. 2734–2752.

[7] Yang, S., Bryant, A., Mawby, P., Xiang, D., Ran, L., and Tavner, P., 'An Industry-Based Survey of Reliability in Power Electronic Converters', *IEEE Transactions on Industry Applications*. 2011; vol. 47(3), pp. 1441–1451.

[8] Hendix, T., 'Industrial and Medical Markets Demand High Efficiency, Too!', *Bodo's Power Systems, Systems Design Motion and Conversion*, 2007; pp. 12.

[9] Wang, H., Liserre, M., and Blaabjerg, F., 'Toward Reliable Power Electronics: Challenges, Design Tools, and Opportunities', *IEEE Industrial Electronics Magazine*, 2013; vol. 7(2), pp. 17–26.

[10] Trehan, N., 'Impact of Failure of Uninterruptible Power Supplies on Nuclear Power Generation Stations', *Proceedings of 35th Intersociety Energy Conversion Engineering Conference and Exhibit*; Las Vegas, USA, July 2000, pp. 741–746.

[11] Panasonic, *Aluminum Electrolytic Capacitors – Technical Guide* [online]. Available from https://industrial.panasonic.com/ww/ds/library/Alumi_TechnicalGuide [Accessed Oct 2017].

[12] Nichicon, *General Descriptions of Aluminum Electrolytic Capacitors – Technical Notes* [online]. Available from http://www.nichicon.co.jp/english/products/pdf/aluminum.pdf [Accessed June 2017].

[13] Morita, G., *Capacitor Selection Guidelines for Analog Devices, Inc., LDOs* [online]. 2010. Available from http://www.analog.com/media/en/technical-documentation/application-notes/AN-1099.pdf [Accessed June 2017].

[14] Barta, M., Pala, S., and Cygan S., *Capacitor for High Temperature Applications* [online]. Available from https://www.avx.com/resources/technical-info-papers/tantalum-niobium-capacitors/ [Accessed June 2017].

[15] CalRamic Technologies LLC, *Capacitor Basics II – Capacitor Types Application Note* [online]. Available from http://www.calramic.com/Design/Assets/PDF_files/AN109-2.pdf [Accessed June 2017].

[16] Vishay Intertechnology, *vPolyTan™ Solid Tantalum Surface-Mount Chip Capacitors, Molded Case, High-Performance Polymer Type* [online]. Available from https://www.vishay.com/docs/40174/t55.pdf [Accessed June 2017].

[17] CDM Cornell Dubilier, *Aluminum Electrolytic Capacitor Application Guide* [online]. Available from http://www.cde.com/resources/catalogs/AEappGUIDE.pdf [Accessed June 2017].

[18] TDK EPCOS, *Aluminum Electrolytic Capacitor – General Technical Information* [online]. Available from https://www.epcoschina.com/download/540988/5f33d2619fa73419e2a4af562122e90c/pdf-generaltechnicalinformation.pdf [Accessed June 2017].

[19] Vishay Intertechnology, *Aluminum Electrolytic Capacitors in Power Applications*, [online]. Available from http://www.chtechnology.com/pdf/AlumElectrolytic%20Caps.pdf [Accessed June 2017].

[20] Bettacchi, P., Montanari, D., Zanarini, D., Orioli, D., Rondelli, G., and Sanua, A., 'Power Film Capacitors for Industrial Applications', *Proceedings CARTS Europe 2010 – Symposium for Passive Electronics*; Munich, Germany, November 2010, pp. 1–10.

[21] Gallay, R., 'Metallized Film Capacitor Lifetime Evaluation and Failure Mode Analysis', *Proceedings of the CAS-CERN Accelerator School: Power Converters*; Baden, Switzerland, May 2014, pp. 45–56.

[22] Pilkor Components, *Film Capacitor* [online]. Available from http://megasan.com/service/pdfhandler.ashx?fileid=4423 [Accessed June 2017].

[23] KEMET, *Introduction to Capacitor Technologies – What is a Capacitor?* [online]. Available from http://www.kemet.com/Lists/TechnicalArticles/Attachments/6/What%20is%20a%20Capacitor.pdf [Accessed June 2017].

[24] CDM Cornell Dubilier, *Power Film Capacitor Application Guide* [online]. Available from http://www.cde.com/resources/catalogs/filmAPPguide.pdf [Accessed June 2017].

[25] Vishay Intertechnology, *Film Capacitors – General Technical Information* [online]. Available from http://www.vishay.com/docs/26033/gentechinfofilm.pdf [Accessed June 2017].

[26] TDK EPCOS, *Film Capacitors for Industrial Applications* [online]. Available from https://en.tdk.eu/download/173546/fc030f0de8e3c68b1d9fe2719d353130/film-capacitors-for-industrial-applications-pp.pdf [Accessed June 2017].

[27] ICEL, *General Technical Information* [online]. Available from http://www.icel.it/wp-content/uploads/2015/09/Generaltechnicalinformation.pdf [Accessed June 2017].

[28] Pan, M. and Randall, C., 'A Brief Introduction to Ceramic Capacitors', *IEEE Electrical Insulation Magazine*. 2010; vol. 26(3), pp. 44–50.

[29] ROHM Semiconductor, *The Important Points of Multi-layer Ceramic Capacitor Used in Buck Converter Circuit* [online]. Available from http://rohmfs.rohm.com/en/products/databook/applinote/ic/power/switching_regulator/cera_cap_appli-e.pdf [Accessed June 2017].

[30] Pithan, C., Hennings, D., and Waser, R., 'Progress in the Synthesis of Nanocrystalline $BaTiO_3$ Powders for MLLC', *International Journal of Applied Ceramic Technology*. 2005; vol. 2(1), pp. 1–14.

[31] CDM Cornell Dubilier, *Multilayer Ceramic Capacitors – Performance Characteristics* [online]. Available from http://www.cde.com/resources/catalogs/ceramperf.pdf [Accessed June 2017].

[32] Ho, J., Jow, T., and Boggs, S., 'Historical Introduction to Capacitor Technology', *IEEE Electrical Insulation Magazine*. 2010; vol. 26(1), pp. 20–25.

[33] Kropp, J. and Bakran, M., 'In-Circuit-Characterization of Ceramic Capacitor with Anti-ferroelectric Material for Voltage Source Inverters', *Proceedings of Power Electronics, Intelligent Motion, Renewable Energy, Energy Management Europe*; Nuremberg, Germany, May 2016, pp. 1400–1407.

[34] Neumayr, D., Bortis, D., Kolar, J., Koini, M., and Konrad, J., 'Comprehensive Large-Signal Performance Analysis of Ceramic Capacitors for Power Pulsation Buffers', *Proceedings of 2016 IEEE 17th Workshop on Control and Modeling for Power Electronics*, Trondheim, Norway, June 2016.

[35] Nippon Chemi-Con, *Technical Note – Judicious use of Aluminum Electrolytic Capacitors* [online]. Available from https://www.chemi-con.co.jp/e/catalog/pdf/al-e/al-sepa-e/001-guide/al-technote-e-170401.pdf [Accessed June 2017].

[36] Gasperi, M., 'Life Prediction Modeling of Bus Capacitors in AC Variable-Frequency Drives', *IEEE Transactions on Industry Applications*. 2005; vol. 41(6), pp. 1430–1435.

[37] Parler, S., 'Improved Spice Models of Aluminum Electrolytic Capacitors for Inverter Applications', *IEEE Transactions on Industry Applications*, 2003; vol. 39(4), pp. 929–935.

[38] Zanobini, A., Iuculano, G., and Falciani, A., 'Automatic-Test Equipment for the Characterization of Aluminum Electrolytic Capacitors', *IEEE Transactions on Instrumentation and Measurement*. 2006; vol. 55(2), pp. 682–688.

[39] IXYS, *Power Capacitors* [online]. Available from http://www.westcode.com/capguide.pdf [Accessed June 2017].

[40] Lackey, J. and Donald, V., 'A Versatile Capacitor Web Coater with Multiple Source Capability', *Product Finishing*. 1992; vol. 45(7), pp. 14–16.

[41] Sarjeant, W., Zirnheld, J., and MacDougall, F., 'Capacitors', *IEEE Transactions on Plasma Science*. 1998; vol. 26(5), pp. 1368–1392.

[42] Brown, R., Electrical and Thermal Modelling of Low Power Metallised Polypropylene Capacitors, [PhD Thesis], Melbourne, Australia, RMIT University, 2007.

[43] Rubycon, *Technical Trend of Power Film Capacitors for Industrial Use* [online]. Available from http://www.rubycon.co.jp/en/products/topics/img/t002_09.pdf [Accessed June 2017].

[44] TDK EPCOS, *Film Capacitors – General Technical Information* [online]. Available from https://en.tdk.eu/download/530754/bb7f3c742f09af6f8ef473fd34f6000e/pdf-generaltechnicalinformation.pdf [Accessed June 2017].

[45] Aitken, N. *Capacitor Failure Modes – Lessons from Industry* [online]. Available from https://escies.org/download/webDocumentFile?id=64630 [Accessed June 2017].

[46] Buatti, G., Ramos, J., Amaral, A., Dworakoski, P., and Cardoso, A., 'Condition Monitoring of Metallized Polypropylene Film Capacitors in Railway Power Trains', *IEEE Transactions on Instrumentation and Measurement*. 2009; vol. 58(10), pp. 3796–3805.

[47] Kyocera AVX, *Medium Power Film Capacitors for Power Applications* [online]. Available from http://catalogs.avx.com/MediumPowerFilm.pdf [Accessed June 2017].

[48] Albertsen, A., *DC-Link Capacitor Technology Comparison – Aluminum Electrolytic vs. Film Capacitors* [online]. Available from http://jianghai-europe.com/wp-content/uploads/6-Jianghai-Europe-DC-Link-Capacitor-Technology-Comparison-Alu-E-Cap-vs.-Film-Capacitor-AAL-2015-02-10-EN.pdf [Accessed June 2017].

[49] COMPOTEC Electronics GmbH, Film Capacitors [online]. Available from http://www.compotec-electronics.com/wordpress/wp-content/uploads/COMPOTEC-Film-Capacitors-Robustness-and-Fail-Safe.pdf [Accessed June 2017].

[50] Rubycon, *Probable Causes of Failure and Failure Mode of Film Capacitors* [online]. Available from http://www.rubycon.co.jp/en/products/film/technote_pdf/filmcapacitor3.pdf [Accessed June 2017].

[51] Bond, J. *A New Mitigation Strategy for Failures in Metallized Polypropylene Capacitors* [online]. Available from http://www.ecicaps.com/wp-content/uploads/New_Mitigation_Strategy_for_Failures.pdf [Accessed June 2017].

[52] Agarwal, N., Ahmad, M., and Anand, S., 'Quasi-Online Technique for Health Monitoring of Capacitor in Single Phase Solar Inverter', *IEEE Transactions on Power Electronics*. 2018; vol. 33(6), pp. 5283–5291.

[53] Amaral, A., Técnicas de Medida para a Caracterizaçáo do Circuito Equivalente de Condensadores Electrolíticos de Alumínio, [PhD Thesis], Coimbra, Portugal, Universidade de Coimbra; 2010.

[54] Amaral, A. and Cardoso, A., 'An ESR Meter for High Frequencies', *Proceedings of IEEE International Conference on Power Electronics and Drives Systems*; Kuala Lumpur, Malaysia, November/December 2005, pp. 1628–1633.

[55] Amaral, A. and Cardoso, A., 'An Experimental Technique for Estimating the Aluminum Electrolytic Capacitor Equivalent Circuit, at High Frequencies', *Proceedings of IEEE International Conference on Industrial Technology*; Hong-Kong, China, December 2005, pp. 86–91.

[56] Amaral, A. and Cardoso, A., 'An Experimental Technique for Estimating the ESR and Reactance Intrinsic Value of Aluminium Electrolytic Capacitors', *Proceedings of the IEEE Instrumentation and Measurement Technology Conference*; Sorrento, Italy, April 2006, pp. 1820–1825.

[57] Amaral, A. and Cardoso, A., 'Using Newton–Raphson Method to Estimate the Condition of Aluminum Electrolytic Capacitors', *Proceedings in IEEE International Symposium on Industrial Electronics*; Vigo, Spain, June 2007, pp. 827–832.

[58] Amaral, A. and Cardoso, A., 'An Automatic Technique to Obtain the Equivalent Circuit of Aluminum Electrolytic Capacitors', *Proceedings of 34th Annual Conference of IEEE Industrial Electronics Society*; Orlando, USA, November 2008, pp. 539–544.

[59] Amaral, A. and Cardoso, A., 'An Economic Off-line Technique for Estimating the Equivalent Circuit of Aluminum Electrolytic Capacitors', *IEEE Transactions on Instrumentation and Measurement*. 2008; vol. 57(12), pp. 2697–2710.

[60] Amaral, A., Buatti, G., Ribeiro, H., and Cardoso, A., 'Using DFT to Obtain the Equivalent Circuit of Aluminum Electrolytic Capacitors', *Proceedings of 7th International Conference on Power Electronics and Drive Systems*; Bangkok, Thailand, November 2007, pp. 434–438.

[61] Amaral, A. and Cardoso, A., 'Using a Simple Charge-Discharge Circuit to Estimate Capacitors Equivalent Circuit at Their Operating Conditions', *Proceedings of IEEE 2009 Instrumentation and Measurement Technology Conference*; Singapore, Singapore, May 2009, pp. 737–742.

[62] Amaral, A. and Cardoso, A., 'Simple Experimental Techniques to Characterize Capacitors in a Wide Range of Frequencies and Temperatures', *IEEE Transactions on Instrumentation and Measurement*. 2010; vol. 59(5), pp. 1258–1267.

[63] Greason, W. and Critchley, J., 'Shelf-Life Evaluation of Aluminum Electrolytic Capacitors', *IEEE Transactions on Components, Hybrids, and Manufacturing Technology*. 1986; vol. CHMT-9(3), pp. 293–299.

[64] Amaral, A. and Cardoso, A., 'Estimating Aluminum Electrolytic Capacitors Condition Using a Low Frequency Transformer Together with a DC Power Supply', *Proceedings of 2010 IEEE International Symposium on Industrial Electronics*, Bari, Italy, 4–7 July.

[65] Amaral, A. and Cardoso, A., 'Condition Monitoring of Electrolytic Capacitors', *International Journal of System Assurance Engineering and Management*. 2010; vol. 2(4), pp. 325–332.

[66] Amaral A. and Cardoso, A., 'Using a Very Simple Capacimeter to Evaluate Aluminum Electrolytic Capacitors Health Status', *International Journal on Engineering Applications*. 2013; vol. 1(4), pp. 234–240.

[67] Amaral, A. and Cardoso, A., 'A Simple Offline Technique for Evaluating the Condition of Aluminum-Electrolytic-Capacitors', *IEEE Transactions on Industrial Electronics*. 2009; vol. 56(8), pp. 3230–3237.

[68] Harada, K. and Katsuki, A., 'Deterioration Diagnosis of Electrolytic Capacitor in a Buck-Boost Converter', *Proceedings of 19th IEEE Power Electronics Specialist Conference*; Kyoto, Japan, April 1988, pp. 1101–1104.

[69] Amaral, A. and Cardoso, A., 'Use of ESR to Predict Failure of Output Filtering Capacitors in Boost Converters', *Proceedings of the IEEE Industrial Electronics Symposium*; Ajaccio, France, May 2004, pp. 1309–1314.

[70] Amaral, A. and Cardoso, A., 'Using Output Voltage and Current to Predict Failures in Switch Mode Power Supplies Operating in Discontinuous Mode', *Proceedings of 25th International Conference for Power Electronics,*

Intelligent Motion and Power Quality; Nuremberg, Germany, May 2004, pp. 563–568.

[71] Amaral, A. and Cardoso, A., 'Using Input Current and Output Voltage Ripple to Estimate the Output Filter Condition of Switch Mode DC/DC Converters', *2009 IEEE International Symposium on Diagnostics for Electrical Machines, Power Electronics and Drives*; Cargèse, Corsica, France, August/September 2009.

[72] Harada, K., Katsuki, A., and Fujiwara, M., 'Use of ESR for Deterioration Diagnosis of Electrolytic Capacitor', *IEEE Transactions on Power Electronics*. 1993; vol. 8(4), pp. 355–360.

[73] Amaral, A. and Cardoso, A., 'Fault Diagnosis on Switch-Mode Power Supplies Operating in Discontinuous Mode', *Proceedings of Second IEE International Conference on Power Electronics Machines and Drives*; Edinburgh, UK, March/April 2004, UK, pp.197–202.

[74] Venet, P., Darnand, H., and Grellet, G., 'Detection of Faults of Filter Capacitors in a Converter: Application to Predict Maintenance', *Proceedings of 15th International Telecommunications Energy Conference*; Paris, France, September 1993, pp. 229–234.

[75] Venet, P., Lahyani, A., Grellet, G., and Jaco, A., 'Influence of Aging on Electrolytic Capacitors Function in Static Converters: Fault Prediction Method', *The European Physical Journal Applied Physics*. 1999; vol. 5(1), pp. 71–83.

[76] Venet, P., Perisse, F., Husseini, H., and Rojat, G., 'Realization of a Smart Electrolytic Capacitor Circuit', *IEEE Industry Applications Magazine*. 2002; vol. 8(1), pp.16–20.

[77] Wlesinger, T. and Ertl, H., 'A Novel Real Time Monitoring Unit for PWM Converter Electrolytic Capacitors', *Proceedings of the 39th Power Electronics Specialist Conference*; Rhodes, Greece, June 2008, pp. 523–528.

[78] Amaral, A. and Cardoso, A., 'Using Spectral Analysis to Determine the Equivalent Circuit of the Output filter of DC–DC Converters at their Operating Frequency', *Proceedings of International Aegean Conference on Electric Machines, Power Electronics and Electromotion*, Bodrum, Turkey, September 2007, pp. 671–675.

[79] Amaral, A. and Cardoso, A., 'State Condition Estimation of Aluminum Electrolytic Capacitors Used on the Primary Side of ATX Power Supplies', *Proceedings of the 35th Annual Conference on the IEEE Industrial Electronics Society*; Porto, Portugal, November 2009, pp. 447–452.

[80] Iman, A., Habetler, T., Harley, R., and Divan, D., 'Failure Prediction of Electrolytic capacitor using DSP Methods', *Proceedings of Applied Power Electronics Conference and Exposition*; Austin, USA, March 2005, pp. 965–970.

[81] Amaral, A. and Cardoso, A., 'A Non-Invasive Technique for Fault Diagnosis of SMPS', *Proceedings of the 39th Power Electronics Specialist Conference*, Rhodes, Greece, June 2008, pp. 2097–2102.

[82] Amaral, A. and Cardoso, A., 'On-line Fault Detection of Aluminum Electrolytic Capacitors, in Step-Down DC–DC Converters, Using Input Current and Output Voltage Ripple', *IET Transaction on Power Electronics*. 2012; vol. 5(3), pp. 315–322.

[83] Chen, Y., Wu, H., Chou, M., and Lee, K., 'Online Failure Prediction of the Electrolytic Capacitor for LC Filter of Switching-Mode Power Converter', *IEEE Transactions on Industrial Electronics*. 2008; vol. 53(1), pp. 400–406.

[84] Aeloíza, E., Kim, J., Ruminot, P., and Enjeti, P., 'A Real Time Method to Estimate Electrolytic Capacitor Condition in PWM Adjustable Speed Drives and Uninterruptible Power Supplies', *Proceedings of IEEE 36th Power Electronics Specialist Conference*; Recife, Brazil, June 2005, pp. 2867–2872.

[85] Vogelsberger, M., Wiesinger, T., and Ertl, H., 'Life-cycle Monitoring and Voltage Managing Unit for DC-Link Electrolytic Capacitors in PWM Converters', *IEEE Transactions on Power Electronics*. 2011; vol. 26(2), pp. 493–503.

[86] Braham, A., Lahyani, A., Venet, P., and Rejeb, N., 'Recent Developments in Fault Detection and Power Loss Estimation of Electrolytic Capacitors', *IEEE Transactions on Power Electronics*. 2010, vol. 25(1), pp. 33–43.

[87] Ahmad, W., Agarwal, N., and Anand, S., 'Online Monitoring Technique for Aluminum Electrolytic Capacitor in Solar PV Based DC System', *IEEE Transactions on Industrial Electronics*. 2016; vol. 63(11), pp. 7059–7066.

[88] Wang, G., Guan, Y., Zhang, J., Wu, L., and Zheng, X., 'ESR Estimation Method for DC–DC Converters Based on Improved EMD Algorithm', *Proceedings of 2012 IEEE Conference on Prognostic and System Health Management*; Beijing, China, May 2012, pp. 1–6.

[89] Farjah, E., Givi, H., and Ghanbari, T., 'Application of an Efficient Rogowski Coil Sensor for Switch Fault Diagnosis and Capacitor ESR Monitoring in Non-Isolated Single Switch DC–DC Converters', *IEEE Transactions on Power Electronics*. 2017; vol. 32(2), pp. 1442–1456.

[90] Ma, H., Mao, X., Zhang, N., and Xu, D., 'Parameter Identification of Power Electronic Circuits based on Hybrid Model', *Proceedings of IEEE 36th Power Electronics Specialist Conference*; Recife, Brazil, June 2005, pp. 2855–2860.

[91] Buatti, G., Amaral, A., and Cardoso, A., 'ESR Estimation Method for DC/DC Converters Through Simplified Regression Models', *Proceedings of the 42nd IEEE Industry Applications Annual Meeting*; New Orleans, USA, September 2007, pp. 2289–2294.

[92] Buatti, G., Amaral, A., and Cardoso, A., 'Parameter Estimation of DC/DC Buck Converter Using a Continuous Time Model', *Proceedings of the 2007 European Conference on Power Electronics and Applications*; Aalborg, Denmark, September 2007.

[93] Buatti, G., Amaral, A., and Cardoso, A., 'An Online Technique for Estimating the Parameters of Passive Components in Non-Isolated DC/DC Converters', *Proceedings of the 2007 IEEE International Symposium on Industrial Electronics*; Vigo, Spain, 4–7 June 2007.

[94] Buatti, G., Amaral, A., and Cardoso, A., 'An Unified Method for Estimating the Parameters of Non-Isolated DC/DC Converters Using Continuous Time Models', *Proceedings of the 29th International Telecommunications Energy Conference*; Rome, Italy, September/October 2007, pp. 334–341.

[95] Imam, A., Divan, D., Harley, R., and Habetler, T., 'A Real-Time Condition Monitoring of the Electrolytic Capacitors for Power Electronics Applications', *Proceedings of the 22nd Annual IEEE Applied Power Electronics Conference*; Anaheim, USA, February/March 2007, pp. 1057–1061.

[96] Abdennadher, K., Venet, P., Rojat, G., Rétif, J., and Rosset, C., 'A Real-Time Predictive-Maintenance System of Aluminum Electrolytic Capacitors Used in Uninterrupted Power Supplies', *IEEE Transactions on Industrial Applications*, 2010, vol. 46(4), pp. 1644–1652.

[97] Buatti, G., Ramos, J., García, C., Amaral, A., and Cardoso, A., 'An Online and Noninvasive Technique for the Condition Monitoring of Capacitors in Boost Converters', *IEEE Transactions on Instrumentation and Measurement*, 2010, vol. 59(8), pp. 2134–2143.

[98] Sepehr, A., Saradarzadeh, M., and Farhangi, S., 'A Noninvasive On-line Failure Prediction Technique for Aluminum Electrolytic Capacitors in Photovoltaic Grid-connected Inverters', *7th Power Electronics and Drive Systems Technologies Conference*; Tehran, Iran, February 2016.

[99] Yao, K., Tang, W., Bi, X., and Lyu, J., 'An Online Monitoring Scheme of DC-Link Capacitor's ESR and C for Boost PFC Converter', *IEEE Transactions on Power Electronics*, 2016, vol. 31(8), pp. 5944–5951.

[100] Yao, K., Tang, W., Hu, W., and Lyu J., 'A Current-Sensorless Online ESR and C Identification Method for Output Capacitor of Buck Converter', *IEEE Transactions on Power Electronics*, 2015, vol. 30(12), pp. 6993–7005.

[101] Laadjal, K., Sahraoui, M., Cardoso, A., and Amaral, A., 'On-line Estimation of Aluminum Electrolytic-Capacitor Parameters Using a Modified Prony's Method', *Proceedings of the 11th International Symposium on Diagnostics for Electric Machines, Power Electronics and Drives*; Tinos, Greece, August/September 2017, pp. 387–393.

[102] Sahraoui, M., Cardoso, A., and Ghoggal, A., 'The Use of a Modified Prony's Method to Track the Broken Rotor Bars Characteristic Frequencies and Amplitudes, in Three-Phase Induction Motors', *IEEE Transactions on Industrial Applications*, 2015, vol. 51(3), pp. 2136–2147.

[103] Celaya, J., Kulkarni, C., Biswas, G., Saha, S., and Goebel, K., 'A Model-based Prognostics Methodology for Electrolytic Capacitors Based on Electrical Overstress Accelerated Aging', *Proceedings of Annual Conference of the Prognostics Health Management Society*; Montreal, Canada, September 2011, pp. 1–9.

[104] Kulkarni, C., Celaya, J., Biswas, G., and Goebel, K., 'Accelerated Aging Experiments for Capacitor Health Monitoring and Prognostics', *Proceedings of IEEE AUTOTESTCON*; Anaheim, USA, September 2012.

[105] Kulkarni, C., Celaya, J., Biswas, G., and Goebel, K., 'Physics Based Electrolytic Capacitor Degradation Models for Prognostic Studies under Thermal Overstress', *1st European Conference of the Prognostics and Health Management Society*; Dresden, Germany, July 2012.

[106] Celaya, J., Kulkarni, C., Saha, S., Biswas, G., and Goebel, K., 'Accelerated Aging in Electrolytic Capacitors for Prognostics', *Proceedings of Annual Reliability and Maintainability Symposium*; Reno, USA, January 2012, pp. 1–6.

[107] Arya, A., Ahmad, M., Agarwal, N., and Anand, S., 'Capacitor Impedance Estimation Utilizing DC-Link Voltage Oscillations in Single Phase Inverter', *IET Power Electronics*. 2017; vol.10(9), pp. 1046–1053.

[108] Sun, P., Gong, C., Du, X., Luo, Q., Wang, H., and Zhou, L., 'Online Condition Monitoring for Both IGBT Module and DC-Link Capacitor of Power Converter Based on Short-Circuit Current Simultaneously', *IEEE Transactions on Industrial Electronics*. 2017; vol. 64(5), pp. 3662–3671.

[109] Lee, K., Kim, M., Yoon, J., Lee, S., and Yoo, J., 'Condition Monitoring of DC-Link Electrolytic Capacitors in Adjustable-Speed Dives', *IEEE Transactions on Industrial Applications*. 2008; vol. 44(5), pp. 1606–1613.

[110] Li, B. and Low, K., 'Low Sampling Rate Online Parameters Monitoring of DC–DC Converters for Predictive Maintenance Using Biogeography-Based Optimization', *IEEE Transactions on Power Electronics*. 2016; vol. 31(4), pp. 2870–2879.

[111] Ahmad, M., Agarwal, N., Kumar, P., and Anand, S., 'Low Frequency Impedance Monitoring and Corresponding Failure Criteria for Aluminum Electrolytic Capacitors', *IEEE Transactions on Industrial Electronics*. 2017; vol. 64(7), pp. 5657–5666.

[112] Nguyen, T. and Lee, D. 'Deterioration Monitoring of DC-Link Capacitors in AC Machine Drives by Current Injection', *IEEE Transactions on Power Electronics*. 2015; vol. 30(3), pp. 1126–1130.

Chapter 6

DC–DC converters

Fernando Bento[1] *and Eunice Ribeiro*[1]

DC–DC converters face an exponential growth in the context of the ever-increasing use of DC electricity (particularly at homes and businesses) driven by the fact that the vast majority of domestic/business loads (computers, communications, LED lighting, home appliances, electric and hybrid vehicles, etc.) operate either in true DC mode or at least involve an intermediate DC-link bus. Due to the increasingly demanding energy efficiency targets, imposed by stringent energy regulations, these DC-compatible loads are now obliged to feature improved capabilities for energy management. DC–DC converters are currently the only power conversion solutions capable of performing such energy management task in DC systems, providing feasible options to solve the bottleneck of energy management in DC-compatible loads, in an efficient manner.

In this context, efficiency and reliability are of major concern. Hence, all efforts taken in the way of fostering the deployment of highly efficient and reliable DC–DC converters aiming real-world applications are valuable and desirable.

Currently, scattered research efforts have been made aiming to develop fault diagnostics algorithms and fault-tolerant DC–DC converter architectures. For that reason, a comprehensive compilation of all the research efforts made so far in this scientific area has a major positive impact. Therefore, both diagnostic techniques – mainly focussed on the occurrence of open-circuit faults and short-circuit faults in the power switches and/or gate drivers – and fault-tolerant strategies – allowing for the continuous operation, even under faulty conditions – are addressed in this final chapter. Most relevant experimental verifications of those strategies, available in the literature, are also presented and evaluated.

Most of the fault detection techniques aim open-circuit faults and short-circuit faults in the power switches, and rely on the analysis of converter parameters, such as converter input current and DC-bus capacitor voltage, and their variation along time. Open-circuit faults do not present, in general, an immediate threat to the DC–DC power converter. Although if they remain undetected for long periods, further faults and damages might be incurred in the converter and, in extreme cases, lead to its total standstill. Reconfiguration strategies used to

[1]CISE – Electromechatronic Systems Research Centre, Universidade da Beira Interior, Portugal

overcome these faults include the use of additional, but cost-effective hardware, and reconfiguration of the gating signals features, when the converter has an inherent fault-tolerant structure.

Nomenclature

AC	Alternating current
CCM	Continuous conduction mode
DAB	Dual active bridge
DC	Direct current
DCM	Discontinuous conduction mode
FB	Full-bridge
FFT	Fast Fourier transform
FPGA	Field-programmable gate array
HB	Half-bridge
IGBT	Insulated gate bipolar transistor
IPOS	Input-parallel output-series
ISOP	Input-series output-parallel
MMC	Modular multilevel converter
MOSFET	Metal oxide semiconductor field effect transistor
MPP	Maximum power point
MPPT	Maximum power point tracking
OC	Open-circuit
PV	Photovoltaic
PWM	Pulse-width modulation
SAB	Single active bridge
SC	Short-circuit
SCR	Silicon-controlled rectifier
SEPIC	Single-ended primary inductance converter
SSR	Solid-state relay
TRIAC	Triode for alternating current
ZVS	Zero voltage switching

6.1 Fault diagnostic algorithms

Generally speaking, switching power converters are susceptible to suffer failures that may impair their smooth operation. Inside power converters, semiconductors and capacitors are the components that are most prone to fail. Semiconductor faults are categorised as open-circuit (OC) faults or short-circuit (SC) faults.

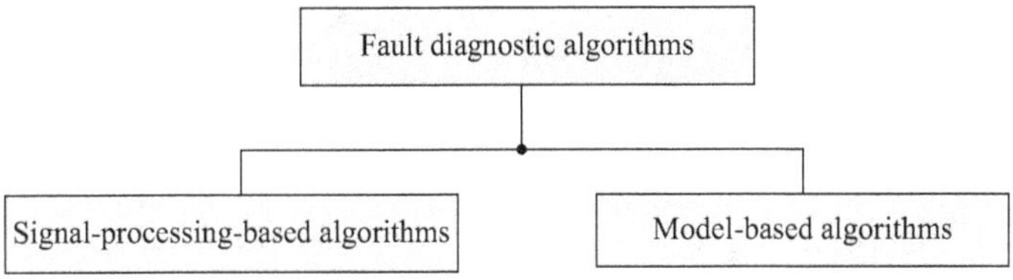

Figure 6.1 Classification of the semiconductor fault diagnostic algorithms

Typically, OC faults are caused by gate drivers' failures, wire lifting or soldering break [1]. Most of the times, OC faults do not represent an immediate threat to the power converter: the energy transfer to the load is maintained, but under degraded conditions (more current and voltage ripple, lower power transfer, etc.). If no proper actions are taken, OC faults in the converter switches may remain undetected for long periods, creating an additional risk factor, in the long term, for the healthy converter components. Therefore, the detection of such failures is fundamental, as it prevents the extension of more severe damages inside a power converter.

On the other hand, SC faults in the converter switches are severe fault events that must be detected and isolated within a few microseconds. Typically, SC faults are isolated from the rest of the converter circuit resorting to hardware protection devices, such as fuses, and usually result in a very limited or, sometimes, impossible operation of the converter. Besides, SC faults require a fast response of the control structure, in order to isolate the fault and avoid further damages in the converter or any other equipment connected to it. Generally, software protections do not achieve the quickness required to overcome the effects of SC faults.

Typically, the diagnostic of semiconductors faults in DC–DC converters takes place in two stages: fault detection and fault identification. During the fault detection stage, a fault alarm is triggered, but the faulty component remains unidentified; during the fault identification stage, the fault mode and the component that has given rise to the fault are identified. In very specific situations, the diagnostic of semiconductor faults might take place in a single stage (i.e. the detection and identification actions are developed concurrently). Such practice mainly depends on the capabilities of the adopted fault diagnostic algorithm and on the selected fault diagnostic variables.

There is no general consensus regarding the classification of the semiconductors fault diagnostic algorithms developed so far. However, the classification of the algorithms as signal-processing-based and model-based algorithms, as represented in Figure 6.1, seems to gather the agreement of most of the scientific community involved in this research field.

Next sections will explain, in detail, the fundamentals of some of the most relevant fault diagnostic algorithms available in the literature.

6.1.1 Signal-processing-based algorithms

Most fault diagnostic algorithms aimed at detecting semiconductor faults in DC–DC converters are based on signal-processing techniques. These algorithms

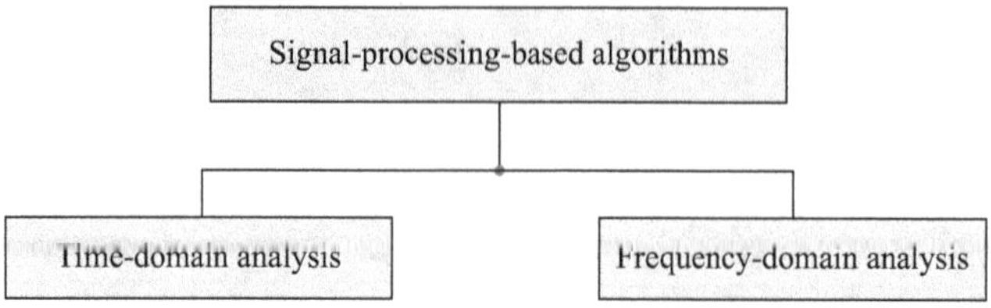

Figure 6.2 Classification of the signal-processing-based fault diagnostic algorithms based on the domain of analysis of the fault diagnostic variables

identify certain fault signatures resorting to an analysis of some of the signals that are commonly used to control the converter, as, for instance, DC-bus current or capacitors voltage.

The success of these algorithms can be justified, among other reasons, by the relatively low computational effort required to implement these algorithms, and by the adoption of a black-box approach, i.e. the need for the determination of the DC–DC converter parameters is obviated, and the analysis is simplified, especially when there is no previous knowledge about the converter model characteristics. As it is possible to witness further ahead, the black-box approach might bring some inconveniences for the fault diagnostic process. The effective fault diagnostic action of some algorithms based on this approach might be compromised by false fault alarms when the converter operation is intended to cover, for example, a wide range of load levels, switching frequencies or conduction modes. In such cases, false fault alarms might be triggered, leading to erroneous diagnostic results and improper implementation of converter reconfiguration strategies.

Signal-processing-based algorithms can be, in turn, classified regarding their analysis of the fault diagnostic variables in either the time or frequency domain (Figure 6.2).

6.1.1.1 Time-domain signal-processing-based algorithms

The fault diagnostic algorithms using signal-processing approaches in the time domain are dominant among the category of fault diagnostic algorithms based on signal-processing techniques.

Next subsections detail the operation principles and highlight the features of some of the most relevant fault diagnostic algorithms that achieve the fault diagnostic through analysis of the fault signatures in the time domain.

Statistical moments

Perturbations in the voltages and currents of switching power converters are commonly correlated to faults in the semiconductors of power converters. The statistical moments of the converters voltages and currents allow to track and evaluate unexpected deviations of these parameters. The algorithm features enable the identification of OC and SC faults in the converter active switches and diodes [2].

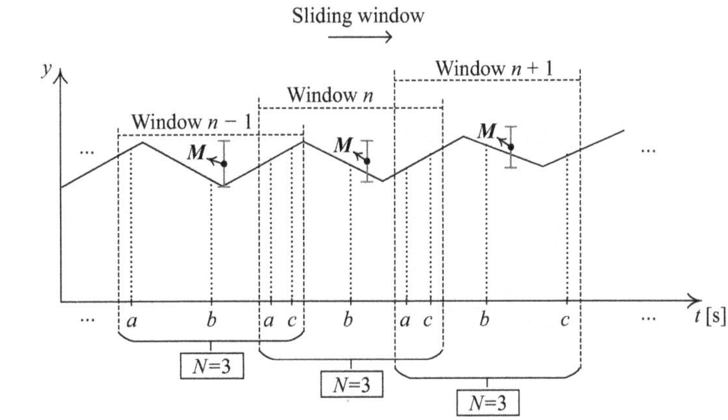

Figure 6.3 Illustrative example showing the elements required for determining the statistical moments of variable y

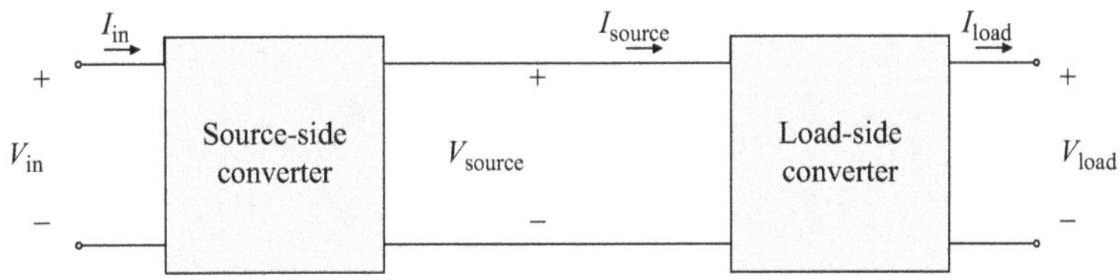

Figure 6.4 Simplification of the two-stage cascaded buck converter

In general terms, the *n*th-order statistical moment of a variable *y* is given by:

$$s^n = \frac{\sum_{k=1}^{N} (y_k - M)^n}{N - 1} \tag{6.1}$$

where y_k is the measured value of variable *y*, for instance, *k*; *M* is the mean value of variable *y* on the moving window; and *N* is the number of samples per moving window. Figure 6.3 depicts a simple example which shows the elements used to determine the statistical moments of a variable *y*.

This strategy is validated in [2] resorting to a two-stage cascaded buck converter prototype, operating at continuous conduction mode (CCM). Nevertheless, and taking into account the algorithm features, it can also be employed in other DC–DC converter topologies. Figure 6.4 depicts a simplification of the converter employed in the study. Each block of Figure 6.4 represents a single-switch buck converter.

In the particular application under analysis, where a cascaded buck converter is considered, the third-order statistical moment of four converter parameters is computed: V_{source}, I_{source}, V_{load}, and I_{load} (refer to Figure 6.4). As an example, the third-order statistical moment of the load-side converter output voltage is obtained as follows:

$$V_{\text{load}}^3 = \frac{1}{N-1}\sum_{k=1}^{N}\left(V_{\text{load},k} - \overline{V_{\text{load}}}\right)^3 \tag{6.2}$$

Fault detection based on this algorithm is achieved resorting to the analysis of the absolute values of the third-order statistical moments of the converter parameters. Under healthy converter operation, all four converter parameters under analysis follow the expected pattern and, consequently, the third-order statistical moments related to all the converter parameters approach to zero. A fault event introduces perturbations in the converter parameters which, in turn, lead to non-zero statistical moments.

Information about the statistical moments of the converter parameters is equally helpful to obtain the cross-variance matrix of the third-order statistical moments of the converter parameters. It is the cross variance between the statistical moments of the converter parameters that allows to identify the fault mode that triggered the fault alarm. The cross-variance matrix is defined as follows:

$$V^3 = \begin{bmatrix} V_{\text{load}}^3 & V_{\text{load}}^1 I_{\text{load}}^2 & V_{\text{load}}^1 V_{\text{source}}^2 & V_{\text{load}}^1 I_{\text{source}}^2 \\ I_{\text{load}}^1 V_{\text{load}}^2 & I_{\text{load}}^3 & I_{\text{load}}^1 V_{\text{source}}^2 & I_{\text{load}}^1 I_{\text{source}}^2 \\ V_{\text{source}}^1 V_{\text{load}}^2 & V_{\text{source}}^1 I_{\text{load}}^2 & V_{\text{source}}^3 & V_{\text{source}}^1 I_{\text{source}}^2 \\ I_{\text{source}}^1 V_{\text{load}}^2 & I_{\text{source}}^1 I_{\text{load}}^2 & I_{\text{source}}^1 V_{\text{source}}^2 & I_{\text{source}}^3 \end{bmatrix} \tag{6.3}$$

Some of the matrix elements are directly related to certain fault modes. The evolution of those matrix elements allows to detect and identify faults in the converter. For instance, a positive value of the third-order statistical moments of V_{load} – element (1,1) of the cross-variance matrix – allows to identify a SC fault in the switch of the load-side converter. Similarly, the cross variance between V_{load} and V_{source} – element (1,3) of the cross-variance matrix – allows to identify the occurrence of an OC fault in the load-side converter switch. Further than the cross variance, this information must be also compared with the information of the two gating signals used to control the entire converter, to clarify the presence (or not) of certain fault modes.

The algorithm is subjected to experimental validation in [2], resorting to a 200 W prototype of the two-stage cascaded buck converter.

This approach does not imply the addition of sensors to the DC–DC converter circuit, covering a wide range of operating conditions. The fault diagnosis does not require the adoption of thresholds for the decision process and, consequently, the fault diagnostic results are feasible for all the converter operation range. On the other hand, the long fault diagnostic time (8 times the switching period) and the significant computation effort required for the algorithm implementation are the main drawbacks.

DC-bus current peak-to-integral ratio

The shape of the DC-bus current contains plenty of information suitable for fault diagnostic purposes. In particular, the information of the peak-to-integral ratio of the DC-bus current proves to be effective for the diagnosis of switch faults in DC–DC converters [3]. Based on the analysis of the converter input current waveform, the ratio between the peak of the waveform and the corresponding integral is observed, as mean to identify both OC and SC faults in the converter active switches. The input current is sensed resorting to a pulse transformer, while the fault diagnostic is implemented through the application of analogue signal processing circuitry.

The DC-bus current i_{in} peak and integral values are computed and sampled resorting to analogue circuitry. Then, the ratio between the current peak and the current integral is computed:

$$\gamma = \frac{i_{pk}}{i_{int}} \tag{6.4}$$

where γ is the ratio between the current peak and the current integral, i_{pk} is the DC-bus current peak value and i_{int} is the current integral, obtained through the following relation:

$$i_{int} = \frac{1}{T}\int_0^T i(t)\,dt \tag{6.5}$$

Figure 6.5 shows a simplification of the building blocks required for the implementation of the algorithm.

The decision process is based on the verification of the current integral i_{int} and the ratio γ. Figure 6.6 depicts the map used for the classification of semiconductor faults. The intervals defined in Figure 6.6, for each fault mode, take into account the thresholds defined empirically in [3]. The selection of these thresholds must consider, among other factors, the technical characteristics of the circuitry used to implement the fault diagnostics algorithm.

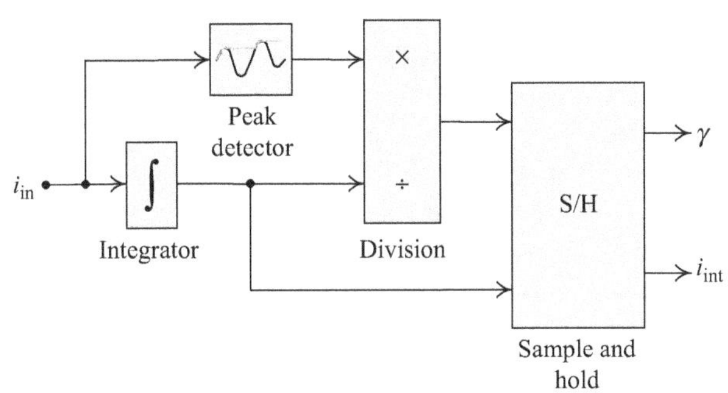

Figure 6.5 Schematic representation of the fault diagnostic algorithm based on the analysis of the peak-to-integral ratio of the DC-bus current [3]

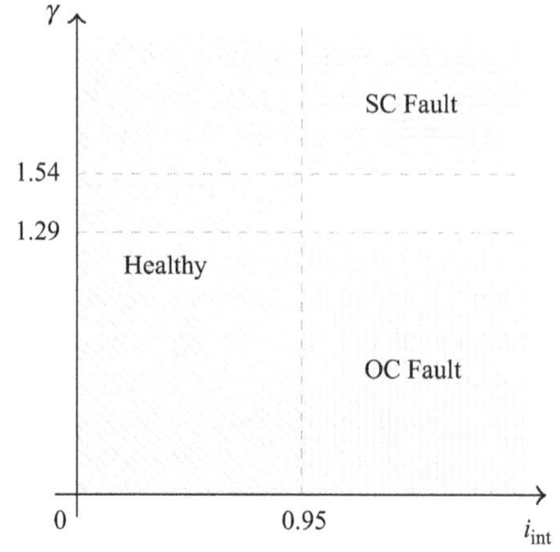

Figure 6.6 Classification of the semiconductor faults taking into account the values of i_{int} and γ, based on the empirical thresholds defined in [3]

The information of i_{int} is important for fault detection purposes. The fault detection is achieved taking into account that large values of i_{int} are related to fault episodes. Looking at Figure 6.6, it is observed that a semiconductor fault alarm is issued when i_{int} surpasses 0.95.

The observation of i_{int}, by itself, does not allow to distinguish OC faults from SC faults. Therefore, the identification of the fault mode is achieved taking into account also the information about ratio γ. High values of ratio γ are related to significant changes in the DC-bus current waveform and, consequently, imply the occurrence of a SC fault. On the other hand, lower values of ratio γ denote a smoother variation of the DC-bus current and, consequently, are associated to an OC fault event. Looking at Figure 6.6, it is observed that the identification of a SC fault occurs when i_{int} surpasses 0.95 and the ratio γ surpasses 1.54; the identification of an OC fault takes place when i_{int} surpasses 0.95 and the ratio γ is lower than 1.29.

The identification of the faulty switch is achieved taking into account the information about the ratio γ and the values of i_{int}. The determination of the range of values of i_{int} and γ related to the different fault scenarios is achieved empirically in [3], based on the results of 50 training experiments.

It should be noted that, despite the fact that the fault diagnostics algorithm includes the ability to distinguish the fault mode (OC or SC fault), the algorithm features do not allow to locate the faulty semiconductor(s), when the converter under analysis contains multiple active switches.

The effectiveness of this fault diagnostic algorithm was confirmed in [3] for an experimental prototype of a 1.1 kW full-bridge (FB) zero voltage switching (ZVS) DC–DC converter. Nevertheless, the algorithm can be applied in other DC–DC converters operating under ZVS condition as well.

As final remarks, the simplicity and low implementation cost appear as two important features of this algorithm. On the other hand, the demand for empirically

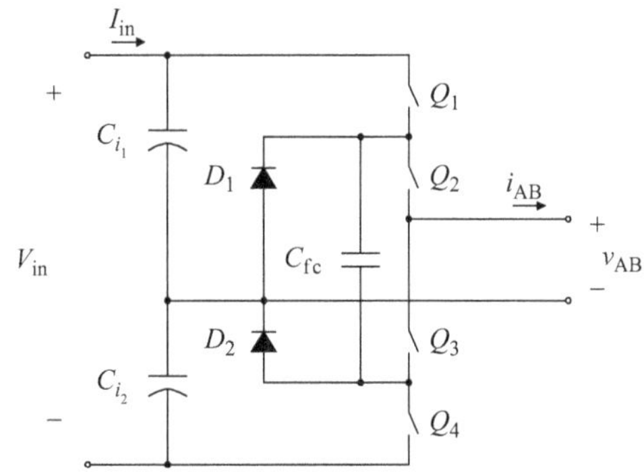

Figure 6.7 Partial representation of the three-level flying capacitor DC–DC converter – transformer primary-side bridge

defined thresholds that vary according to the converter and fault diagnostic system designs constitutes the main disadvantage of this approach. The algorithm features lead us to infer that there is a remote risk of activation of false fault alarms due to transients in the diagnostic variable.

Flying capacitor voltage

In multilevel converters, maintaining the balance of the capacitors voltage is a vital condition that should be continuously met. However, switch faults preclude such criterion, leading to unbalances in the capacitors voltage. Based on this principle, it is possible to obtain fault signatures through the analysis of the flying capacitor voltage to detect switch faults [4]. The fault diagnostics algorithm based on this principle is suitable for fault detection in three-level flying capacitor DC–DC converters. For ease of comprehension of the fault diagnostic algorithm, let us consider the information available in Figure 6.7, which provides a schematic representation of the transformer primary-side bridge of a three-level flying capacitor DC–DC converter.

The fault diagnostics algorithm based on this principle detects the presence of a faulty semiconductor, in either the primary-side or secondary-side bridge of the isolation transformer, taking into account the fact that any semiconductor fault (OC or SC fault) introduces an unbalance in the voltage of the flying capacitor. Let us consider the converter structure depicted in Figure 6.7. Under healthy converter operating conditions, the voltages of the two DC-bus capacitors (C_{i_1} and C_{i_2}) are similar and, consequently, the flying capacitor voltage is constant and equal to half of the DC-bus voltage:

$$V_{fc} = \frac{V_{in}}{2} \tag{6.6}$$

where V_{fc} is the flying capacitor voltage and V_{in} is the DC-bus voltage.

In a fault event, V_{fc} is not balanced anymore, and its value evolves to zero or V_{in}, depending on the fault mode.

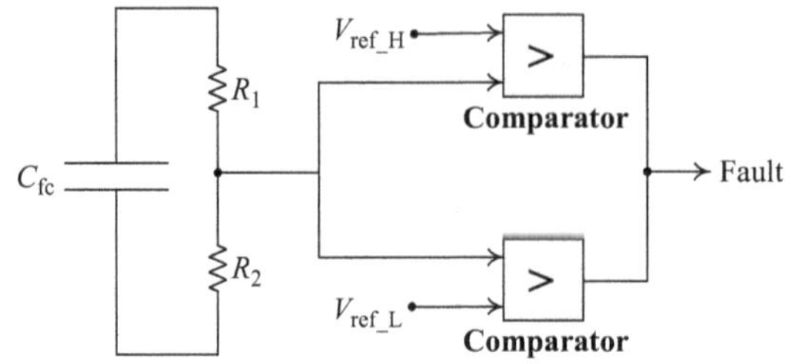

Figure 6.8 Schematic representation of the circuitry used for fault diagnostic purposes. The algorithm action is based on the observation of the flying capacitor C_{fc} voltage [4]

This fault diagnostic algorithm can be implemented resorting to simple analogue circuitry, while maintaining a small fault detection time [4]. Figure 6.8 depicts a simplified sketch of the fault diagnostic algorithm.

This circuit comprises a voltage divider, responsible for sensing V_{fc}, and two comparators, whose function is to determine when V_{fc} decreases below V_{ref_L}, or increases above V_{ref_H}. V_{ref_L} and V_{ref_H} are the low reference and high reference thresholds, respectively, and are computed as follows:

$$V_{ref\_L} = \frac{V_{in\_L}}{2n} \tag{6.7}$$

$$V_{ref\_H} = \frac{1+k}{2n} V_{in\_H} \tag{6.8}$$

In (6.7) and (6.8), V_{in_L} is the minimum input voltage, V_{in_H} is the maximum input voltage, k is the voltage ripple at the input capacitor, and n is the ratio between resistors R_1 and R_2 (refer to Figure 6.8), that constitute the voltage divider used to sense V_{fc}.

The algorithm performance is confirmed in [4], on a 30 kW experimental prototype of the three-level flying capacitor DC–DC converter.

The main constraint of the algorithm is related to the limited applicability, since it can only detect semiconductor faults in three-level DC–DC converters.

Transformer voltage

In isolated DC–DC converters and, particularly, in FB DC–DC converters, it is possible to obtain meaningful fault signatures resorting to the information of the isolation transformer voltage [5]. This algorithm is suitable for the diagnosis of OC faults in FB DC–DC converters. Due to the algorithm nature, this approach can only be applied to DC–DC power converters featuring galvanic isolation.

The fault diagnostic algorithm uses the AC voltages measured at the terminals of the transformer windings as diagnostic variables. Again, and for ease of comprehension of the fault diagnostic algorithm, let us consider the information available in Figure 6.9, which provides a schematic representation of the transformer primary-side bridge of a dual-active bridge (DAB) DC–DC converter.

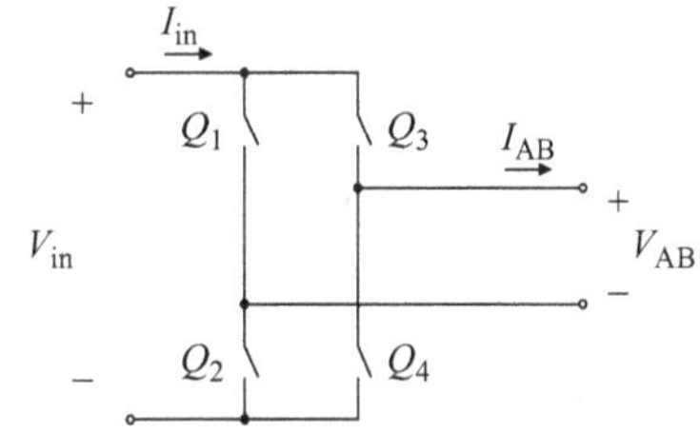

Figure 6.9 Partial representation of the DAB DC–DC converter – transformer primary-side bridge

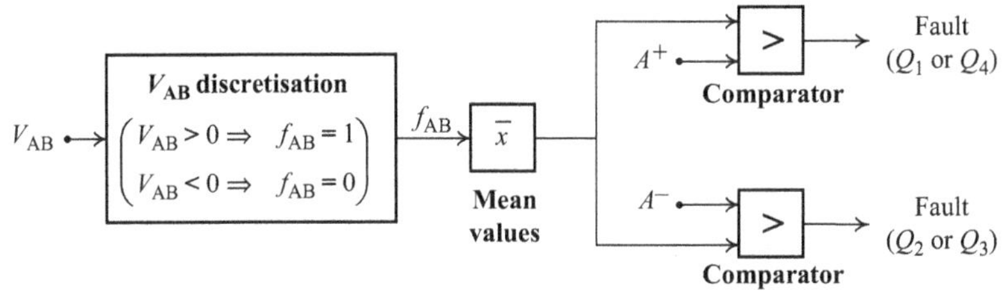

Figure 6.10 Schematic representation of the fault diagnostic algorithm, based on the observation of the isolation transformer voltage V_{AB} [5]

The primary-side winding voltage is used to diagnose faults in the primary-side bridge, while the secondary-side winding voltage is used on the diagnostic of faults in the secondary-side bridge.

The algorithm allows to identify a set of two switches whose functions might have been compromised by an OC fault. The symmetric structure of the converter limits the fault identification capability, as the occurrence of faults in diagonally opposed semiconductors produces similar fault signatures. For this reason, the identification of the faulty switch, among the group of two switches that might have been compromised, is not possible with this algorithm. Figure 6.10 provides a simple and schematic representation of the algorithm.

The fault detection algorithm performs an evaluation, in the time domain, of the mean value of the transformer windings voltage. Under normal converter operation, the mean value of the AC transformer winding voltage is zero. A failure in any of the converter switches produces an unbalance in the windings voltage, as confirmed in Figure 6.11. As a consequence of such unbalance in the windings voltage, the mean value of that voltage deviates from zero. The direction of the deviation in the mean windings voltage depends on the location of the faulty switch. As stated in Figure 6.11(a), an OC fault in switch Q_1 causes a temporary decrement of voltage V_{AB}. Similarly, an OC fault in switch Q_2 leads to a temporary increment of voltage V_{AB}, as highlighted in Figure 6.11(b).

The definition of empirical thresholds will be fundamental to identify the pair of switches where the faulty switch is included. These thresholds are denoted as A^+ and A^- in Figure 6.10.

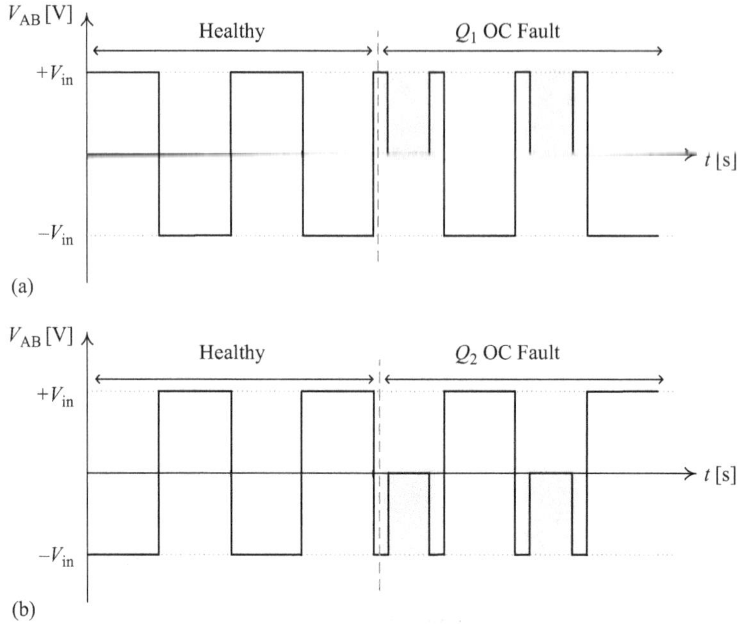

Figure 6.11 Partial representation of the DAB DC–DC converter transformer primary-side voltage evolution, considering an OC fault scenario: (a) fault in switch Q_1; (b) fault in switch Q_2

So far, the algorithm effectiveness was solely tested resorting to a simulation model of a DAB converter, without any experimental validation [5].

To overcome some of the pitfalls of the previous diagnostic algorithm, alternative approaches for the analysis of the transformer winding voltage were developed. The comparison of the transformer winding voltage with predefined thresholds provides a feasible diagnostic tool aimed at FB DC–DC converters [6]. An auxiliary transformer winding, added to the original converter, is required in order to sense the voltage of the transformer primary auxiliary winding $v_{aux}(t)$.

The diagnostic process comprises two distinctive stages. During the first stage, the information about the transformer primary voltage is compared, using simple analogue circuitry, to a predefined threshold V_{ref_1}, allowing to detect the presence of an OC fault. Threshold V_{ref_1} is a function of the converter input current V_{DC} and of the turns ratio between the transformer primary winding and the auxiliary winding K_{aux}. Threshold V_{ref_2} is defined taking into account the phase-shift and technical specifications of the fault detection circuit. Figure 6.12 depicts a simplification of the fault detection component of this algorithm.

This stage is unable to identify the faulty switch and, consequently, a fault identification stage must be deployed. Accordingly, the fault identification stage can take place by forcing the converter states. The phase shift between gating signals applied to diagonally opposed switches of the SAB is reduced, and the

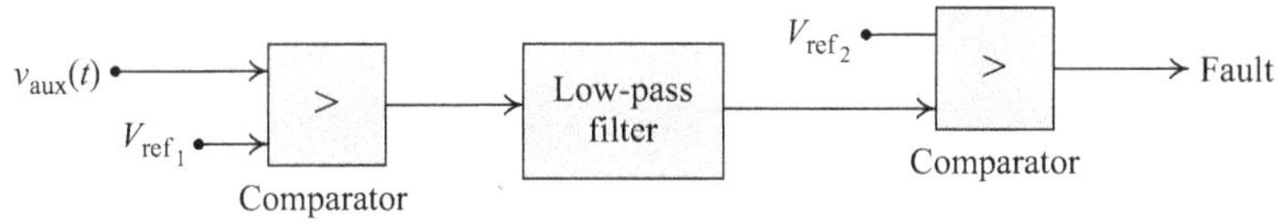

Figure 6.12 Simplification of the fault detection algorithm based on the observation of the transformer primary winding voltage [6]

influence of this adaption on the transformer primary winding voltage is studied. With this adaption in the control strategy, the transformer primary winding voltage becomes particularly high or low, depending on the position of the faulty switch. Based on the converter response to that stimulus, the determination of the faulty switch is achieved.

In the literature, the validation of this fault diagnostic algorithm is conducted on a single active bridge (SAB) converter [6].

It should be also noted that the fault diagnostic algorithm is not effective for all the range of converter operation points. Experimental results conducted in [6] demonstrated that the fault diagnostic algorithm is effective between 20% and 100% of the converter rated load. Along with this limitation, there is a relevant increase in the converter cost, due to the inclusion of an auxiliary transformer winding. Despite its small size, the auxiliary transformer winding increases the converter bill of material.

Converter input current slope

Selection of the input current of DC–DC converters as fault diagnostic variable is a common practice within the scientific community, which has inspired the development of plenty alternative fault diagnostic algorithms aimed at many DC–DC converter topologies.

To achieve the fault diagnostics objective, one of those algorithms relies on an analysis, in the time domain, of the converter inductor current i_L slope. Among the reasons for selecting the inductor current i_L as fault diagnostic variable is the fact that i_L is commonly used for control purposes in most DC–DC converters and, consequently, it is already available in the converter controller. Additionally, the oscillation pattern of i_L is similar for all single-ended DC–DC converters, which confer a relevant advantage to this fault diagnostics approach [7].

The algorithm can detect both OC and SC faults in single-ended converters (buck, boost, buck-boost, Ćuk, SEPIC, and dual SEPIC topologies), as long as the CCM conduction mode is adopted.

Progressive improvements have been attained while developing this algorithm. The preliminary version of the algorithm has relevant pitfalls related to it, due to the algorithm inability to distinguish whether the faulty switch has suffered an OC or SC fault [7].

Two modules complement their fault diagnostic actions, by operating in parallel to ensure the effective fault detection. While the first module ensures a fast fault response to detect faults, the second is used as backup, providing robust

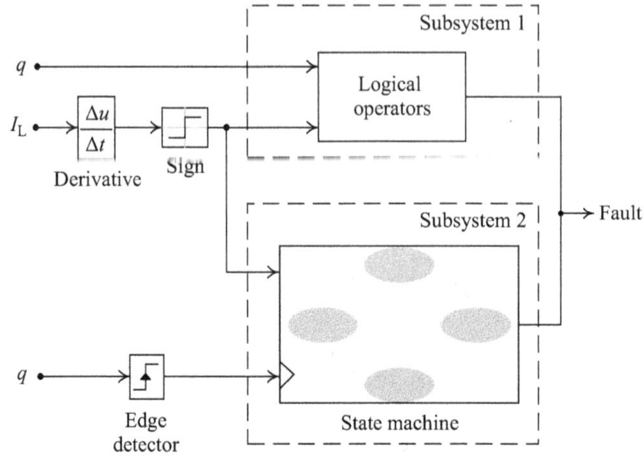

Figure 6.13 Simplification of the fault diagnostic algorithm based on the analysis of the inductor current slope [7]

results while detecting both OC and SC faults, over a wide range of operating conditions (namely, higher switching frequency and wider range of duty cycle).

Subsystem 1 (refer to Figure 6.13) monitors the sign of the inductor current derivative sgn(di_L/dt).

Under healthy converter operating conditions, there are two distinctive periods that can be distinguished. The first one occurs when a high level of gating signal q is applied to the converter switch: the converter inductor is storing energy and, therefore, the current i_L increases, resulting in a positive value of sgn(di_L/dt):

$$q = 1 \quad \Rightarrow \quad \frac{di_L}{dt} > 0 \Rightarrow \mathrm{sgn}\left(\frac{di_L}{dt}\right) = 1 \tag{6.9}$$

The second period concerns the switch off-state: gating signal q is at low level and the inductor current i_L decreases linearly, as a result of its discharging process, resulting in a negative value of sgn(di_L/dt):

$$q = 0 \quad \Rightarrow \quad \frac{di_L}{dt} < 0 \Rightarrow \mathrm{sgn}\left(\frac{di_L}{dt}\right) = -1 \tag{6.10}$$

Under OC fault conditions, sgn(di_L/dt) remains either positive (in case of SC fault) or negative (in case of OC fault), regardless of the expected switch state, defined by gating signal q. To evaluate the presence of a switch fault, an error signal compares sgn(di_L/dt) and the expected sign of the inductor current derivative S_q, estimated resorting to the information of the switch gating signal q sign:

$$\begin{cases} q = 1 \Rightarrow S_q = 1 \\ q = 0 \Rightarrow S_q = -1 \end{cases} \tag{6.11}$$

The error signal is obtained as follows:

$$\text{error} = \text{sgn}\left(\frac{di_L}{dt}\right) - S_q \tag{6.12}$$

It is considered that a fault event happens when the error diverges from zero. To avoid false fault alarms resulting from the non-ideal behaviour of the converter, a margin is considered, and the fault alarm is triggered only when the error signal remains at high level for a reasonably long period, longer than N sampling periods.

The fault diagnostic effectiveness of the first subsystem (refer to Figure 6.13) is not granted for all circumstances. The fault diagnostic might reveal unsuccessful when the duty cycle or the switching frequency are high, or if the fault occurs close to the moments of transition between switch states [7]. To overcome these hurdles, an auxiliary fault diagnostic subsystem, based on a state machine, is implemented. The state machine has two inputs (sign of the inductor current derivative sgn(di_L/dt) and gating signal q), and four states. The rising edge of gating signal q triggers the transition between states of the machine. The principle of implementation of the state machine is quite similar to the principle adopted for the first subsystem: the sign of the inductor current derivative is checked, to confirm whether the measured sign of the inductor current derivative at that moment matches with the expected sign for the present machine state. If the expected sign of the inductor current derivative does not match with the expected derivative sign during a certain period, the state machine follows a sequence of states that leads to the activation of a fault alarm. The fault alarm issued by subsystem 2 (refer to Figure 6.13) does not provide enough information to conclude on the verified fault mode (OC or SC fault mode).

To overcome some of the pitfalls of the initial version of the algorithm, several derivations of the initial fault diagnostic algorithm are known [8,9].

In one of those derivations [8], subsystem 1, represented in Figure 6.13, is replaced by a state machine, also with four states. Thus, two state machines, with a similar structure, are operated in parallel to realise the fault diagnostic action. The principles used to develop the new state machine remain unchanged. The adaptation of subsystem 1 into a state machine enables the fault identification feature, which was not available in the preliminary version of the algorithm.

A second derivation of the initial fault diagnostic algorithm aims to provide a meaningful reduction of the fault diagnostics time and, at the same time, improve the diagnostic effectiveness [9]. Figure 6.14 depicts a schematic view of the algorithm. The new feature of the algorithm is highlighted with the dark grey background.

The fault diagnostic action is also based on the comparison of the sign of the inductor current slope sgn(di_L/dt) and the state of gating signal q. The delay between the rising edge of gating signal q and the sign of the inductor current slope sgn(di_L/dt) is compensated, resorting to an estimation of that delay. This means that the algorithm compares sgn(di_L/dt) and the delayed version of gating signal (q_d). In practice, this action allows to reduce the time span required to issue a more reliable fault alarm. The algorithm limitations for very low or very high duty cycle values remain as the most important pitfalls of this derivation of the fault diagnostics algorithm.

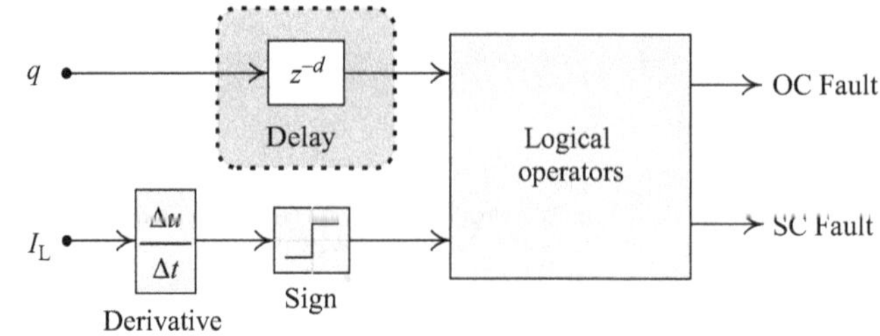

Figure 6.14 Schematic representation of the fault diagnostic algorithm based on the analysis of the converter inductor current I_L slope. A delay compensation (highlighted with a dark grey background) allows to reduce the fault diagnostic time and, at the same time, improve the diagnostic effectiveness [9]

The adoption of a fault diagnostic algorithm based on two subsystems which execute the fault diagnostic function concurrently [7] demands a fairly high computational effort. The adoption of alternative algorithms, demanding less computational effort, is feasible. Based on the same fault diagnostic variable (inductor current waveform), a single module can correctly detect a fault [10,11]. The observation of the switch states, through the analysis of the inductor current slope sign sgn(di_L/dt) during one switching period, is performed in this case resorting to a single state machine. The state machine comprises, as inputs, the inductor current slope sign sgn(di_L/dt) and the state machine command signals, responsible for controlling the transition between states. The selection of the right moments for the transition between states within the state machine is vital to ensure fast diagnostic response, on the one hand, and reliable diagnostic results, on the other hand. In simpler implementations of the algorithm, the transition between machine states is solely controlled by the two edges of the converter gating signal q [10]. Feasible diagnostic results are achieved, but the algorithm is unable to distinguish OC from SC fault episodes, as demonstrated in the flowchart of Figure 6.15(a). Additionally, the diagnostic response may be somehow delayed. The transition between states, denoted as x_1 and x_2 in Figure 6.15(a), takes into account the information provided by gating signal q. Each transition is triggered by the edges of the gating signal q, as depicted in Figure 6.15(b).

The limitations of the simpler version of the algorithm are easily overcome by simply 'splitting' one switching period of gating signal q into four commands for the state machine, faster diagnostic response of the state machine and improved capability to distinguish OC from SC fault episodes are attainable [11]. Figure 6.16(a) depicts a simplification of the flowchart followed by the state machine, while Figure 6.16(b) shows a timeline which illustrates the moments in which the most meaningful transition between states takes place within the state machine.

Note that the flowcharts depicted in Figures 6.15(a) and 6.16(a) are a simplification of the state machine, that does not make mention to all the states and transitions of the state machine [10,11]. In Figure 6.16(a), the transition between

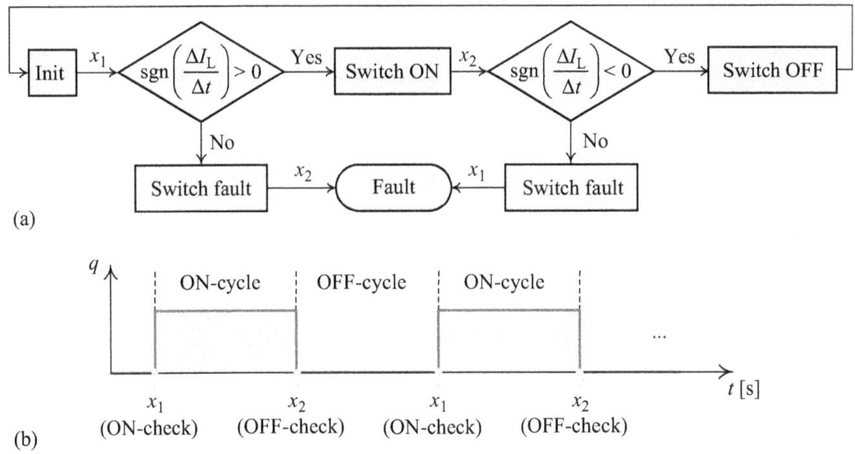

Figure 6.15 (a) Simplified flowchart of the state machine used for fault diagnostics [10]; (b) gating signal q and corresponding instants used to control the transition between states within the state machine developed in [10]

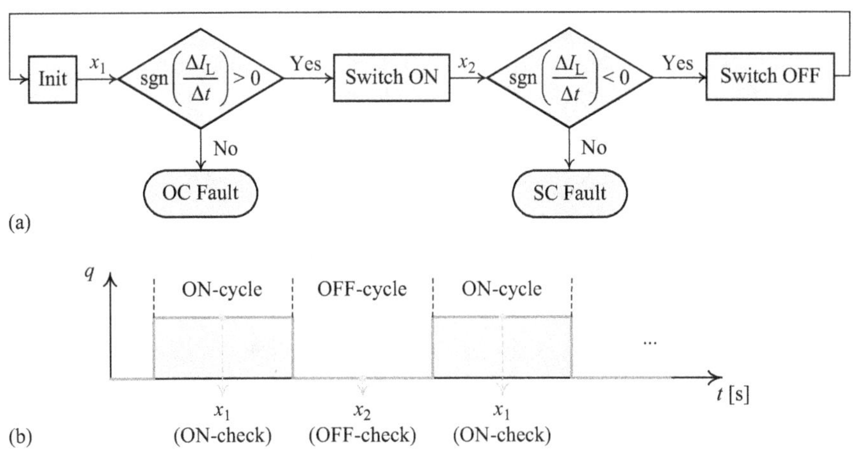

Figure 6.16 (a) Simplified flowchart of the state machine used for fault diagnostics [11]; (b) gating signal q and corresponding instants used to control the transition between states within the state machine developed in [11]

states, denoted as x_1 and x_2, takes into account the information provided by gating signal q. As depicted in Figure 6.16(b), each transition takes place at very well-defined moments. Along with the two edges of the converter gating signal q, two additional command signals are used to control the transition between machine states: x_1 and x_2 (refer to Figure 6.16(b)). Transition x_1 triggers the verification of the converter switch ON-state, and occurs in the middle of the switch ON-cycle. On the other hand, transition x_2 triggers the verification of the converter switch OFF-state, and occurs in the middle of the switch OFF-cycle. The selection of these instants for the control of the transition between machine states allows to obtain the best compromise between effective diagnostic results and fast response to fault events.

The performance of these fault diagnostic algorithms [7–11] was evaluated resorting to a laboratory test bench, comprising the prototypes of single-ended DC–DC boost converters, delivering a few hundreds of watts to the load.

Unlike other fault diagnostic algorithms based on simple analogue circuits, the implementation of these algorithms takes advantage from the capabilities of a field-programmable gate array (FPGA) to control the converter and implement a fault diagnostic algorithm, in a single platform. Another advantage arises when fault diagnostic algorithms are implemented in a FPGA: the fault diagnostic time is small, allowing a fast response to faults. Obviously, such approach implies a higher implementation cost, as the hardware complexity also increases. Additionally, the effectiveness of this algorithm is limited to single-ended converters operating at CCM. Adoption of two distinct approaches to detect a fault, as mean to overcome the difficulties of the primary algorithm to diagnose faults under certain circumstances, is also seen as a disadvantage.

An alternative and simpler approach, equally based on the inductor current slope, can also detect OC faults in single-switch converters in an effective manner, requiring less computational effort [12]. Figure 6.17 depicts a typical evolution pattern of the inductor current I_L in a simple single-switch boost converter, considering both healthy and faulty switch conditions. As shown in Figure 6.17, the algorithm samples current I_L at three distinctive moments during one switching period T_{sw} to obtain three values of current I_L: $I_L(k-1)$, $I_L(k)$, and $I_L(k+1)$.

The acquisition of data for fault diagnostic purposes must obey some rules. The acquisition of the first value – $I_L(k-1)$ – may occur at any moment of the switching pattern. As shown in Figure 6.17, the three moments used to sample current I_L must be separated from each other by half of the switching period ($0.5T_{sw}$). Then, a logical relation is established between those three values. Under healthy converter operation, the absolute values of $I_L(k-1)$ and $I_L(k+1)$ are quite similar. On the other hand, a switch OC fault forces a descendent trend in current I_L that is transversal to all three sampled values.

The algorithm simplicity and low cost of implementation are the advantages of the approach. On the other hand, the algorithm resiliency is somehow limited, especially in case of load transients or any other events that may perturb the stability of the inductor current I_L. For that reason, the adoption of this algorithm for diagnostic purposes should be carefully considered, based on the converter application requirements.

Output DC-bus capacitors voltage

Diagnostic of semiconductor faults in DC–DC converters used for PV applications featuring maximum power point tracking (MPPT) control capabilities may take advantage of the power-conditioning system intrinsic features [13]. Accordingly,

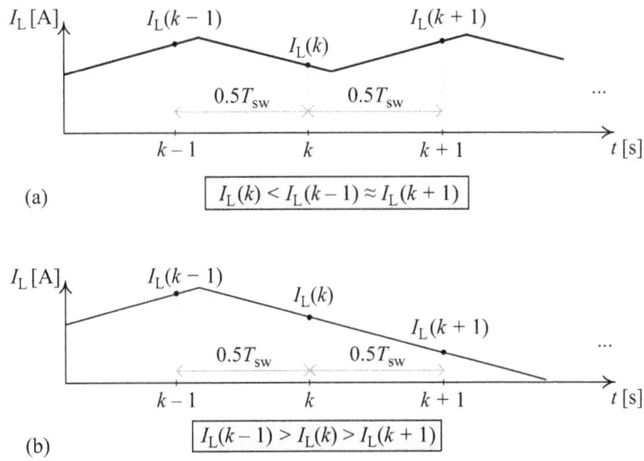

Figure 6.17 Inductor current evolution under: (a) healthy converter operation; (b) faulty switch condition [12]

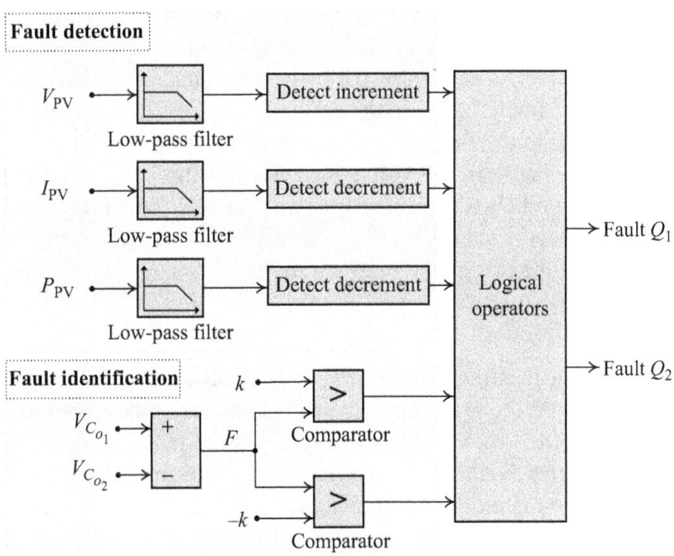

Figure 6.18 Schematic representation of the fault diagnostic algorithm based on the analysis of the output DC-bus capacitors voltage [13]

a fault diagnostics algorithm based on the PV panel variables – voltage, current and power – and on the output DC-bus capacitor(s) voltage(s), provides an effective and relevant application-oriented diagnostics tool.

Figure 6.18 depicts a simplified representation of this fault diagnostic algorithm, suitable for multilevel DC–DC converters.

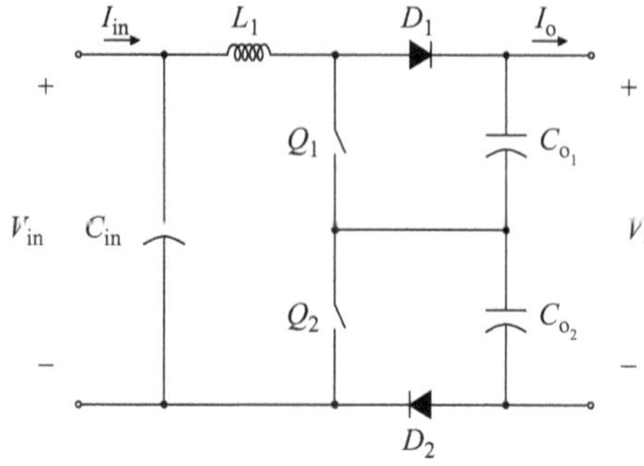

Figure 6.19 Three-level non-isolated DC–DC converter

A fault is detected by taking into account the low-frequency oscillations of the PV panel variables. The increment of the PV panel voltage, along with the decrement of the PV panel current and power, put into evidence the occurrence of an OC fault in the power converter responsible for the implementation of the MPPT control strategy. However, the simple action of monitoring the PV panel variables does not allow, by itself, to properly identify the faulty switch(es).

Taking into account the behaviour of the selected converter topology (three-level converter), depicted in Figure 6.19, the algorithm takes advantage of the converter intrinsic characteristics. In the particular case of the three-level converter, the algorithm resorts to the fact that any OC fault in the converter power switches disturbs the balance between the voltages of the converter DC-bus capacitors C_{o_1} and C_{o_2}. Such voltage balance is mandatory in a healthy converter condition. The fault identification step tracks the DC-bus capacitors voltages. A diagnostic variable F is defined as the difference between the DC-bus capacitors voltages:

$$F = V_{C_{o_1}} - V_{C_{o_2}} \tag{6.13}$$

When F surpasses a positive threshold k, it is concluded that the upper DC-bus switch Q_1 is faulty; on the other hand, when F assumes values that are lower than $-k$, it is stated that the lower DC-bus switch Q_2 is the faulty switch. There are no specific methodologies available for the selection of threshold k, the reason why its selection is based on the empirical knowledge.

The algorithm was validated experimentally on a three-level boost converter connected to two PV modules of 175 W, each [13].

Simplicity, low implementation cost, and the possibility to cover a wide range of converter operating conditions seem to be important features of the algorithm. Again, the demand for empirically defined thresholds appears as the main drawback of this algorithm.

Converter inductor current evolution

The analysis of the absolute values of the converter inductor current waveform, along specific periods, presents a simple yet an effective way to diagnose switch faults in simpler converter topologies [14].

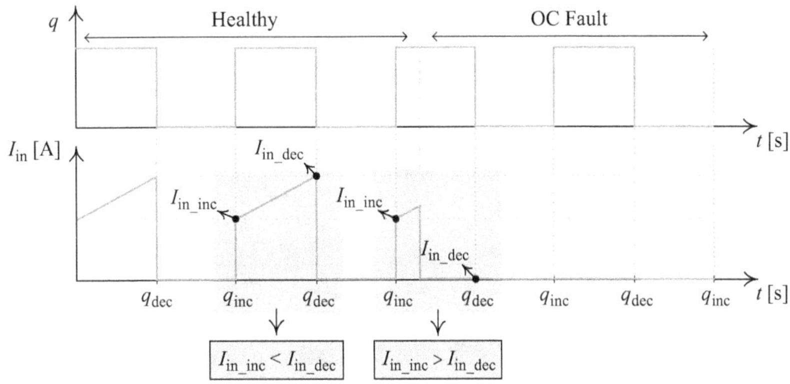

Figure 6.20 Typical evolution of the buck converter input current under healthy and faulty conditions [14]

Among other converter topologies, the fault diagnostic algorithm based on this principle can be implemented on a multi-input DC–DC converter, used as the interface system between a telecommunication system and power generation system, comprising a PV and a wind generator [14]. Two approaches are used to detect OC faults in the power switches of each conversion stage.

For the buck converter, connected to the wind generator, this algorithm compares the converter input current at the rising and falling edges of the switch gating signal. Under healthy operation of the buck converter, the input current at the rising edge I_{in_inc} should be smaller than the input current at the falling edge I_{in_dec}, due to the switching action of the power switch that allows the inductor to charge:

$$I_{in\_inc} < I_{in\_dec} \tag{6.14}$$

Under faulty condition, the input current does not increase between the rising edge and falling edge of the command signal, as a result of the loss of the converter switching function:

$$I_{in\_inc} \geq I_{in\_dec} \tag{6.15}$$

Figure 6.20 depicts the evolution of the buck converter input current during the pre-fault and post-fault periods. As highlighted in Figure 6.20, the converter input current increases between the rising and falling edges of gating signal q, as long as the switch action maintains intact. Right after the OC fault, the converter input current drops to 0 A. Under such circumstances, the relation expressed in (6.14) is not met and, consequently, an OC fault is detected.

The confirmation of the OC fault in the buck converter is complemented with the verification of two additional conditions.

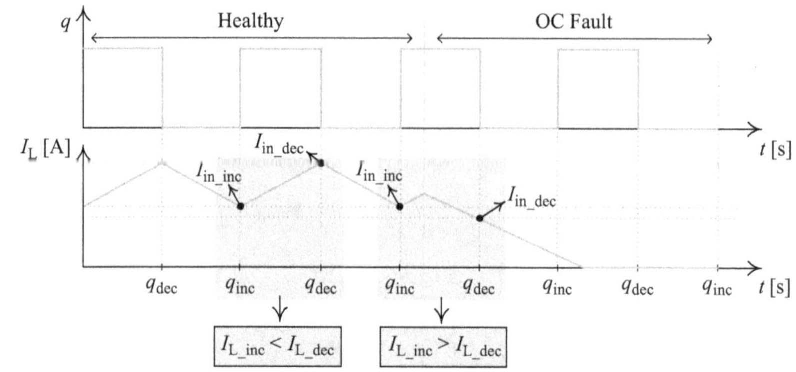

Figure 6.21 Typical evolution of the converter inductor current under healthy and faulty conditions [15,16]

The second diagnostic strategy, applied to the Ćuk converter connected to a PV system, establishes a mathematical derivation of the MPPT algorithm to obtain a fault diagnostic variable F_{PV}:

$$F_{\text{PV}} = \frac{dV_{\text{PV}}/dI_{\text{PV}}}{V_{\text{PV}}/I_{\text{PV}}} \tag{6.16}$$

After an OC fault in the power switch of the Ćuk converter, the maximum power point is lost and, consequently, F_{PV} approaches zero.

Similar fault diagnostic approach, equally based on the analysis of the absolute values of the converter current, can be followed in other converter topologies, attaining feasible diagnostic results. Effectiveness of the algorithm has been confirmed for non-isolated bidirectional DC–DC converters [15], and non-isolated unidirectional buck converters [16]. The detection and identification of OC faults is also achieved by comparing the absolute values of the inductor current at the rising and falling edges of the gating signals. Referring to Figure 6.21, the inductor current of non-isolated bidirectional DC–DC converters and non-isolated unidirectional buck converters increases between the rising and falling edges of gating signal q, due to the switching action of the converter switch(es). After the OC fault, the inductor charging cycle is interrupted and the current decreases linearly until it fully extinguishes.

To obtain fault diagnostic results with higher degree of confidence, additional conditions are tested, to ascertain that load transients or other non-linearities do not trigger false fault alarms. For both the non-isolated bidirectional DC–DC converters [15] and non-isolated unidirectional buck converters [16], an OC fault event is confirmed by checking that the converter voltage in the high-side V_{HV} is higher than the converter voltage in the low-side V_{LV}:

$$V_{\text{HV}} > V_{\text{LV}} \tag{6.17}$$

This condition must be observed to obtain feasible fault diagnostic results; otherwise, dubious diagnostic results are obtained, precluding any convincing conclusions.

These fault diagnostic algorithms have a simple structure, allowing their implementation even when implementation cost is a concern. Unlike most fault diagnostic algorithms performing an analysis in the time domain, whose effectiveness is typically limited to power converters operating at CCM, these algorithms can identify OC faults when the power converter operates in DCM. Additionally, these strategies do not require the establishment of thresholds. As main drawback of these approaches, it can be pointed out the limited range of switching frequencies in which the algorithms effectiveness is ensured. Despite the remote possibility of false fault alarms caused by transients or noise in the diagnostic variable, the probability of such events is higher for these algorithms, if compared to their peers. It should be also highlighted that the validation of the fault diagnostic algorithms was limited to the simulation studies in [14] and [15]. The fault diagnostic algorithm proposed in [16] was validated on an experimental setup of a wind energy conversion system supplying a set of batteries.

SAB converter output current

This fault diagnostic algorithm, specifically developed for the parallel-connected SAB DC–DC converter, uses the converter output current I_o as diagnostic variable to identify OC faults in any of the converter modules [17]. The converter architecture enables the adoption of a phase-shift control strategy. Due to the adoption of this switching control strategy, the converter output current I_o is interleaved, allowing a significant reduction of the converter output current ripple.

For fault diagnostic purposes, the converter output current I_o is sampled at precisely selected instants. The algorithm takes advantage from the fact that the switches' turnoff moment is overlapped with the peak of the output current I_o, as shown in Figure 6.22. In Figure 6.22, labels of the x-axis identify the converter module responsible for each peak of current I_o, i.e. module 1, …, module 4.

Under healthy converter operation, the converter output current I_o follows a uniform pattern, where all current peaks are aligned, as shown in Figure 6.22(a).

When an OC fault occurs in a single switch, one of the peaks of the converter output current I_o stands out from the other current peaks, due to its abnormally lower value, as shown in Figure 6.22(b). The position of the current peak with abnormal behaviour allows to identify the faulty module. In the example shown in Figure 6.22(b), the peak of current I_o related to module 2 (denoted as m_2) is abnormally low.

When two or more converter switches from one of the converter modules are compromised, the operation of that module is not feasible any longer, due to the inability to transfer energy to the corresponding isolation transformer. In that case, the frequency of the output current ripple reduces after the fault, and one of the peaks is extinguished. Again, the position of the missing current peak allows to localise the module with faulty switches.

Experimental tests were conducted on a 1 kW prototype of the converter. The results presented in [17] confirm the algorithm resiliency.

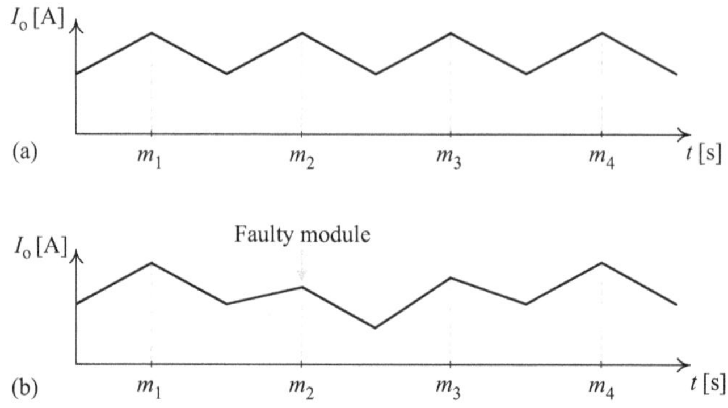

Figure 6.22 Converter output current evolution pattern under: (a) healthy converter operation; (b) faulty switch condition. Labels of the x-axis identify the converter module responsible for each peak of current I_o [17]

The algorithm features include low computational effort and low implementation cost. On the other hand, the limited applicability of the algorithm and the high risk of false fault alarms, due to load transients, are the negative aspects of the algorithm.

Converter input current derivative sign

The diagnostic of semiconductor faults based on the analysis of the derivative sign of the converter input current I_{in} has given proofs of effectiveness [18]. The algorithm is capable of detecting OC faults of the converter switches. The algorithm was tested on an interleaved DC–DC boost converter [18], but its effectiveness is extensible to other converter topologies. Unlike other fault diagnostic algorithms that use the same diagnostic variable, such as those presented in [7] and [10], this algorithm can effectively diagnose switch faults on both DCM and CCM.

Along with the converter input current I_{in}, the gating signals and the duty cycle D are the inputs of the fault diagnostic algorithm. To obtain a reliable fault diagnostic action, it must be ensured that the selected intervals of I_{in} have the same number of ON commands. Figure 6.23 provides a clear view of this rule.

Labels of the x-axis in bold highlight the intervals which have a single ON command. As those intervals have the same number of ON commands, they can be selected for the comparative analysis required to identify OC faults. Further ahead, each one of these intervals will be denoted as interval X.

Figure 6.24 depicts a simple overview on the architecture of the fault diagnostics algorithm based on the analysis of the input current derivative sign.

A sampling time T_c is applied to sample the sign of the I_{in} derivative. For each interval X, the number of positive (N_p) and negative (N_n) derivative values is determined. The sum of these two values ($N_p + N_n = N_s$) is also computed. The information of N_p, N_n, and N_s allows to conclude, for each interval, whether the derivative of the input current I_{in} is predominantly positive or negative. This information is

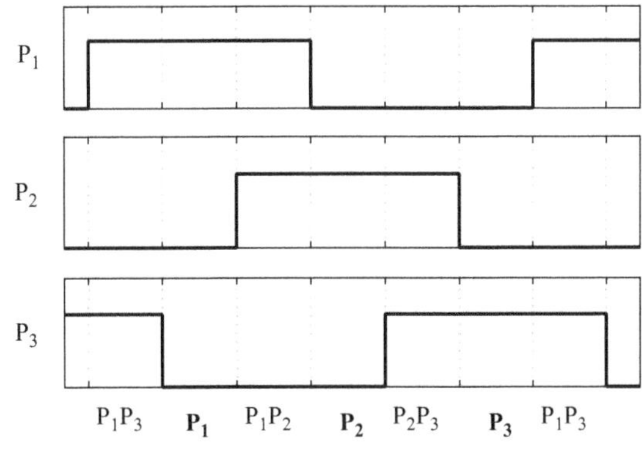

Figure 6.23 Typical switching pattern of a three-phase interleaved DC–DC converter. The x-axis labels identify the active gating signals for each interval

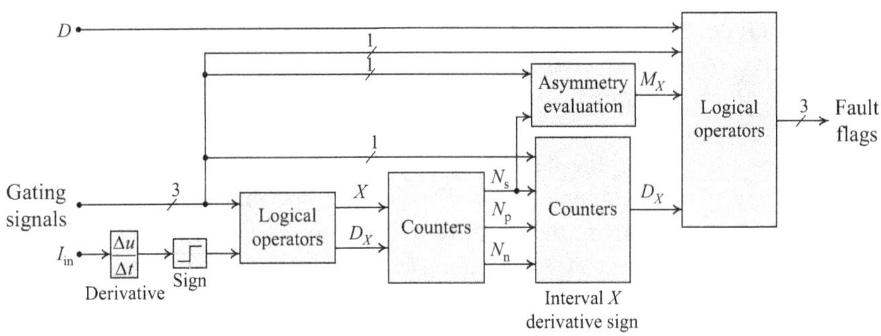

Figure 6.24 Schematic representation of the fault diagnostic algorithm based on the analysis of the input current derivative sign [18]

transmitted through variable D_X, as depicted in Figure 6.24. The presence of potential asymmetries in the switching pattern is also checked, resorting to the 'Asymmetry Evaluation' block, shown in Figure 6.24. This operation aims to certify that all intervals under evaluation have the same number of samples of the I_{in} derivative. Albeit seldom, these asymmetries impair the effective action of the algorithm.

After identifying the prevailing sign of the input current I_{in} derivative for each interval X, the horizon of the analysis is extended to an entire switching period, and the measured sign of the input current derivative is confronted with the expected sign of the input current derivative, on each interval. The expected derivative sign for each interval X is deduced taking into account the information of the duty cycle D and one of the converter control signals.

Figure 6.25 shows a generic representation of the input current of a three-phase interleaved DC–DC converter. Dashed lines in light grey colour underline the intervals used for the analysis of the derivative of the converter input current. The extension of the two switching periods, depicted in Figure 6.25, is marked off

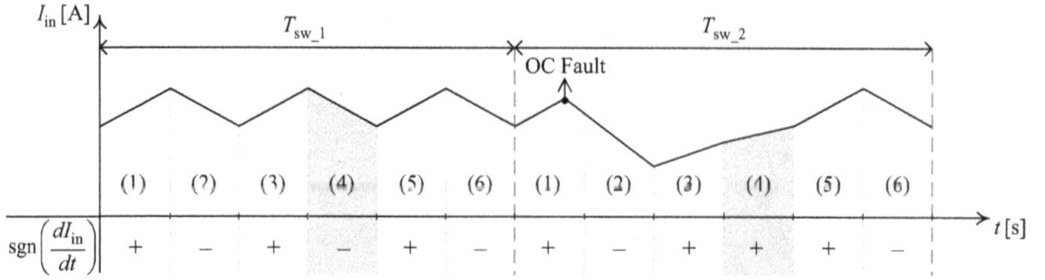

Figure 6.25 Three-phase interleaved DC–DC converter input current evolution along two switching periods. Shadowed intervals highlight the differences in the sign of derivative of I_{in} [18]

by arrows T_{sw_1} and T_{sw_2}. The six intervals that comprise one switching period are numbered for ease of comparison between the two switching periods. An OC fault occurs in one of the converter switches during interval (1) of the second switching period T_{sw_2}, as represented in Figure 6.25.

A careful comparison allows to verify that the OC fault introduces changes in the derivative of the converter input current in interval (4) of switching period T_{sw_2}, highlighted in Figure 6.25. Under healthy converter operation, the derivative of the converter input current is positive during interval (4) of switching period T_{sw_1}; on the other hand, the derivative of the converter input current is negative during interval (4) of switching period T_{sw_2}, as a consequence of the switch fault.

The implementation of the algorithm does not imply the adoption of any thresholds, ensuring that its effectiveness is independent of the converter load level. Fast diagnostic action is attained. In addition, the algorithm action covers a wide range of operating conditions and, though not mentioned in [18], a relevant range of converter topologies. Addition of sensors solely dedicated to fault diagnostic actions is also avoided, configuring another relevant advantage to this algorithm.

Magnetic components voltage

The fault diagnostic algorithm performs a time-domain analysis of the voltage measured in any of the magnetic components (inductors or transformers) of the DC–DC converter. By cross-checking the information available in the voltage waveform, denoted as v_m, and the gating signal(s) applied to the converter switch(es), it is possible to identify both OC and SC faults in the converter switches [19]. To measure voltage v_m, an auxiliary winding has to be introduced next to the magnetic component of the converter. Then, the remaining fault diagnostic structure can be easily implemented resorting to simple analogue circuitry, namely, logic gates and comparators.

The algorithm features the possibility for implementation in a wide range of DC–DC converter topologies, on either single-switch or multi-switch DC–DC converters. In the case of power converter topologies with multiple magnetic components, the selected magnetic component should have a direct link to the switch(es) to be monitored.

When the fault diagnostic action aims single-switch converter topologies, the algorithm compares the measured sign of voltage v_m with the expected sign of this

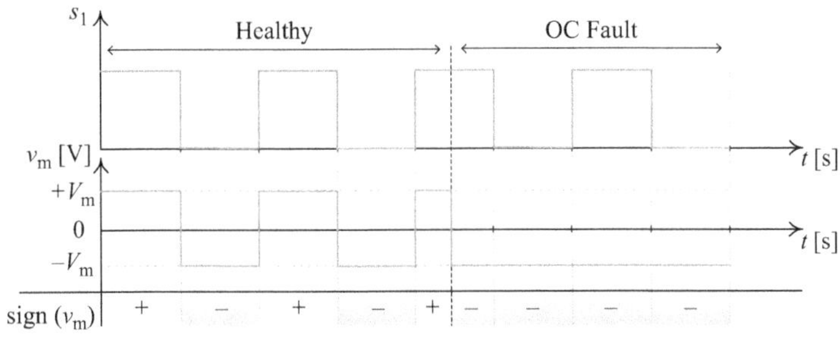

Figure 6.26 Evolution of the magnetic components voltage, in single-switch converter topologies, before and after a switch OC fault [19]

waveform. To identify the expected sign of voltage v_m, information of the gating signal s_1 is used, in order to determine the instants when the ON and OFF switch conditions occur. Considering healthy converter operating conditions, the following statements are met: during an OFF-period of the gating signal s_1, voltage v_m is negative; an ON-period of gating signal s_1 leads to a positive voltage v_m:

$$\begin{cases} s_1 = 0 \Rightarrow v_m < 0 \\ s_1 = 1 \Rightarrow v_m > 0 \end{cases} \tag{6.18}$$

If a fault compromises the normal operation of the converter switch, voltage v_m evolves in different ways, depending on the switch fault mode. Hence, voltage v_m is always negative after an OC fault, independently of the gating signal s_1 state:

$$\begin{cases} s_1 = 0 \Rightarrow v_m < 0 \\ s_1 = 1 \Rightarrow v_m < 0 \end{cases} \tag{6.19}$$

Referring to Figure 6.26, it is easily stated that the voltage across the terminals of the magnetic components of a single-switch converter remains negative right after the occurrence of an OC fault, as the switching function was lost.

On the other hand, a SC fault leads to a positive voltage v_m, independently of the gating signal s_1 state:

$$\begin{cases} s_1 = 0 \Rightarrow v_m > 0 \\ s_1 = 1 \Rightarrow v_m > 0 \end{cases} \tag{6.20}$$

Looking at Figure 6.27, it can be seen that the voltage across the terminals of the magnetic components of a single-switch converter remains positive right after the occurrence of a SC fault, independently of the gating signal state.

In conclusion, an OC fault alarm is issued if voltage v_m remains negative while $s_1 = 1$, and a SC fault alarm is issued if voltage v_m remains positive while $s_1 = 0$.

Figure 6.28 depicts a schematic representation of the fault diagnostic algorithm used to identify switch faults in single-switch DC–DC converter topologies.

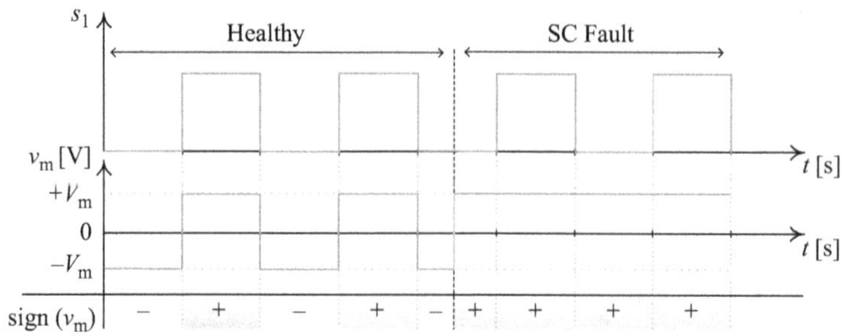

Figure 6.27 Evolution of the magnetic components voltage, in single-switch converter topologies, before and after a switch SC fault [19]

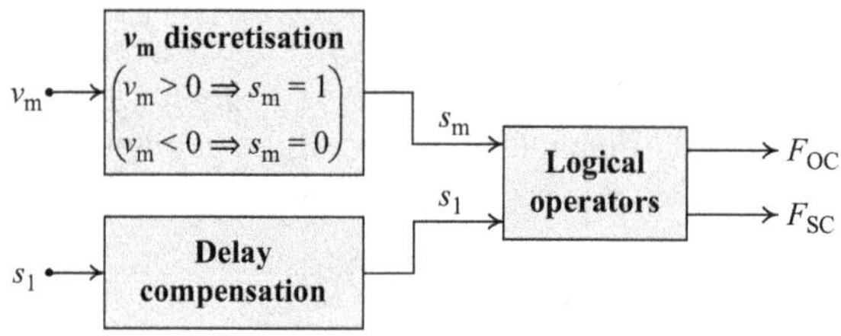

Figure 6.28 Simplification of the fault diagnostic algorithm, based on the analysis of the converter magnetic components voltage, suitable for single-switch DC–DC converter topologies [19]

In Figure 6.28, F_{OC} and F_{SC} denote the flags for OC and SC fault events, respectively. Additional capabilities, such as delay compensation and correct fault identification, are also included. These features accommodate non-ideal behaviours, such as delays while triggering the converter switch, or temporary changes in the converter operation mode.

First, voltage v_m is discretised, using the relations shown in Figure 6.28, to create a binary diagnostic variable s_m. This diagnostic variable is then compared to the delayed version of gating signal s_1. Two variables (F_{OC} and F_{SC}) show the results of the diagnostic process: F_{OC} is at low level to flag an OC fault, while a low level of variable F_{SC} confirms a SC fault.

Pertaining to dual-switch converter topologies, as it is the case of push-pull and HB converters, the same algorithm can be also implemented, considering now small changes. Figure 6.29 depicts a schematic view of the algorithm implemented on a two-switch DC–DC converter.

As shown in Figure 6.29, additional fault diagnostic blocks must be used for each switch of the converter. The principle of the fault diagnostic action remains unchanged: voltage v_m is compared with its expected value. Gating signals s_1 and s_2, used to control the converter switches, are quite useful to determine the expected voltage level (positive, negative or zero voltage).

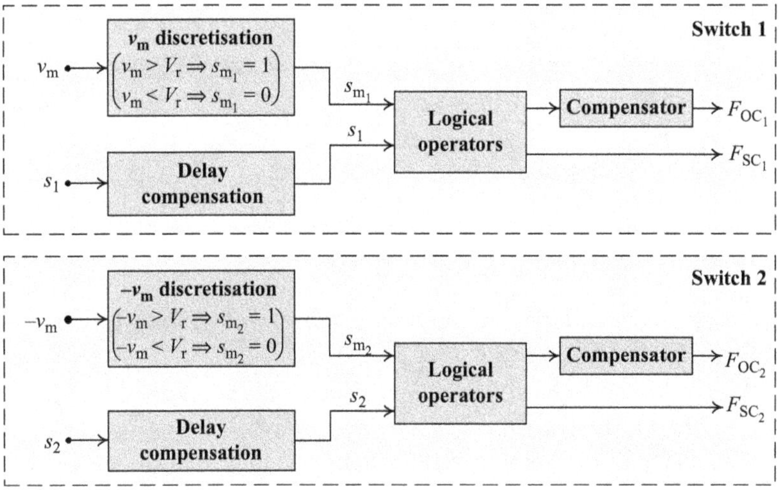

Figure 6.29 Simplification of the fault diagnostic algorithm, based on the analysis of the converter magnetic components voltage, suitable for dual-switch DC–DC converter topologies [19]

For the discretisation of voltage v_m, a threshold V_r, different from zero, must be defined for the implementation of the algorithm in dual-switch converters. In [19], the threshold is defined based on the empirical experience. Additionally, a compensator, comprising a delay unit and an OR logic gate, is applied to the OC fault flags F_{OC_1} and F_{OC_2}, to avoid any false fault alarms.

Simplicity, low implementation cost, modular structure, and fast diagnostic response are the main merits of this approach. On the other hand, the addition of an extra sensing component, which increases the overall cost of implementation and, most of all, reduces the reliability of the converter, seems to be the main drawback. The algorithm effectiveness restriction to DC–DC converters operating at CCM, leaving behind any cases where converter operation at DCM is desired, also presents an adverse impact.

An alternative fault diagnostics algorithm allows to override some of the pitfalls manifested by the original algorithm [20]. The fault signatures are also extracted from the magnetic components voltage waveform v_m, and gating commands $s_{1\ldots n}$ are also used as auxiliary variables. The modular structure of the algorithm is also adopted. Small changes are introduced in the logical operators used for fault diagnostic purposes, allowing to overcome some of the pitfalls manifested by the original algorithm, by improving the fault diagnostic capabilities for converters operating at DCM. Two low-pass filters and two thresholds, included in the fault diagnostic scheme, must be dimensioned and defined to obtain a good response of the decision-making process. The thresholds must be continuously updated due to their dependence on the duty cycle, switching period, and delay time. The filters' cut-off frequency, whose selection does not obey to any particular rule, seems to be the main drawback of the algorithm.

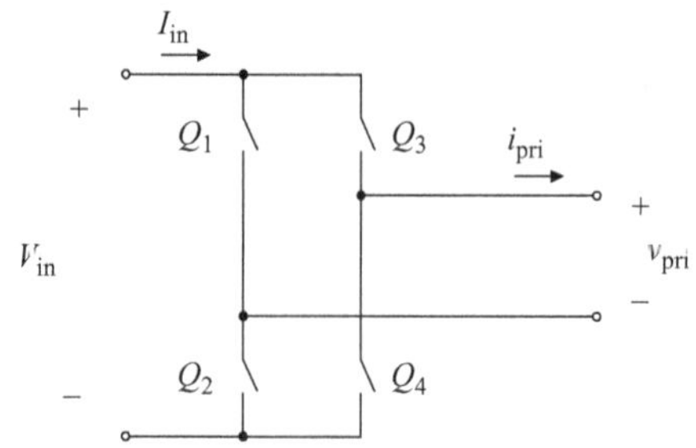

Figure 6.30 Representation of one of the bridges of the phase-shift FB DC–DC converter

DC-bus current and transformer primary voltage

This fault diagnostic algorithm results from the improvement of a previously established strategy, proposed in [6], where an OC fault diagnostic strategy was developed, aiming the same DC–DC converter topology, i.e. the phase-shift FB DC–DC converter. The algorithm aims at the diagnostic of SC faults in the primary-side bridge of the converter [21]. With this aim, the algorithm extracts information from two variables to obtain the fault signatures: the DC-bus current I_{in} and the auxiliary transformer winding voltage v_m, which is proportional to the transformer primary voltage v_{pri}. Again, and for ease of comprehension of the fault diagnostic algorithm, let us consider the information available in Figure 6.30, which provides a schematic representation of the transformer primary-side bridge of a phase-shift FB DC–DC converter.

A SC fault in any of the bridge switches entails a significant increment in current I_{in}, far exceeding a predefined threshold I_{th}, selected empirically. Based on this statement, a fault flag s_{id} is defined as follows:

$$\begin{cases} I_{in} > I_{th} \Rightarrow s_{id} = 1 \\ I_{in} < I_{th} \Rightarrow s_{id} = 0 \end{cases} \tag{6.21}$$

$$I_{th} = 3I_{in\_max} = 3\frac{I_o}{K} \tag{6.22}$$

where I_{in_max} is the maximum DC-bus current, I_o is the rated load current of the converter, and K is the transformer turns ratio. To better understand the principles of action of the algorithm, Figure 6.31 shows a draft of the algorithm responsible for the detection of switch faults.

However, the simple analysis of current I_{in} does not provide enough information to locate the faulty switch. Therefore, the fault identification stage must be implemented separately. Fault identification involves two stages. During the first stage, information from two of the four converter gating signals, pertaining to different legs of the primary-side bridge (for instance s_1 and s_3, used to trigger switches Q_1 and Q_3 of Figure 6.30), and from the fault flag s_{id} is cross-checked to determine, based on a predefined table, a combination of two possible faulty

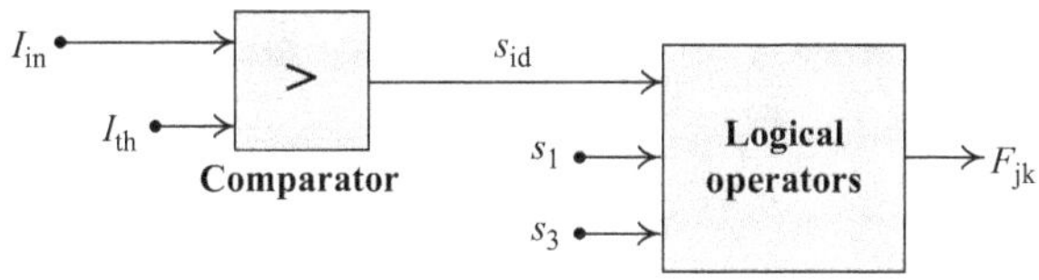

Figure 6.31 Schematic representation of the fault detection algorithm based on the observation of the converter input current I_{in} [21]

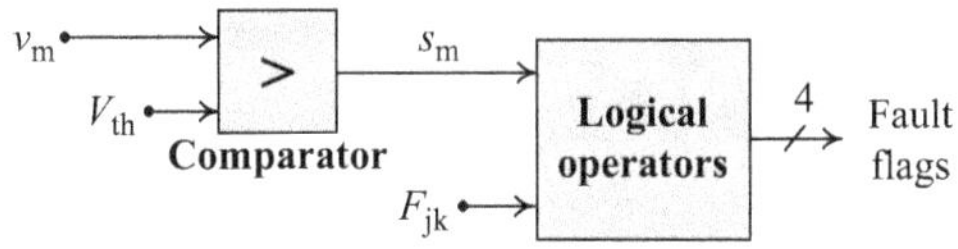

Figure 6.32 Schematic representation of the fault identification algorithm based on the observation of the transformer auxiliary winding voltage v_m [21]

switches, one switch from each of the converter legs. Information about the set of possibly faulty switches is carried by variable F_{jk} (refer to Figure 6.31).

To identify which one of those two switches is the one that is effectively faulty, the algorithm checks the midpoint voltages of both converter legs, which depend on the transformer primary voltage waveform v_{pri} and, consequently, on the auxiliary winding voltage v_m. More particularly, this observation aims at finding the sign of voltage v_m, as the position of the faulty switch affects the sign of v_m. A predefined threshold V_{th} is selected empirically, and defined as:

$$\begin{cases} v_m > V_{th} \Rightarrow s_m = 1 \\ v_m < V_{th} \Rightarrow s_m = 0 \end{cases} \tag{6.23}$$

$$V_{th} = \pm \frac{V_{in}}{4k_{aux}} \tag{6.24}$$

where V_{in} is the DC-bus voltage and k_{aux} is the transformer primary-auxiliary winding turns ratio (n_{pri}/n_{aux}). Figure 6.32 shows a draft of the algorithm responsible for the identification of the faulty switch within the phase-shift FB DC–DC converter.

In most cases, a simple analysis of the voltage v_m sign provides enough information to determine the faulty switch. On the other hand, when the analysis of voltage v_m does not allow to distinguish the faulty switch among the pair of possible faulty switches, an alternative approach must be followed. For such sporadic cases, the identification of the faulty switch takes place by forcing converter states.

The fast diagnostic action appears as the main merit of the algorithm. The main drawbacks of the fault diagnostic algorithm include requirement of two additional sensors (DC-bus current sensor and transformer auxiliary winding) and other hardware to implement the algorithm, restricted applicability of the algorithm (only

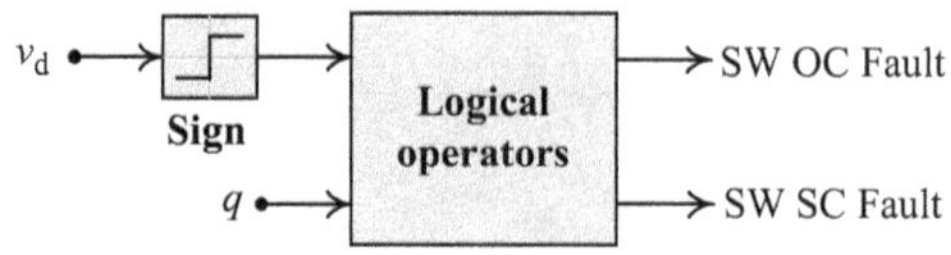

Figure 6.33 Schematic representation of the strategy used to diagnose switch faults at non-isolated DC–DC converters, based on the observation of the diode voltage v_d [22]

phase-shift FB DC–DC converters), and requirement of empirically established thresholds.

Diode voltage

This fault diagnostic algorithm, which uses the diode voltage as diagnostic variable, is aimed at non-isolated DC–DC converters. The algorithm effectiveness is confirmed for three single-switch non-isolated DC–DC converter topologies: buck, boost and buck-boost converter [22]. To obtain the fault signatures, the algorithm resorts to the voltage waveform measured at the terminals of the converter diode. Those signatures enable the diagnostic of OC and SC faults, not only in the converter switch but also in the diode.

For the diagnostic of faults in the converter switch SW, the algorithm action aims to identify abnormal conditions during particular periods of time, resorting to the information of the diode voltage v_{d} and the gating signal q. Therefore, the effects of an OC fault in switch SW become evident due to the positive value of voltage v_{d}, observed during the on-period of the gating signal q:

$$q = 1 \text{ and } v_{\mathrm{d}} > 0 \Rightarrow \text{SW OC Fault} \tag{6.25}$$

Similarly, the effects of a SC fault in the switch are expressed by a negative value of voltage v_{d}, observed in the off-period of the gating signal q:

$$q = 0 \text{ and } v_{\mathrm{d}} < 0 \Rightarrow \text{SW SC Fault} \tag{6.26}$$

Figure 6.33 shows a draft of the building blocks required to deploy the logical operations described in (6.25) and (6.26).

For the diagnostic of faults in the converter diode D, the diode voltage is compared to the predefined thresholds v_{th_1} and v_{th_2}. A diode OC fault alarm is triggered if the diode voltage surpasses the threshold v_{th_1} during the period in which the diode is directly polarised:

$$q = 0 \text{ and } v_{\mathrm{d}} > v_{\mathrm{th}_1} \Rightarrow \text{D OC Fault} \tag{6.27}$$

In this relation, $v_{\mathrm{th}_1} = 4v_{\mathrm{f}_{\max}}$, where $v_{\mathrm{f}_{\max}}$ corresponds to the maximum forward voltage drop of the diode. On the other hand, a diode SC fault condition is identified when the diode voltage v_{d} is negative, but higher than threshold v_{th_2}, during the period of inverse polarisation of the diode:

$$q = 1 \text{ and } v_{\mathrm{th}_2} < v_{\mathrm{d}} < 0 \Rightarrow \text{D SC Fault} \tag{6.28}$$

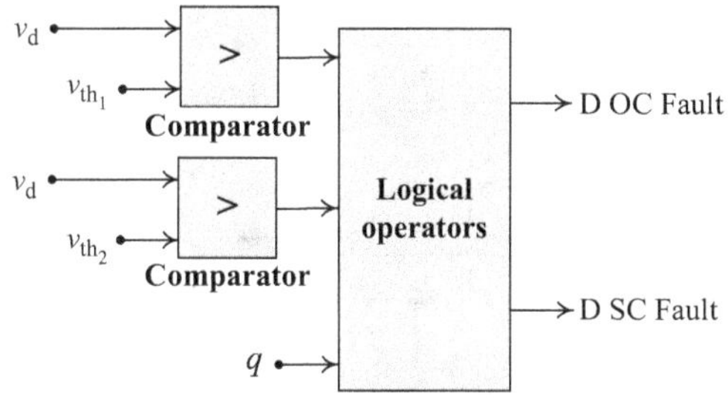

Figure 6.34 Schematic representation of the strategy used to diagnose diode faults at non-isolated DC–DC converters, based on the observation of the diode voltage v_d [22]

In this expression, $v_{th_2} = -v_{in} + v_{S_{max}}$, where v_{in} denotes the converter input voltage and $v_{S_{max}}$ defines the switch maximum on-state voltage drop.

Figure 6.34 shows a draft of the building blocks required to implement part of the algorithm, dedicated to the diagnostic of diode faults.

Versatility and fast fault diagnostic are the main attributes of this algorithm. The algorithm is implemented resorting to an additional voltage sensor and other simple analogue circuitry to generate the fault flags. A general increment of the implementation costs is expected, conferring a significant disadvantage to the algorithm.

MMC sub-module output voltage

The diagnostic of OC faults in MMC DC–DC converters is a challenging task which requires a careful selection of the fault diagnostic variables. In this algorithm, the voltage at the output of each MMC converter sub-module is used as the fault diagnostic variable [23]. The algorithm is specifically devoted to the diagnostic of OC faults in the switches of MMC DC–DC converters.

To assess the health state of the switches that compose each sub-module of an MMC DC–DC converter, the voltage measured at the output of each sub-module (V_{out}) is compared to the voltage measured at the sub-module input (V_{in}). Each voltage is sensed and compared resorting to simple analogue circuitry, namely, a voltage divider and a comparator. The implementation of the voltage dividers also aims to ensure that the following two conditions are met:

$$\begin{cases} V_{out} > V_{in}, & \text{if } q_1 = 1 \\ V_{out} < V_{in}, & \text{if } q_1 = 0 \end{cases} \tag{6.29}$$

where q_1 is the gating signal related to the sub-module upper switch Q_1, and q_2 represents the gating signal related to the sub-module lower switch Q_2 (refer to Figure 6.35).

To attain a fault diagnostic action that covers all converter switches, each sub-module of the MMC converter must contain the voltage dividers and comparators previously described.

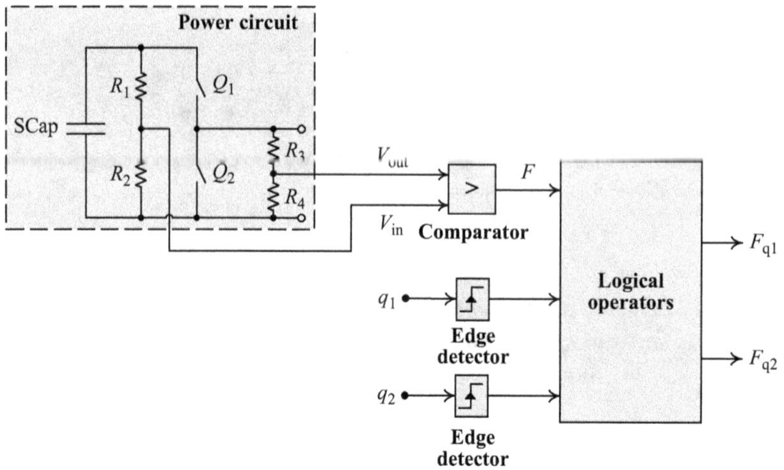

Figure 6.35 *Simplification of the fault diagnostic algorithm implemented in one of the modules of an MMC [23]*

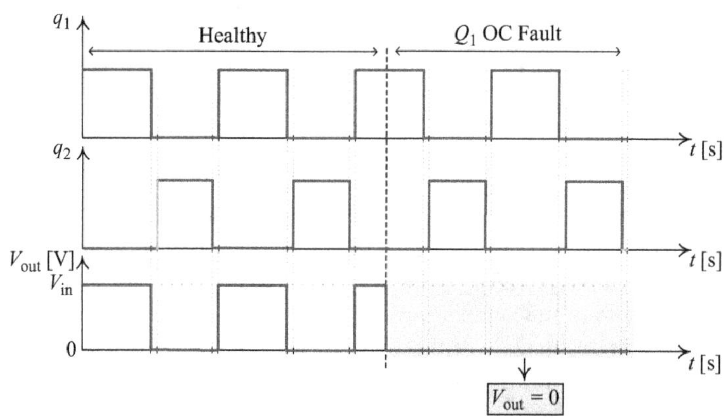

Figure 6.36 *Typical switching pattern and MMC module output voltage evolution considering supercapacitor charging operation. A scenario where an OC fault occurs in switch Q_1 is considered [23]*

The result of the comparison between voltage V_{out} and voltage V_{in} is reflected on variable F (refer to Figure 6.35). Briefly, it is possible to conclude that variable F remains at high level as long as the sub-module output voltage V_{out} is fairly higher than zero.

If an OC fault impacts the sub-module upper switch Q_1, the fault effects will be noticed if the switch is activated while the energy storage system module discharges. Under such condition, the sub-module output voltage V_{out} will be null, as stated in Figure 6.36.

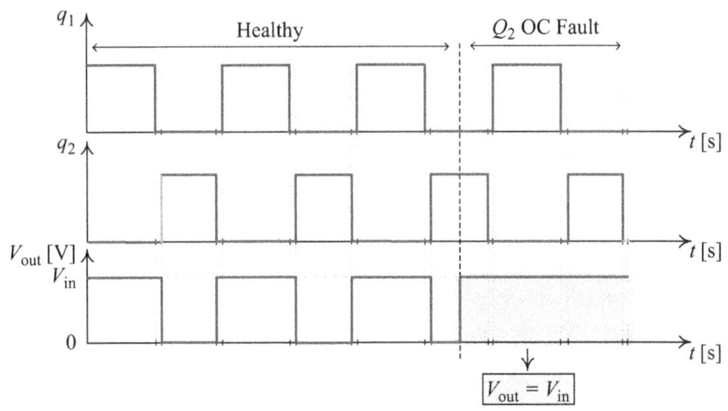

Figure 6.37 Typical switching pattern and MMC module output voltage evolution during SC discharging operation. A scenario where an OC fault occurs in switch Q_2 is considered [23]

Similarly, when an OC fault impacts the sub-module lower switch Q_2, it is possible to identify the fault if the energy storage system is charging. For such condition, voltage V_{out} will be equal to the voltage V_{in}, as shown in Figure 6.37.

The remainder of the fault diagnostic algorithm is implemented on an FPGA, responsible for identifying the faulty switch(es), by cross-checking the information of variable F with gating signals q_1 and q_2.

The algorithm can achieve fast fault diagnostic in all the sub-modules of an MMC converter. Its simplicity and modular structure are additional merits of the algorithm. The reduced applicability of the algorithm (only MMC converters), the possibility – albeit quite remote – of fault misdiagnosis due to delays in the system, and the medium to high implementation cost are the main drawbacks of this algorithm.

6.1.1.2 Frequency-domain signal-processing-based algorithms

As referred in previous sections, fault diagnostic algorithms based on a frequency-domain analysis require a significantly higher sampling rate of the diagnostic variables and, hence, lead to an increased computational effort. Still, the resiliency and performance of fault diagnostic algorithms based on analysis in the frequency domain should be highlighted.

Magnetic near field

Despite the classification as a signal-processing-based algorithm, this algorithm proposed in [24] can be also framed in the group of algorithms based on artificial intelligence, as its implementation resort to neural networks to identify the faulty component.

Despite the similarity of this approach to the one used in [19], where the fault diagnostic algorithm performs a time-domain analysis of the voltage measured in any of the magnetic components (inductors or transformers), the two fault diagnostic algorithms differ on the way of processing such diagnostic variables. While

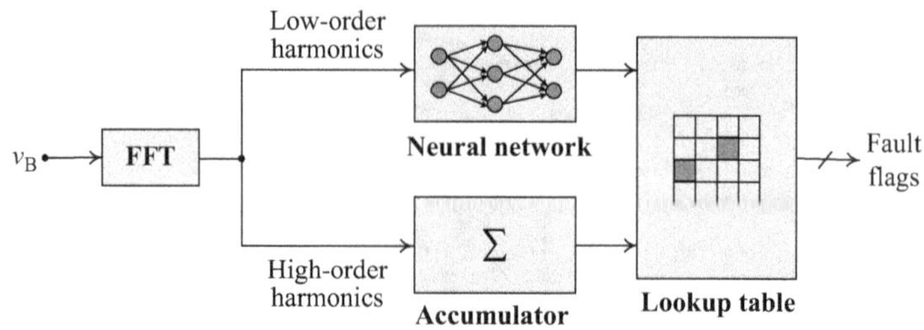

Figure 6.38 Simplification of the fault diagnostic algorithm based on the spectral analysis of the magnetic near field of the converter magnetic components [24]

this algorithm performs a spectral analysis of the fault diagnostic variables, a time-domain analysis of the fault diagnostic variables is adopted in reference [19].

This algorithm uses the magnetic near field of the converter magnetic components (inductors or transformer) as fault diagnostic variable, measuring it resorting to a dedicated probe. The magnetic near field waveform should be captured with a high sampling rate, to preserve the high-frequency components of the waveform. The magnetic near field waveform can be captured resorting, for instance, to a high-speed digital oscilloscope [24].

The fault diagnostic is based on the fact that both OC and SC faults of the converter switching semiconductors introduce perturbations in the converter currents and voltages that will, ultimately, conduct to additional electromagnetic noise.

The algorithm comprises the extraction of the information, in the frequency domain, of the magnetic near field, through the computation of the fast Fourier transform (FFT) of the magnetic near field waveform. Then, the low-order frequency components of the FFT are analysed by a previously trained neural network, while the high-frequency components are introduced in the accumulator. The results obtained from the neural network and accumulator are then used to conclude about the converter components state, by comparing the results of the neural network and accumulator with the data available in a look-up table.

Figure 6.38 contains a simplification of the building blocks required for the implementation of the fault diagnostics algorithm.

Due to its nature, this fault diagnostic algorithm is well known for its versatility. The algorithm was tested in two converter topologies: buck converter and phase-shift FB converter. Despite this, its implementation is equally feasible in most switching power converters. The algorithm sensitivity against variations in the probe position seems to be the main pitfall. The probe position should be identical during both training and fault diagnostic processes, to ensure that the measured waveforms while carrying the fault diagnostic follow the fault signatures that were obtained previously, for training purposes. The significant computational effort and the large number of training sets, required to recognise all switch fault events, constitute additional limitations of the algorithm.

6.1.1.3 Main features of the signal-processing-based algorithms

Table 6.1 summarises the most relevant features and limitations of the fault diagnostic algorithms available in the literature, as well as the conditions used to test the algorithms. The evaluation of the implementation cost takes into account factors such as the number of additional components required to implement the fault diagnostic algorithm and the corresponding complexity. It should be noted that some fault diagnostic algorithms resort to analogue circuitry to deploy the fault diagnostic strategies. For such cases, the sampling time criterion is not valid and, consequently, it is not presented.

6.1.2 Model-based algorithms

Model-based fault diagnostic algorithms have become particularly popular among the scientific community in the last few years, with many researchers proposing new fault diagnostic algorithms based on the switching models of the DC–DC converters. Last-generation converter controllers, with high processing capability, helped triggering this evolution, as they allowed the implementation of fault diagnostic algorithms requesting high computational effort, a common characteristic of model-based algorithms.

These algorithms are well known for their resiliency and effectiveness while detecting OC and/or SC faults, independently of the DC–DC converter operating conditions (e.g. switching frequency, load level, conduction mode, etc.). The robustness against non-linearities of the fault signatures, such as noise or transients, is also much higher in model-based fault diagnostic algorithms.

To deploy these algorithms, a state-space model of the DC–DC converter must be established. Previous knowledge of the DC–DC converter parameters is, therefore, a premise for the development of model-based fault diagnostic algorithms.

Like model-free algorithms, model-based fault diagnostic algorithms comprise two stages: fault detection and fault identification.

During the fault detection stage, the measured converter response is compared to the estimated converter response, through the definition of residuals. Under healthy converter operation, both measured converter states and estimated converter states converge, leading to residuals that approach to zero. A fault in the converter switches introduce deviations between the measured converter response and the estimated converter response, thus leading to non-zero residuals.

After triggering a fault alarm, the fault identification stage performs a thinner analysis of the residuals, allowing to locate the faulty component(s).

Next subsections present the most relevant fault diagnostic algorithms based on models, highlighting their operation principles and main features.

6.1.2.1 Sliding mode observer

The first fault diagnostic algorithm based on a sliding mode observer is presented in [25], where it is implemented on an MMC DC–DC converter, for the diagnosis of switch OC faults. Validation of the algorithm effectiveness is solely confirmed on a simulation environment [25].

Table 6.1 Features of some of the most relevant signal-processing-based fault diagnostic algorithms

Ref.	Converter topologies	Diagnostic variable	Faults	Switching frequency	Sampling time	Detection time	Cost
[2]	Cascaded buck converter, most DC–DC converters	Statistical moments	OC SC	20 kHz	20 μs	400 μs ($8T_{sw}$)	Low to medium
[3]	FB ZVS converter, other ZVS converters	DC-bus current	OC SC	80 kHz	–[a]	–[b]	Low
[4]	Three-level parallel resonant converter	Flying capacitor voltage	OC SC	200 kHz	–[a]	<1 μs	Low
[5]	DAB converter	Isolation transformer voltages	OC	20 kHz	–[b]	$<T_{sw}$ (Simulation)	Low to medium
[6]	FB converter	Transformer primary voltage	OC	50 kHz	–[a]	2 ms ($100T_{sw}$)	Low
[7]	Non-isolated single-switch converters operating at CCM	Inductor current derivative sign	OC SC	15 kHz	1 μs	$<2T_{sw}$	Medium to high
[8] [9]	Non-isolated single-switch converters operating at CCM	Inductor current derivative sign	OC SC	–[b]	–[b]	$<T_{sw}$	Medium to high
[10]	Non-isolated single-switch converters operating at CCM	Inductor current derivative sign	OC	15 kHz	1 μs	$<T_{sw}$	Medium
[11]	Non-isolated single-switch converters operating at CCM	Inductor current derivative sign	OC SC	15 kHz	–[b]	$<T_{sw}$	Medium to high
[12]	Non-isolated single-switch converters operating at CCM	Inductor current evolution	OC SC	40 kHz	–[b]	–[b]	Low
[13]	Three-level non-isolated boost converter, other multilevel converters	Output DC-bus capacitors voltage	OC	5 kHz	50 μs	–[b]	Low

Table 6.1 (Continued)

Ref.	Converter topologies	Diagnostic variable	Faults	Switching frequency	Sampling time	Detection time	Cost
[14]	Multi-input converter	Inductor current evolution	OC	–[b]	–[b]	<2 ms (Simulation)	Low
[15]	Non-isolated bidirectional converter	Inductor current evolution	OC	–[b]	–[b]	<2 ms (Simulation)	Low
[16]	Unidirectional non-isolated converters	Inductor current evolution	OC	–[b]	–[b]	<0.5 ms	Low
[17]	Parallel-connected SAB converter	Converter output current	OC	10 kHz	–[b]	$<2T_{sw}$	Low
[18]	Interleaved boost converter	DC-bus current derivative sign	OC	1 kHz	25 μs, 50 μs	$<2T_{sw}$	Low
[19] [20]	Buck converter [19], HB converter [20], most DC–DC converters	Magnetic component voltage	OC SC	48 kHz [19] 45 kHz [20]	–[a]	$<T_{sw}$ [19] $<2T_{sw}$ [20]	Low
[21]	Phase-shift FB converter	DC-bus current, transformer primary voltage	SC	50 kHz	–[a]	$<T_{sw}$	High
[22]	Buck, boost, buck-boost converters, non-isolated DC–DC converters	Diode voltage	OC SC	50 kHz	–[a]	–[b]	Low to Medium
[23]	MMC DC–DC converter	Sub-module output voltage	OC	4 kHz	–[b]	$<T_{sw}$	Medium
[24]	Buck and phase-shift FB (PSFB) converters, most switching converters	Magnetic near field	OC SC	24 kHz (Buck) 135 kHz (PSFB)	40 ns	–[b]	High

[a]Not applicable
[b]Not specified

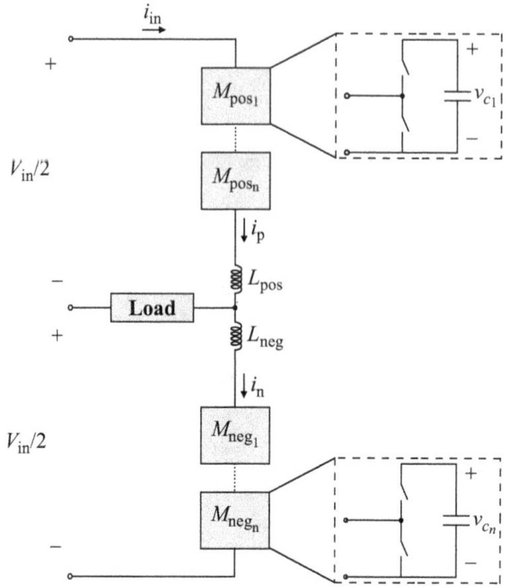

Figure 6.39 Simplified structure of an MMC, as implemented in [25]

Fault detection and, most of all, fault identification on MMC converters are particularly challenging tasks, due to the significant number of components of the converter. In the MMC converter under analysis, each module is composed of a simple HB converter, as depicted in Figure 6.39.

As any other fault diagnostic algorithm based on observers, this algorithm establishes a comparison between the estimated states and the observed states. Information about the observed states is acquired through the measurement of the converter variables whose evolution allows to extract meaningful fault signatures, as, for instance, currents or voltages.

Any linear first-order system can be described by the following condition:

$$x' = Ax + Bu \tag{6.30}$$

To obtain the estimation of the converter state using a sliding mode observer, the mathematical conditions that define the converter model are combined with the observer vectors:

$$\hat{x}' = A\hat{x} + Bu + \mathrm{L}\,\mathrm{sgn}(x - \hat{x}) \tag{6.31}$$

where $\hat{x}$ is the estimated state of variable x, L denotes the sliding mode observer gains, and $\mathrm{sgn}(x - \hat{x})$ refers to the sign of the error between the measured and the estimated states:

$$\mathrm{sgn}(x - \hat{x}) = \begin{cases} 1, x - \hat{x} > 0 \\ 0, x - \hat{x} = 0 \\ -1, x - \hat{x} < 0 \end{cases} \tag{6.32}$$

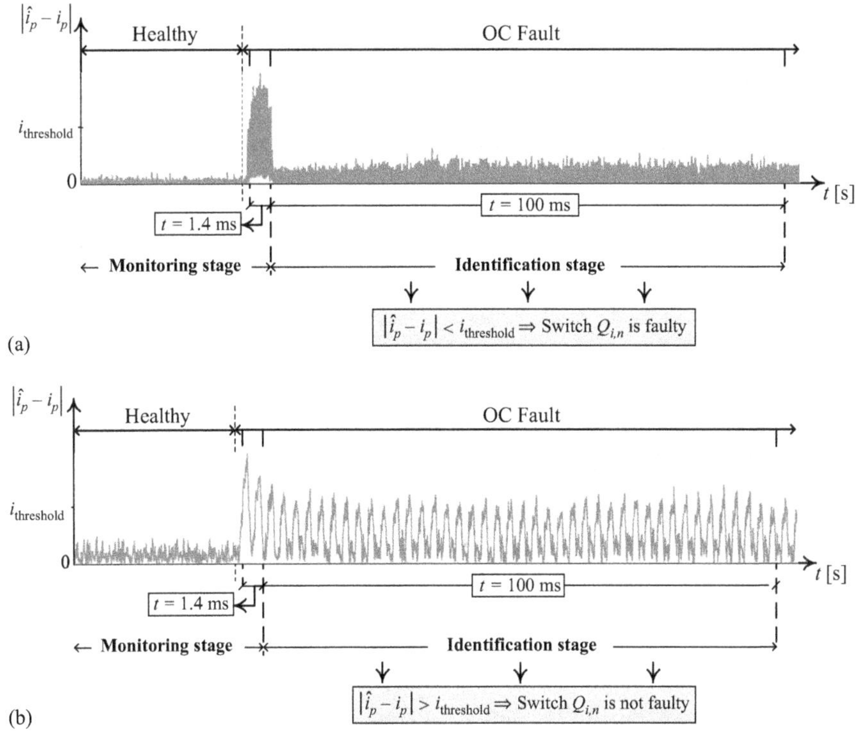

Figure 6.40 Sequence of events verified along the fault diagnostic procedure, showing the results of the assumption made after detecting a fault: (a) switch $SW_{i,n}$ is effectively faulty; (b) switch $Q_{i,n}$, assumed to be faulty is, in reality, healthy

Sliding mode observers can be defined to estimate any meaningful converter parameters, suitable for the fault diagnostic purposes. In [25], sliding mode observers are defined to estimate the converter arm current $\hat{i}_p$ and the voltage of each DC-bus capacitor $\hat{v}_{c_{1\ldots n}}$, placed at the terminals of each converter module (refer to Figure 6.39).

The fault diagnostic process comprises two distinctive stages, identified in Figure 6.40: monitoring stage and identification stage. Please note that the waveforms of the arm current error are illustrative only.

In the monitoring stage, a period of observation is defined, with the aim of comparing the arm current error with a predefined arm current error threshold $i_{\text{threshold}}$:

$$|\hat{i}_p - i_p| \geq i_{\text{threshold}} \tag{6.33}$$

where $\hat{i}_p$ is the estimated arm current and i_p is the measured arm current. The arm current error threshold $i_{\text{threshold}}$ is defined, empirically, as the converter arm current peak ($i_{\text{p\_peak}}$). If the current error remains higher than $i_{\text{threshold}}$ for a long period of time (in [25], this period is defined empirically as $t = 1.4$ ms), it means that a switch fault is perturbing the proper converter operation. The second stage of the fault diagnostic process is then activated, to locate the faulty switch.

In the fault identification stage, an assumption approach is adopted, i.e. it is assumed that a certain converter module contains the faulty switch. It is assumed that the faulty switch is located in switch number n of module number i ($Q_{i,n}$), and a mathematical model of the converter arm is built based on that assumption. Changes in the estimated values $\hat{i}_p$ and $\hat{v}_{c_{1\ldots n}}$ are witnessed, as the assumed converter model is different from the original converter model. Then, the arm current error and the DC-bus capacitor voltage error are compared to the corresponding predefined thresholds:

$$|\hat{i}_p - i_p| \geq i_{\text{threshold}} \tag{6.34}$$

$$|\hat{v}_{ci} - v_{ci}| \geq v_{\text{threshold}} \tag{6.35}$$

The voltage error threshold $v_{\text{threshold}}$ is also defined empirically, and it is equal to 1/10 of the rated capacitor voltage.

Both arm current error and capacitor voltage errors are compared to the corresponding thresholds, during an observation time of 100 ms (period is defined empirically). If the arm current error and capacitor voltage error do not surpass the corresponding thresholds during the entire observation period, it means that the assumed converter model matches with the measured converter response. Therefore, the location of the fault is confirmed, and the assumption taken before is confirmed. Otherwise, the location of the faulty switch is not confirmed, and another assumption must be taken, by repeating the identification process as many times as necessary, until the faulty switch is found.

Also based on sliding mode observers, further improvements to the fault diagnostic process are introduced in [26]. In comparison to [25], the fault diagnostic procedure proposed in [26] takes half of the time to be completed. The computational effort is also reduced, due to the elimination of one observer. Only the converter circulating current i_{in} (refer to Figure 6.39), defined as the median of the upper arm and lower arm currents, is estimated resorting to a sliding mode observer.

Like sliding mode control, this algorithm features robustness against noise and low sensitivity against uncertainty of the converter parameters. The algorithm features allow its application in many other DC–DC converter topologies. High fault diagnostic times, complexity of the algorithm, high implementation cost, and the requirement of empirically defined thresholds seem to be the main disadvantages of fault diagnostic algorithms based on sliding mode observers.

6.1.2.2 Inductor current emulation

The emulation of the inductor current is one of the most recent strategies used to diagnose semiconductor faults within DC–DC converters [27,28]. This algorithm is able to detect both OC and SC fault events in the active switches of any non-isolated DC–DC converter topology. It uses a detailed model of the converter to predict the inductor current evolution, allowing to detect abnormal deviations of

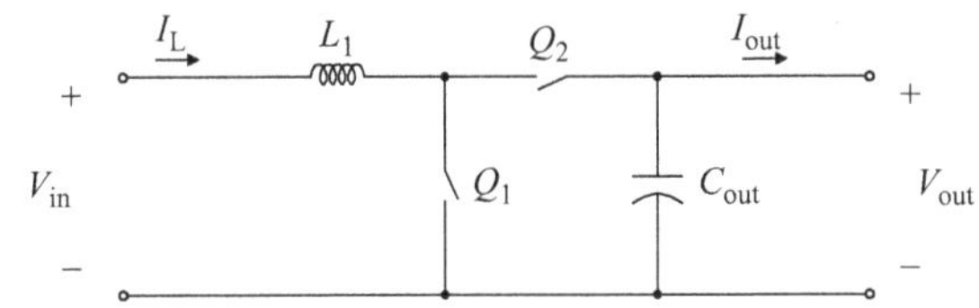

Figure 6.41 Simplification of the synchronous boost converter

that parameter. Along with a detailed converter model, this algorithm uses the information of the converter input and output voltages (V_{in} and V_{out}, respectively), and the inductor current I_L, thus obviating the requirement of additional sensors.

Considering an ideal model of the synchronous boost converter, as depicted in Figure 6.41, it is possible to derive a mathematical relation that allows to compute the predicted inductor current $\hat{I}_L$, for a moment n [27]:

$$\hat{I}_L(n) = I_L(n-N) + \frac{V_{in}}{L}T_{sw} - \frac{V_{out}}{L}(1-D)T_{sw} \tag{6.36}$$

In this equation, I_L denotes the inductor current, V_{in} corresponds to the converter input voltage, L is the inductance, T_{sw} denotes the switching period, V_{out} is the converter output voltage, and D specifies the switching duty cycle. It must be emphasised that (6.36) is valid only if the switching period T_{sw} is an integer multiple of the sampling period T_s, i.e. $T_{sw} = NT_s$.

The fault diagnostic action takes place in two steps. Faults are detected by comparing the expected inductor current for a moment n, $\hat{I}_L(n)$, and the measured inductor current at the same moment n, $I_L(n)$, to obtain the inductor current error $I_L(n) - \hat{I}_L(n)$. A fault is detected if the absolute inductor current error surpasses a predefined threshold $I_{threshold}$:

$$|I_L(n) - \hat{I}_L(n)| \geq I_{threshold} \tag{6.37}$$

This threshold is computed taking into account the internal resistances of the converter components, namely, voltage source (R_{in}), active switches (R_{sw}), and inductor (R_L) [27]:

$$I_{threshold} = \frac{I_L(R_{in} + R_{sw} + R_L)}{L}T_{sw} \tag{6.38}$$

The identification of the faulty component is deduced resorting to the sign of inductor current error $\text{sgn}(I_L(n) - \hat{I}_L(n))$. In the case of an OC fault, the inductor current error is negative:

$$\text{sgn}\left(I_L(n) - \hat{I}_L(n)\right) < 0 \Rightarrow Q_1 \text{ OC Fault} \tag{6.39}$$

On the other hand, a SC fault introduces a positive error of the inductor current. Further information must be obtained if the SC fault alarm is triggered. To this end, information about the load voltage V_{out} is assessed. While a SC fault in the bottom

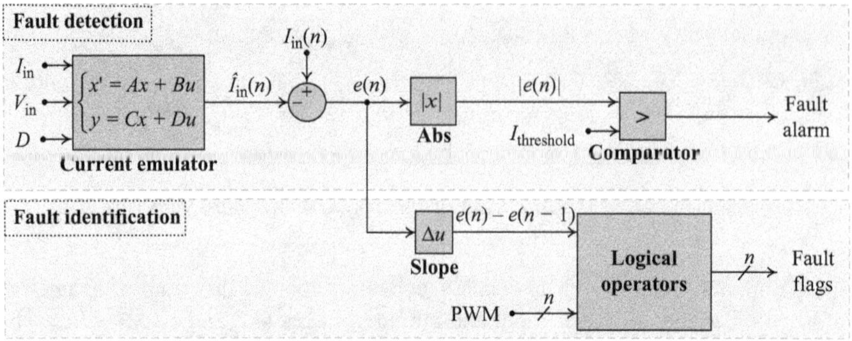

Figure 6.42 Simplification of the fault diagnostic algorithm based on the converter current emulation, suitable for an n-phase interleaved DC–DC converter [28]

switch implies a reduction of the converter output voltage V_{out} to zero, a SC fault in the synchronous switch leads to a significant increment in voltage V_{out}:

$$\mathrm{sgn}\big(I_L(n) - \hat{I}_L(n)\big) > 0 \quad \text{and} \quad V_{out} = 0 \quad \Rightarrow Q_1 \text{ SC Fault} \tag{6.40}$$

$$\mathrm{sgn}\big(I_L(n) - \hat{I}_L(n)\big) > 0 \quad \text{and} \quad V_{out} \gg 0 \quad \Rightarrow Q_2 \text{ SC Fault} \tag{6.41}$$

The same fault diagnostic algorithm can be easily implemented in other converter topologies, such as the interleaved DC–DC boost converter [28]. The mathematical model used to predict the converter input current has to be adapted according to the converter topology. Along with the parameters of the converter model, the fault diagnostic algorithm inputs are the converter input voltage V_{in}, the converter input current I_{in}, and the switching duty cycle D. The fault detection is based on the same principle: a fault alarm is issued when the absolute value of the current error e surpasses the predefined threshold $I_{threshold}$, as shown in Figure 6.42. Due to the inherent characteristics of interleaved DC–DC converters, the fault identification procedure developed for the non-isolated synchronous boost converter cannot be replicated for interleaved DC–DC converters. Therefore, a new approach is developed for the identification of faults in interleaved DC–DC converters. Information of the slope of the error is cross-checked with the converter gating signals. The slope of the current error is analysed during the OFF-state of the switches.

This algorithm provides a solid alternative to diagnose faults in the switches of non-isolated DC–DC converters, allowing to localise those faults in a very short time (less than one switching period). No additional sensors are introduced in the converter, thus increasing the overall converter reliability. The significant number of converter model parameters required to implement the algorithm is the only drawback of the fault diagnostic algorithm.

6.1.2.3 State estimation

The state estimation of DC–DC converters configures itself as an excellent and meaningful model-based fault diagnostic algorithm [29]. It extends the range of detectable fault modes, allowing to diagnose not only OC and SC faults in the switching devices of power converters, but also in the passive components (inductors and capacitors), and sensors of power converters. Also, the range of converter topologies embraced by this algorithm is widely extended, allowing the diagnostic of faults in most switching power converters.

This fault diagnostic algorithm consists on a state estimator that allows to generate residuals, by comparing the estimated converter response to the measured one. When non-zero residuals occur, it is possible to detect the presence of a fault. The direction of evolution of the generated residuals allows to identify not only the fault mode but also the faulty component.

The general expression for the state-space representation of a linear first-order system, operating at healthy conditions, is:

$$\begin{cases} x' = Ax + Bu \\ y = Cx \end{cases} \tag{6.42}$$

Faults in the converter switches and passive components express themselves through deviations in the parameters of the state matrices A and B. Hence, a new state-space representation of the linear first-order system containing a faulty component is obtained:

$$x' = Ax + Bu + \phi f \tag{6.43}$$

where ϕ is a scalar component fault magnitude function, and f is the vector containing the library of fault signatures related to each converter component.

On the other hand, faults in sensors are distinguished by the introduction of deviations in the parameters of matrix C. The state-space representation of the linear first-order system containing faulty sensor(s) is:

$$y = Cx + \theta g \tag{6.44}$$

where θ is a scalar sensor fault magnitude function and g is the vector containing the library of fault signatures related to each sensor.

This algorithm follows the typical sequence of events of fault diagnostic algorithms, consisting on a fault detection stage and fault identification stage.

During the fault detection stage, the residual of the difference between the measured converter state and the estimated converter state is determined:

$$\gamma = y - C\hat{x} \tag{6.45}$$

where γ is the residual, y is the measured converter state, and $C\hat{x}$ denotes the estimated converter state. A fault detection alarm is triggered if the Euclidean norm of residual γ surpasses a predefined fault detection threshold. That threshold is defined empirically, based on a worst-case scenario, where the transients on voltage and current considerably increase the Euclidean norm of residual γ.

Table 6.2 Features of some of the most relevant model-based fault diagnostic algorithms

Ref.	Converter topology	Diagnostic variable	Faults	Switching frequency	Sampling time	Detection time	Cost
[25] [26]	MMC	Sliding-mode observers	OC	–[a]	–[a]	100 ms [25], 50 ms [26]	Low
[27] [28]	Synchronous boost converter, other non-isolated DC–DC converters	Inductor current emulation	OC SC [27], SC [28]	20 kHz [27], 10 kHz [28]	10 μs	$<T_{sw}$	Low to medium
[29]	Switching power converters	State estimation	OC SC	10–20 kHz	100 μs	<10 ms	Medium to high

[a]Not specified

The fault identification stage comprises the computation of the inner product between residual γ and each one of the fault signatures, available on a pre-established library. The inner product between the two arrays allows to find out which fault signature is closely aligned to the residual γ. To this end, the magnitude of the inner product between the residual γ and the fault signatures is compared to a predefined fault identification threshold. The magnitude of the inner product surpasses the defined fault identification threshold when the fault signature selected from the library matches with the fault occurring in the converter.

The fault diagnostic algorithm based on the state estimation of DC–DC converters is validated experimentally resorting to a nanogrid prototype, consisting on a common DC bus with four different switching power conversion solutions plugged in that bus [29].

The wide applicability, high robustness and fast fault diagnosis are some of the major virtues of this fault diagnostic algorithm.

The model estimator must be executed in real time, at a very high sampling rate, demanding a significant computational effort. For this reason, a fast and powerful FPGA platform must be used. This can be considered the main disadvantage of the algorithm.

6.1.2.4 Main features of model-based fault diagnostic algorithms

Table 6.2 provides some of the most important indicators about the fault diagnostic algorithms, available in the literature, based on the models of the DC–DC converters. The implementation of cost weighting takes into account the number of additional components used to implement the fault diagnostic algorithm, the algorithm complexity, and the details required to define the converter model.

6.2 Fault-tolerant strategies

Although fault diagnostics assumes a major importance, converter reconfiguration after a fault also plays an important role, as it enables converter operation continuity, with acceptable quality levels.

It is relevant to maintain the hardware structure of a fault-tolerant converter as simple as possible, considering a fault-tolerant converter without any additional hardware the most attractive solution.

Obviously, the adoption of fault-tolerant strategies does not allow, in most cases, to fully recover the power conversion capabilities of a DC–DC converter. Power quality degradation and derating of the power transferred to the load are some of the most common effects that are still observed after the reconfiguration of the faulty DC–DC converter. Additional side effects might be experienced, due to the implementation of the fault-tolerant control strategy, such as higher conduction and switching losses in the converter switching devices (IGBTs, MOSFETs, etc.).

Next subsections present and detail the most relevant fault-tolerant architectures and control strategies developed so far to overcome the negative effects of OC and/or SC faults in the active switches and/or diodes of DC–DC converters. The applicability, main merits and drawbacks of such strategies are enumerated. All the reconfiguration strategies described in the next subsections were validated in laboratory environment by their proposers, resorting to experimental prototypes of the DC–DC converters.

6.2.1 Bypass of faulty module(s)

In DC–DC converter topologies with modular structure, it is quite common to develop bypass schemes that allow to isolate the faulty module(s), while maintaining the operation of the healthy ones. Due to the inherently fault-tolerant structure of most modular DC–DC converters, most of the fault-tolerant schemes proposed on the literature concerning modular DC–DC converters do not include any additional redundant modules, restricting the reconfiguration action to the isolation of the faulty module and adaption of the control applied to the healthy modules, whenever required. Depending on the converter topology, additional discrete components, such as thyristors or solid-state relays (SSRs), might be required to bypass the faulty module(s).

The input-series output-parallel (ISOP) modular converter consists of an arrangement of simpler DC–DC converter topologies, connected in series on the input side, and connected in parallel on the output side, as shown in Figure 6.43. In Figure 6.43, each module symbolises a simple DC–DC converter, while switches Q_1, Q_2, and Q_n, covered by shaded areas, represent the devices responsible for bypassing the faulty module(s).

Any faulty module of an ISOP converter is bypassed, resorting to a silicon-controlled rectifier (SCR) [30] or a thyristor [31], when an OC or SC fault impairs the proper operation of the module switches. Let us take the example of a fault in Converter 1 (refer to Figure 6.43). After detecting a fault in Converter 1, the converter control acts in the way of activating switch Q_1 permanently, to ensure the proper operation of the remaining healthy converter modules.

Fault tolerance based on bypass schemes is also attainable in cascaded DC–DC converters. Due to the cascaded configuration, a single switch fault may preclude the entire converter operation. Hence, the isolation of the module that contains the

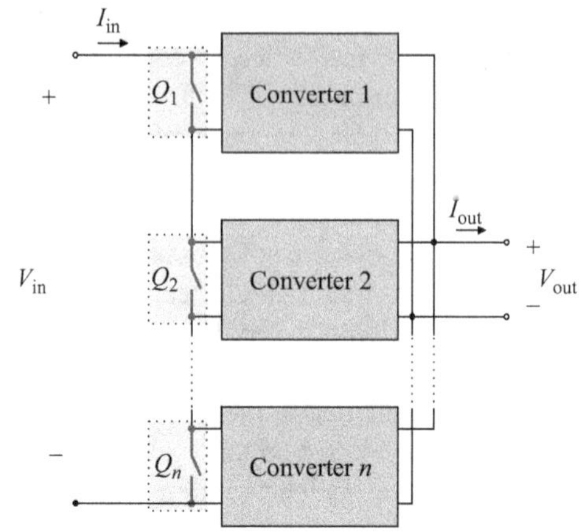

Figure 6.43 General structure of a fault-tolerant ISOP converter [30,31]. Each module symbolises a simpler DC–DC converter. Bypass switches Q_1, Q_2, and Q_n are highlighted in the shaded areas

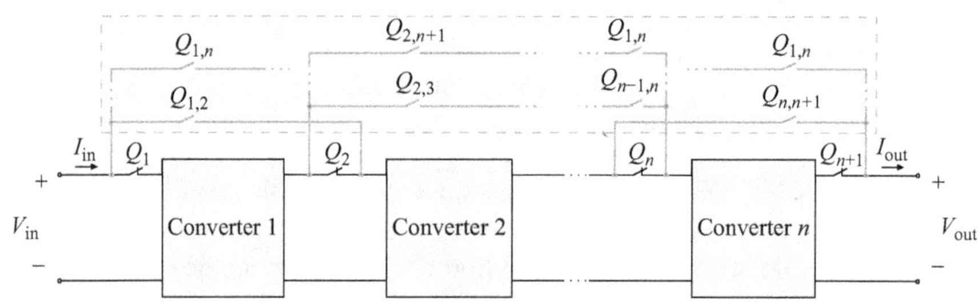

Figure 6.44 General structure of a fault-tolerant cascaded converter, as proposed in [32]. Each block symbolises a simpler DC–DC converter. Bypass switches are highlighted in the light grey area

faulty switch(es) has utmost importance. The isolation of the faulty module(s) can be obtained through the addition of SSRs to the original converter circuit [32], following the scheme depicted in Figure 6.44.

When one module gets damaged due to switch OC or SC fault, the additional SSRs are activated/deactivated in a way that allows to isolate and bypass the faulty module. Let us take the example of a fault in Converter 1 (refer to Figure 6.44). After detecting a fault in Converter 1, the converter control isolates the faulty module, by deactivating SSRs Q_1 and Q_2, while activating SSR $Q_{1,2}$ permanently, to bypass the faulty module and ensure the proper operation of the remaining healthy converter modules.

The implementation of this reconfiguration strategy implies a significant number of additional components, which significantly increase the total cost of the

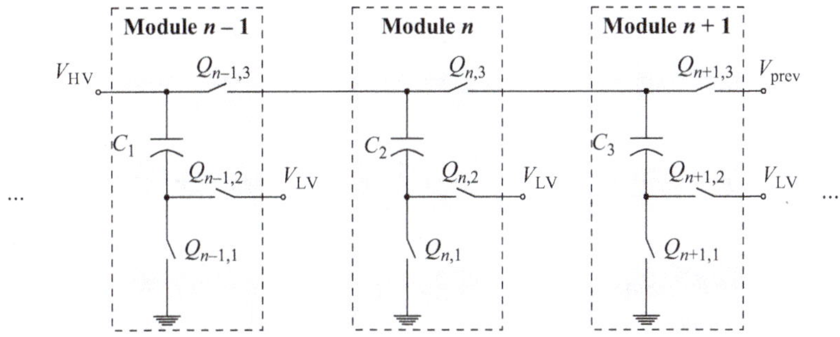

Figure 6.45 Structure of the fault-tolerant bidirectional MMC [33,34]

fault-tolerant converter. In the case of a converter with three modules, nine additional SSRs are required to implement the proposed fault-tolerant architecture.

MMC converters are increasingly popular converters for applications where high-power requirements must be met. A single faulty switch may also impair the entire operation of a MMC. Reconfiguration techniques based on the bypass of faulty modules are also available for bidirectional MMCs [33,34]. Figure 6.45 shows part of the structure of the fault-tolerant converter.

The converter can be built with as many modules as desired, by simply connecting additional modules in series with terminals V_{HV} and V_{prev}, identified in Figure 6.45. In Figure 6.45, V_{HV} denotes the voltage in the converter high-voltage side, V_{LV} is the voltage in the converter low-voltage side, and V_{prev} is the voltage of module $n+2$. Each one of the converter modules, shown in Figure 6.45, comprises three active switches: two of those switches are connected in a configuration that enables the bidirectional power flow (switches $Q_{n,1}$ and $Q_{n,2}$ of module n, as depicted in Figure 6.45); a third switch, which also plays an active role on the converter operation during healthy condition, allows to bypass that module (switch $Q_{n,3}$ of module n, as depicted in Figure 6.45). The structure of the module itself includes all the components required to perform the bypass function. The fault-tolerant operation of the converter implies changes in the control of the switches related to the bypassed module. In the event of an OC fault in either $Q_{n,1}$ or $Q_{n,2}$, switch $Q_{n,3}$ is continuously conducting to bypass that module, while switches $Q_{n,1}$ and $Q_{n,2}$ are deactivated, by simply clearing the gating signals applied to them.

It should be emphasised that certain fault modes totally preclude the operation of the entire converter, namely, an OC fault in any of the converter switches responsible for the bypass of the faulty module, or a SC fault in any $Q_{\ldots,1}$ or $Q_{\ldots,2}$ switch.

Bypassing the faulty module(s) of a modular DC–DC converter topology presents some advantages over other reconfiguration strategies. Typically, the implementation cost is quite reasonable, due to the reduced number of additional components required for the implementation of the fault-tolerant converter architecture. Additionally, the transition to the fault-tolerant control strategy does not imply relevant changes in the pre-fault control scheme. On the other hand, the loss

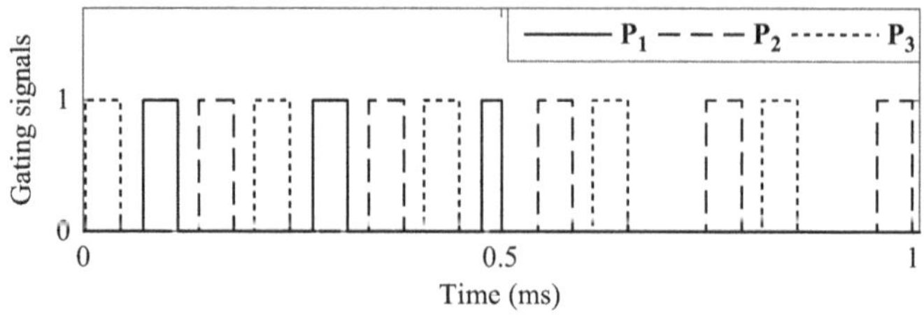

Figure 6.46 Evolution in time of the gating signals applied to a three-phase interleaved DC–DC converter. After an OC fault in P_1, at $t = 0.5$ ms, no reconfiguration strategy is adopted and, consequently, the phase-shift between the gating signals related to healthy switches (P_2 and P_3) remains unchanged

of power conversion capabilities, which might be significant in some cases, is one of the barriers presented by these fault-tolerant converter architectures.

6.2.2 Phase-shift adjustment

This fault-tolerant control strategy introduces changes in the features of the gating signals used to control the DC–DC converter switches. It does not imply the addition of redundant components to the converter, reason why it is commonly employed in certain DC–DC converter topologies featuring an inherent fault-tolerant architecture.

After a fault (either OC or SC fault) in any of the converter switches, the faulty phase is isolated, and the phase-shift between the remaining healthy phases is corrected. In the case of a three-phase interleaved converter, the phase-shift between gating signals associated to healthy phases is changed from $2\pi/3$ rad to π rad after a switch fault; in other words, the control strategy of a three-phase converter is reformulated and reconfigured into a two-phase control strategy. To better understand the effects of this reconfiguration strategy in the typical switching pattern of a three-phase interleaved DC–DC converter, Figure 6.46 depicts a switching pattern without any reconfiguration after an OC fault, while Figure 6.47 depicts the same switching pattern subjected to the described reconfiguration strategy.

In Figure 6.46, it is clearly observable that the OC fault introduced at $t = 0.5$ ms creates a significant unbalance in the switching pattern. This asymmetry in the switching pattern is easily compensated by simply introducing the reconfiguration strategy, as shown in Figure 6.47.

The most relevant effects of the adoption of such fault-tolerant control strategy are the mitigation of the converter input and output current ripple. The reduction of the stress imposed to the converter output capacitor is also attained.

In the literature, this fault-tolerant control strategy is commonly employed in faulty multi-phase converters, also known as interleaved DC–DC converters [18,35–37].

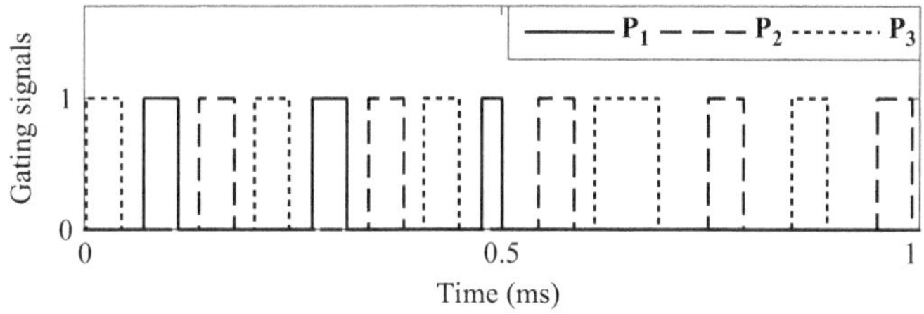

Figure 6.47 Evolution in time of the gating signals applied to a three-phase interleaved DC–DC converter. After an OC fault in P_1, at $t = 0.5$ ms, the reconfiguration strategy is implemented and, consequently, gating signals P_2 and P_3 are shifted by π rad between each other, after $t = 0.5$ ms

Even though it is equally possible to implement this fault-tolerant control strategy in other converter topologies that employ a phase-shift modulation strategy, this modulation strategy is implemented in DC–DC converters containing multiple modules or phases connected in parallel in the converter input side.

Phase-shift adjustment is a valid option to minimise the effects of OC or SC faults in parallel-connected SAB DC–DC converters [17]. The converter module containing a faulty switch is isolated, by simply deactivating all the switches of the faulty module. Concurrently, the phase-shift between the control signals applied to the healthy converter modules is corrected.

A fault-tolerant scheme of an input-parallel output-series (IPOS) modular converter based on the correction of the control signals phase-shift is equally feasible [35]. In this converter topology, each module is composed of a secondary phase-shift FB converter. The fault-tolerant scheme based on the phase-shift correction is able to overcome the effects of OC and SC faults in any of the converter switches. In the case of SC faults in the transformer primary-side bridge, the faulty module is bypassed, by simply deactivating the gating signals related to that module. As the modulation strategy used to control each module consists on applying a phase shift between the gating signals of each module, the bypass of one module implies the correction of the phase shift, to maintain a symmetrical switching pattern.

Simplicity, low or null implementation cost, and effectiveness are some of the major advantages of this fault-tolerant control strategy. On the other hand, higher conduction and switching losses during the post-fault converter operation are among some of the secondary effects arising from the implementation of this fault-tolerant control strategy.

6.2.3 Inclusion of additional components

The adoption of fault-tolerant converter architectures based on additional components is the dominant strategy used to develop fault-tolerant DC–DC converters. As the name itself suggests, additional semiconductors are included in the original

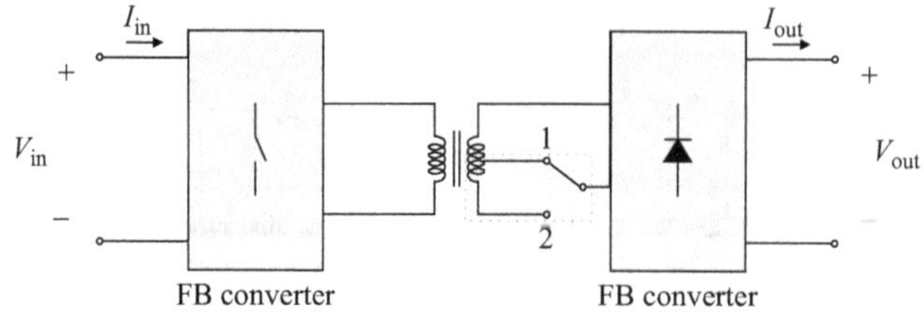

Figure 6.48 Implementation of an auxiliary winding in the transformer secondary side (highlighted in light grey background), as mean to compensate the converter output voltage sag resulting from faults in the active switches of the transformer primary-side bridge [6]

converter architecture, to mitigate the effects of OC and/or SC faults. The inclusion of redundant components can be performed at the device level (e.g. single TRIACs, IGBTs, MOSFETs), or at the leg level (e.g. redundant HB), depending on the DC–DC converter topology.

6.2.3.1 Fault-tolerant architectures based on additional discrete components

In this category, it is possible to frame the fault-tolerant converter architectures that resort to redundant components which might be different from the ones used in the original DC–DC converter, such as TRIACs. This approach might present some advantages over others that resort to redundant legs or modules: the inclusion of a redundant leg is avoided, thus reducing the cost of the fault-tolerant architecture. For ease of comprehension, the next subsections describe these fault-tolerant converter architectures, considering the different target converter topologies.

FB converters

Several fault-tolerant schemes of the phase-shift FB DC–DC converter have been developed so far, allowing to overcome the effects of OC faults in the switches of the transformer primary-side bridge. The proposed fault-tolerant schemes are extensible to other FB DC–DC converter topologies as well.

The successful implementation of one of those fault-tolerant architectures requires an additional redundant transformer winding, placed in the transformer secondary winding [6]. Figure 6.48 depicts a simplification of the fault-tolerant phase-shift FB DC–DC converter.

In general terms, the implementation of the reconfiguration strategy consists on changing the transformer primary-side FB configuration into a HB configuration. This reconfiguration of the transformer primary-side bridge has a negative effect on the converter output voltage [6]. The output voltage of an isolated DC–DC converter whose transformer primary-side bridge consists on a FB configuration is given by:

$$V_o = nV_i \tag{6.46}$$

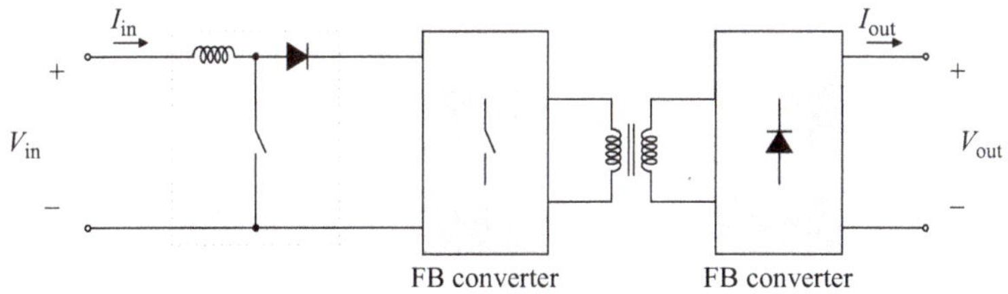

Figure 6.49 Implementation of a boost converter in the input DC-bus (highlighted in light grey background), as mean to compensate the converter voltage sag resulting from faults in the active switches of the transformer primary-side bridge [36]

where V_o is the converter output voltage, n is the transformer turns ratio and V_i is the converter input voltage. On the other hand, the output voltage of an isolated DC–DC converter consisting on a HB configuration at the transformer primary side is given by:

$$V_o = n\frac{V_i}{2} \tag{6.47}$$

These relations confirm that the reconfiguration of the transformer primary-side bridge from FB to HB leads to a reduction of the converter output voltage, to half of its original value.

To recover the pre-fault voltage level at the converter output, the redundant transformer winding is activated, after reconfiguring the transformer primary-side bridge into a HB. The main results of the converter reconfiguration are the restitution of the converter output voltage to pre-fault levels, the reduction in the power transferred to the load and an additional cost of implementation.

As mentioned before, the insertion of an auxiliary transformer winding as mean to compensate the reduction of the converter output voltage has a fairly high implementation cost. To reduce the implementation cost of the aforementioned reconfiguration strategy, the insertion of a simple boost converter proves to be a feasible alternative approach to compensate the converter output voltage in the post-fault period [36,37]. Figures 6.49 and 6.50 depict two effective fault-tolerant converter architectures based on the insertion of an additional boost converter. In those fault-tolerant converter architectures, the additional boost converter is inserted into the original FB converter in a cascaded configuration, in either of the transformer primary side (refer to Figure 6.49), or the transformer secondary side (refer to Figure 6.50).

Non-isolated multilevel converters

In most situations, semiconductor faults in multilevel DC–DC converters determine the total loss of power conversion functions of the converter. A fault-tolerant three-level DC–DC boost converter is proposed to implement a MPPT algorithm for a PV system, thus overcoming the problem of loss of conversion capabilities [13]. The

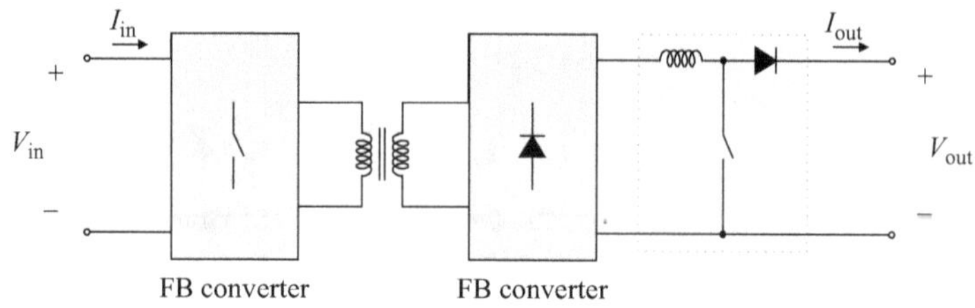

Figure 6.50 Implementation of a boost converter in the output DC-bus (highlighted in light grey background), as mean to compensate the converter voltage sag resulting from faults in the active switches of the transformer primary-side bridge [37]

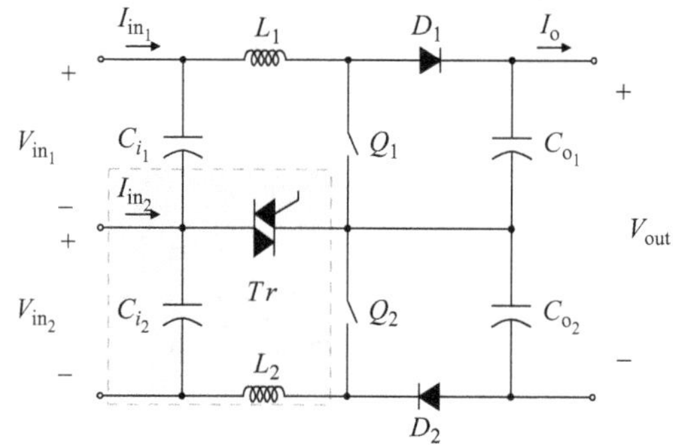

Figure 6.51 Fault-tolerant non-isolated three-level DC–DC converter [13]. Additional components introduced in the fault-tolerant converter architecture are highlighted in the grey background

fault-tolerant architecture consists of a rearrangement of the converter input, realised through the connection of multiple PV modules in series, at the input of the multilevel converter; in [13], an association of two PV modules connected in series is used. As shown in Figure 6.51, a TRIAC is introduced between the midpoint of the input DC-bus capacitors and the midpoint of the two converter switches. Some additional passive components (capacitor and inductor) are included as well.

In healthy converter operation, one of the converter switches is controlled to realise the MPPT, while the other switch is controlled to ensure the balance of the output DC-bus capacitor voltages.

In case of a switch OC fault, the converter is reconfigured into a two-level DC–DC boost converter. In order to achieve this configuration, the TRIAC is activated, allowing power flow on the respective branch. The converter control is also rearranged: the remaining healthy switch is controlled to maximise the power extracted from one of the PV modules. This shift in the control strategy implies the loss of MPPT in one of the PV modules, and the loss of the voltage-balancing feature.

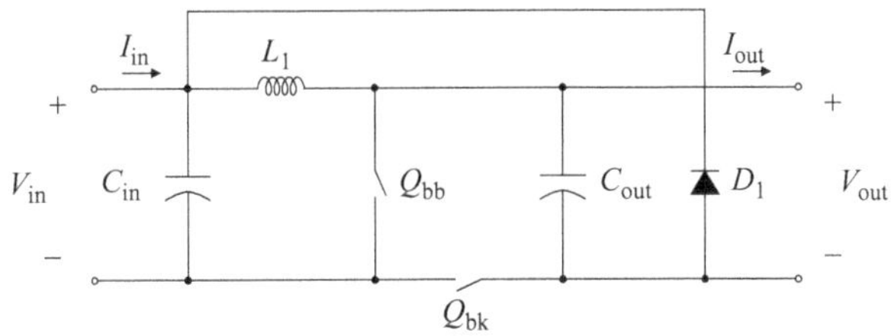

Figure 6.52 Fault-tolerant architecture of a DC–DC buck converter [38]

In the original architecture of the multilevel DC–DC boost converter, an OC fault in any of the converter switches implies total loss of the power conversion capabilities, meaning that the original multilevel converter is not able to deliver power to the output after a switch fault (OC or SC fault). In opposition, the fault-tolerant configuration ensures the extension of the converter operation, despite the partial loss of power conversion capability [13].

Buck converter

The fault-tolerant architecture of a buck converter is derived from the equivalent circuit of two distinctive DC–DC converter topologies. By superimposing the equivalent DC circuit of the original buck converter configuration and the equivalent DC circuit of the single-switch buck-boost converter, it is possible to derive a new buck converter configuration, featuring fault tolerance against OC faults [38]. In comparison to the traditional buck converter, this topology includes one additional active switch.

This converter configuration has two important advantages: (1) it possesses the fault-tolerant feature; and (2) it allows to select the desired operation mode: buck mode or buck-boost mode.

Under healthy converter operation, one of the switches is in active mode, while the other is kept in idle mode. As shown in Figure 6.52, to operate the converter in buck (or step-down) mode, switch Q_{bk} is activated and switch Q_{bb} is kept in idle mode; to operate the converter in buck-boost mode, switch Q_{bb} is active and switch Q_{bk} is at idle mode.

If an OC fault impacts the converter active switch, the redundant switch, which was previously in idle mode, is activated to ensure the converter operation afterwards.

It is important to make two remarks that configure some of the pitfalls of this fault-tolerant converter configuration: as the post-fault operation is based on a converter that is different from the initial configuration, the duty cycle of the PWM used to control the converter must be updated in the post-fault period and, consequently, a closed-loop control strategy is required; this converter configuration has a floating ground at the output, reason why is not suitable for applications that require a common ground in both the input and output of the converter.

Series-resonant DC–DC converters

In [39], a fault-tolerant architecture of the FB series-resonant converter is proposed. It is extensible to the group of series-resonant DC–DC converters [40], where the

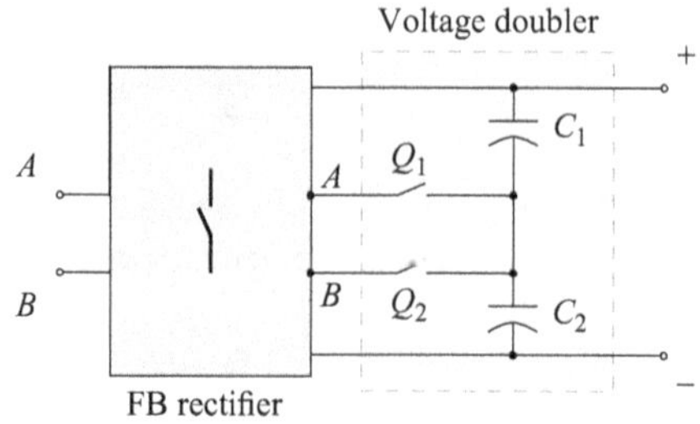

Figure 6.53 Implementation of a voltage doubler in the transformer secondary-side FB rectifier, as mean to compensate the converter voltage sag resulting from faults in the active switches of the transformer primary-side bridge [39,40]

following configurations can be framed: HB, FB and multilevel series-resonant DC–DC converters. In this group of DC–DC converters, the main concern arises from the occurrence of OC or SC faults in the active switches of the transformer primary-side bridge. Those faults imply the reconfiguration of the transformer primary-side bridge into a HB configuration. The configuration of the transformer primary-side bridge influences the series-resonant DC–DC converter output voltage [40]. The adaptation of the transformer primary-side FB configuration into a HB configuration reduces the converter output voltage to half of its original value.

Such significant reduction in the converter output voltage does not allow to maintain the proper operation of the loads connected to converter, particularly those that require a very stable power supply. To solve this problem, two capacitors and two switches are added to the FB rectifier connected in the transformer secondary side, to obtain a voltage doubler rectifier configuration, as depicted in Figure 6.53.

Under healthy converter operation, switches Q_1 and Q_2 are deactivated, and the voltage doubler function is disabled. If a fault (either OC or SC fault) compromises any of the transformer primary-side switches, that bridge is reconfigured into a HB configuration, and the voltage doubler rectifier of the transformer secondary side is put to work, by activating the switches Q_1 and Q_2.

The implementation of this fault-tolerant architecture requires neither any fuses nor isolation switches, but only two switches and two capacitors, used to build a voltage doubler rectifier. In addition, the post-fault operation does not imply any changes in the control strategy of the HB switches.

6.2.3.2 Fault-tolerant architectures based on redundant legs

In this category, it is possible to frame the fault-tolerant architectures that apply redundant legs, aiming the replacement of the original components, when they become faulty. Typically, fault-tolerant architectures based on redundant legs imply several components and, consequently, the overall cost of implementation increases, in comparison to fault-tolerant architectures that just add discrete components.

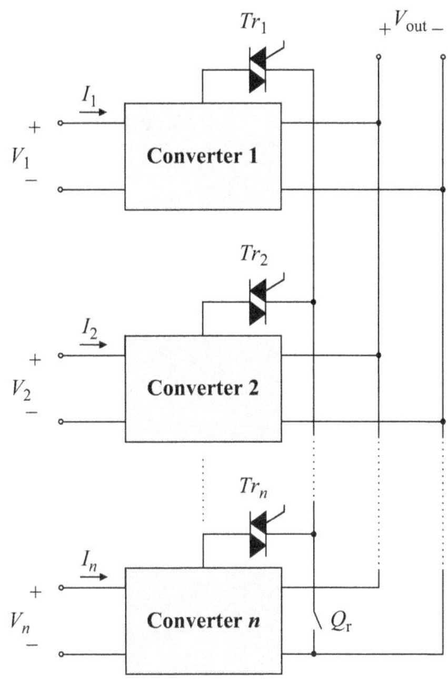

Figure 6.54 Implementation of a redundant converter leg based on a single redundant switch (Q_r) [8,11]

It is the case of the fault-tolerant architecture of a single-switch DC–DC boost converter [8,11]. A redundant leg, comprising one active switch and one TRIAC [8], or just a single active switch [11], is connected in parallel with the original converter switch. The faulty switch is isolated resorting to a fuse, and the redundant leg operation is triggered. To this end, the TRIAC is activated, to enable the operation of the redundant switch, that is controlled in the same way as the original switch. This converter reconfiguration strategy is not a particularly interesting and cost-effective solution when it is employed in a single-switch DC–DC boost converter. However, this solution becomes more cost-effective and attractive if the outputs of several single-switch boost DC–DC converters are connected in parallel to a single DC-bus [8,11]. In that case, the redundant leg replaces the faulty switch of any of the converters being paralleled, as shown in Figure 6.54.

6.2.4 Comparative analysis of the fault-tolerant strategies

Table 6.3 provides a simplified overview on the features of the most relevant fault-tolerant strategies aimed at DC–DC converters. Special attention should be devoted to the column 'Control reconfiguration?' of Table 6.3. It is considered that a reconfiguration of the converter control is conducted when the control strategy of the healthy converter modules, legs, or switches is adapted for the post-fault operation. Changes in the control of switches that aim the isolation of faulty

Table 6.3 Features of some of the most relevant fault-tolerant strategies

Ref.	Converter topology	Reconfiguration strategy	Additional components?	Control reconfiguration?	Cost
[6]	Phase shift FB converter	Inclusion of redundant components	Yes (2)	Yes	High
[8] [11]	Single-switch boost converter	Inclusion of redundant legs	Yes (2)	No	Medium
[13]	Non-isolated multilevel converter	Inclusion of redundant components	Yes (1)	Yes	Low
[17]	Parallel-connected SAB converter	Phase-shift adjustment	No	Yes	–[a]
[18] [41] [42] [43]	Interleaved converters	Phase-shift adjustment	No	Yes	–[a]
[30] [31]	ISOP	Bypass of faulty module	Yes (1 per module)	No	Low
[32]	Cascaded DC–DC converters	Bypass of faulty module	Yes (at least 5)	No	High
[33] [34]	MMC	Bypass of faulty module	No	No	–[a]
[35]	IPOS	Phase-shift adjustment	No	Yes	–[a]
[38]	Single-switch buck converter	Inclusion of redundant components	Yes (1)	No	Low
[36] [37]	FB converters	Inclusion of redundant components	Yes (3)	Yes	Medium
[39] [40]	Series-resonant DC–DC converter	Inclusion of redundant components	Yes (4)	No	Medium

[a]Not applicable

modules are not classified as control reconfiguration. Once again, the implementation cost metric is based on the number of additional components required to implement the fault-tolerant strategy.

6.3 Conclusions

This chapter outlined some of the most recent and important advances achieved in the development of fault diagnostic tools and fault-tolerant strategies aimed at DC–DC converters. It is a research topic that has attracted much attention in the last few years, and will certainly continue to attract attentions, due to the compelling need for highly reliable and efficient power conversion solutions for the emerging topic of DC grids.

The algorithms available in the literature provide effective fault diagnostic and fault-tolerant solutions for the wide range of DC–DC converter topologies and their typical end-users, allowing to establish a solid framework for the development of reliable DC grids. Still, further developments are required to obtain cheaper and highly reliable DC–DC power conversion solutions.

Acknowledgement

This work was supported by the European Regional Development Fund (ERDF) through the Operational Programme for Competitiveness and Internationalization (COMPETE 2020), under Project POCI-01-0145-FEDER-029494, and by National Funds through the FCT – Portuguese Foundation for Science and Technology, under Projects PTDC/EEI-EEE/29494/2017, UID/EEA/04131/2013 and SFRH/BD/131002/2017.

References

[1] S. Yang, A. Bryant, P. Mawby, D. Xiang, L. Ran, and P. Tavner, "An Industry-Based Survey of Reliability in Power Electronic Converters", *IEEE Transactions on Industry Applications*, vol. 47, no. 3, pp. 1441–1451, 2011.

[2] R. Jayabalan and B. Fahimi, "Monitoring and Fault Diagnosis of Multi-converter Systems in Hybrid Electric Vehicles", *IEEE Transactions on Vehicular Technology*, vol. 55, no. 5, pp. 1475–1484, 2006.

[3] S. Y. Kim, K. Nam, H. S. Song, and H. G. Kim, "Fault Diagnosis of a ZVS DC–DC Converter Based on DC-Link Current Pulse Shapes", *IEEE Transactions on Industrial Electronics*, vol. 55, no. 3, pp. 1491–1494, 2008.

[4] H. Sheng, F. Wang, and C. W. Tipton IV, "A Fault Detection and Protection Scheme for Three-Level DC–DC Converters Based on Monitoring Flying Capacitor Voltage", *IEEE Transactions on Power Electronics*, vol. 27, no. 2, pp. 685–697, 2012.

[5] E. Ribeiro, A. J. M. Cardoso, and C. Boccaletti, "Fault Analysis of Dual Active Bridge Converters", in *IECON 2012 – 38th Annual Conference on IEEE Industrial Electronics Society*, Montreal, QC, 2012.

[6] X. Pei, S. Nie, Y. Chen, and Y. Kang, "Open-Circuit Fault Diagnosis and Fault-Tolerant Strategies for Full-Bridge DC–DC Converters", *IEEE Transactions on Power Electronics*, vol. 27, no. 5, pp. 2550–2565, 2012.

[7] M. Shahbazi, E. Jamshidpour, P. Poure, S. Saadate, and M. R. Zolghadri, "Open- and Short-Circuit Switch Fault Diagnosis for Nonisolated DC–DC Converters Using Field Programmable Gate Array", *IEEE Transactions on Industrial Electronics*, vol. 60, no. 9, pp. 4136–4146, 2013.

[8] E. Jamshidpour, P. Poure, E. Gholipour, and S. Saadate, "Single-Switch DC–DC Converter With Fault-Tolerant Capability Under Open- and Short-Circuit Switch Failures", *IEEE Transactions on Power Electronics*, vol. 30, no. 5, pp. 2703–2712, 2015.

[9] E. Jamshidpour, P. Poure, and S. Saadate, "Switch Failure Diagnosis Based on Inductor Current Observation for Boost Converters", *International Journal of Electronics*, vol. 103, no. 9, pp. 1498–1509, 2016.

[10] E. Jamshidpour, M. Shahbazi, S. Saadate, P. Poure, and E. Gholipour, "FPGA Based Fault Detection and Fault Tolerance Operation in DC–DC Converters", in *2014 International Symposium on Power Electronics, Electrical Drives, Automation and Motion*, Ischia, 2014.

[11] E. Jamshidpour, P. Poure, and S. Saadate, "Photovoltaic Systems Reliability Improvement by Real-Time FPGA-Based Switch Failure Diagnosis and Fault-Tolerant DC–DC Converter", *IEEE Transactions on Industrial Electronics*, vol. 62, no. 11, pp. 7247–7255, 2015.

[12] T. Park and T. Kim, "Novel Fault Tolerant Power Conversion System for Hybrid Electric Vehicles", in *2011 IEEE Vehicle Power and Propulsion Conference*, Chicago, IL, 2011.

[13] E. Ribeiro, A. J. M. Cardoso, and C. Boccaletti, "Fault-Tolerant Strategy for a Photovoltaic DC–DC Converter", *IEEE Transactions on Power Electronics*, vol. 28, no. 6, pp. 3008–3018, 2013.

[14] E. Ribeiro, A. J. M. Cardoso, and C. Boccaletti, "Fault Diagnosis in a Multi-Input Power Interface for a Photovoltaic Wind Supply System for Telecommunications", in *Intelec 2013; 35th International Telecommunications Energy Conference, Smart Power and Efficiency*, Hamburg, Germany, 2013.

[15] E. Ribeiro, A. J. M. Cardoso, and C. Boccaletti, "Fault Diagnosis in Non-Isolated Bidirectional Half-Bridge DC–DC Converters", in *IECON 2014 – 40th Annual Conference of the IEEE Industrial Electronics Society*, Dallas, TX, 2014.

[16] E. Ribeiro, A. J. M. Cardoso, and C. Boccaletti, "Fault Diagnosis in Unidirectional Non-Isolated DC–DC Converters", in *2014 IEEE Energy Conversion Congress and Exposition (ECCE)*, Pittsburgh, PA, 2014.

[17] K. Park and Z. Chen, "Open-Circuit Fault Detection and Tolerant Operation for a Parallel-Connected SAB DC–DC Converter", in *2014 IEEE Applied Power Electronics Conference and Exposition – APEC 2014*, Fort Worth, TX, 2014.

[18] E. Ribeiro, A. J. M. Cardoso, and C. Boccaletti, "Open-Circuit Fault Diagnosis in Interleaved DC–DC Converters", *IEEE Transactions on Power Electronics*, vol. 29, no. 6, pp. 3091–3102, 2014.

[19] S. Nie, X. Pei, Y. Chen, and Y. Kang, "Fault Diagnosis of PWM DC–DC Converters Based on Magnetic Component Voltages Equation", *IEEE Transactions on Power Electronics*, vol. 29, no. 9, pp. 4978–4988, 2014.

[20] H. K. Cho, S. S. Kwak, and S. H. Lee, "Fault Diagnosis Algorithm Based on Switching Function for Boost Converters", *International Journal of Electronics*, vol. 102, no. 7, pp. 1229–1243, 2015.

[21] X. Pei, S. Nie, and Y. Kang, "Switch Short-Circuit Fault Diagnosis and Remedial Strategy for Full-Bridge DC–DC Converters", *IEEE Transactions on Power Electronics*, vol. 30, no. 2, pp. 996–1004, 2015.

[22] H. Givi, E. Farjah, and T. Ghanbari, "Switch and Diode Fault Diagnosis in Non-isolated DC–DC Converters Using Diode Voltage Signature", *IEEE Transactions on Industrial Electronics*, vol. 65, no. 2, pp. 1606–1615, 2018.

[23] K. Bi, Q. An, J. Duan, L. Sun, and K. Gai, "Fast Diagnostic Method of Open-Circuit Fault for Modular Multilevel DC–DC Converter Applied in Energy Storage System", *IEEE Transactions on Power Electronics*, vol. 32, no. 5, pp. 3292–3296, 2017.

[24] Y. Chen, X. Pei, S. Nie, and Y. Kang, "Monitoring and Diagnosis for the DC–DC Converter Using the Magnetic Near Field Waveform", *IEEE Transactions on Industrial Electronics*, vol. 58, no. 5, pp. 1634–1647, 2011.

[25] S. Shao, P. W. Wheeler, J. C. Clare, and A. J. Watson, "Fault Detection for Modular Multilevel Converters Based on Sliding Mode Observer", *IEEE Transactions on Power Electronics*, vol. 28, no. 11, pp. 4867–4872, 2013.

[26] S. Shao, A. J. Watson, J. C. Clare, and P. W. Wheeler, "Robustness Analysis and Experimental Validation of a Fault Detection and Isolation Method for the Modular Multilevel Converter", *IEEE Transactions on Power Electronics*, vol. 31, no. 5, pp. 3794–3805, 2016.

[27] E. Pazouki, Y. Sozer, and J. A. D. Abreu-Garcia, "Fault Diagnosis and Fault-Tolerant Control Operation of Nonisolated DC–DC Converters", *IEEE Transactions on Industry Applications*, vol. 54, no. 1, pp. 310–320, 2018.

[28] E. Pazouki, J. A. D. Abreu-Garcia, and Y. Sozer, "Short-Circuit Fault Diagnosis for Interleaved DC–DC Converter Using DC-link Current Emulator", in *2017 IEEE Applied Power Electronics Conference and Exposition (APEC)*, Tampa, FL, 2017.

[29] J. Poon, P. Jain, I. C. Konstantakopoulos, C. Spanos, S. K. Panda, and S. R. Sanders, "Model-Based Fault Detection and Identification for Switching Power Converters", *IEEE Transactions on Power Electronics*, vol. 32, no. 2, pp. 1419–1430, 2017.

[30] V. Choudhary, E. Ledezma, R. Ayyanar, and R. M. Button, "Fault Tolerant Circuit Topology and Control Method for Input-Series and Output-Parallel Modular DC–DC Converters", *IEEE Transactions on Power Electronics*, vol. 23, no. 1, pp. 402–411, 2008.

[31] Y. Hayashi, Y. Matsugaki, T. Ninomiya, and H. Ohash, "Active Gate Controlled SiC Transfer Switch for Fault Tolerant Operation of ISOP Multicellular DC–DC Converter", in *2016 IEEE International Conference on Power Electronics, Drives and Energy Systems (PEDES)*, Trivandrum, 2016.

[32] M. M. Haji-Esmaeili, M. Naseri, H. Khoun-Jahan, and M. Abapour, "Fault-Tolerant and Reliable Structure for a Cascaded Quasi-Z-Source DC–DC Converter", *IEEE Transactions on Power Electronics*, vol. 32, no. 8, pp. 6455–6467, 2017.

[33] F. H. Khan and L. M. Tolbert, “Multiple Load-Source Integration in a Multilevel Modular Capacitor Clamped DC–DC Converter Featuring Fault Tolerant Capability”, *IEEE Transactions on Power Electronics*, vol. 24, pp. 14–24, 2009.

[34] F. H. Khan and L. M. Tolbert, “Bi-Directional Power Management and Fault Tolerant Feature in a 5-kW Multilevel DC–DC Converter with Modular Architecture”, *IET Power Electronics*, vol. 2, no. 5, pp. 595–604, 2009.

[35] T. Li and L. Parsa, “Design, Control and Analysis of a Fault-Tolerant Soft-Switching DC–DC Converter for High Power High Voltage Applications”, *IEEE Transactions on Power Electronics*, vol. 33, no. 2, pp. 1094–1104, 2018.

[36] J. Y. Lee, Y. S. Jeong, and B. M. Han, “An Isolated DC/DC Converter Using High-Frequency Unregulated LLC Resonant Converter for Fuel Cell Applications”, *IEEE Transactions on Industrial Electronics*, vol. 58, no. 7, pp. 2926–2934, 2011.

[37] C. Yao, X. Ruan, X. Wang, and C. K. Tse, “Isolated Buck–Boost DC/DC Converters Suitable for Wide Input-Voltage Range”, *IEEE Transactions on Power Electronics*, vol. 26, no. 9, pp. 2599–2613, 2011.

[38] D. D. C. Lu, J. L. Soon, and D. Verstraete, “Derivation of Dual-Switch Step-Down DC–DC Converters With Fault-Tolerant Capability”, *IEEE Transactions on Power Electronics*, vol. 31, no. 9, pp. 6064–6068, 2016.

[39] L. Costa, G. Buticchi, and M. Liserre, “A Fault-Tolerant Series-Resonant DC–DC Converter”, *IEEE Transactions on Power Electronics*, vol. 32, no. 2, pp. 900–905, 2017.

[40] L. Costa, G. Buticchi, and M. Liserre, “A Family of Series-Resonant DC–DC Converter with Fault-Tolerance Capability”, *IEEE Transactions on Industry Applications*, vol. 54, no. 1, pp. 335–344, 2018.

[41] Z. Lukic, C. Blake, S. C. Huerta, and A. Prodic, “Universal and Fault-Tolerant Multiphase Digital PWM Controller IC for High-Frequency DC–DC Converters”, in *APEC 07 – Twenty-Second Annual IEEE Applied Power Electronics Conference and Exposition*, Anaheim, CA, USA, 2007.

[42] E. Ribeiro, A. Monteiro, A. J. M. Cardoso, and C. Boccaletti, “Fault Tolerant Small Wind Power System for Telecommunications with Maximum Power Extraction”, in *2014 IEEE 36th International Telecommunications Energy Conference (INTELEC)*, Vancouver, BC, 2014.

[43] M. Gleissner and M. M. Bakran, “Design and Control of Fault-Tolerant Nonisolated Multiphase Multilevel DC–DC Converters for Automotive Power Systems”, *IEEE Transactions on Industry Applications*, vol. 52, no. 2, pp. 1785–1795, 2016.

Index